STUDENT'S SOLUTIONS MANUAL

to accompany

ELEMENTARY STATISTICS
SEVENTH EDITION

STUDENT'S SOLUTIONS MANUAL

MILTON LOYER
Pennsylvania State University

to accompany

ELEMENTARY STATISTICS
SEVENTH EDITION
MARIO F. TRIOLA

 **ADDISON-WESLEY**

An imprint of Addison Wesley Longman, Inc.

Reading, Massachusetts • Menlo Park, California • New York • Harlow, England
Don Mills, Ontario • Sydney • Mexico City • Madrid • Amsterdam

ISBN   0-201-85923-8

6 7 8 9 10-CRS-99 98

PREFACE

This manual contains the solutions to the odd-numbered exercises for each section of the textbook <u>Elementary Statistics</u>, Seventh Edition, by Mario Triola, and the solutions for all end-of-chapter review and cumulative review exercises of that text. In the worked problems, intermediate steps are provided for the calculations. When appropriate, additional hints and comments are included and prefaced by NOTE.

Many statistical problems are best solved using particular formats. Recognizing and following these patterns promotes understanding and develops the capacity to apply the concepts to other problems. This manual identifies and employs such formats whenever practicable.

For best results, read the text carefully <u>before</u> attempting the exercises, and attempt the exercises <u>before</u> consulting the solutions. This manual has been prepared to provide a check and extra insights for exercises that have already been completed and to provide guidance for solving exercises that have already been attempted but have not been successfully completed.

I would like to thank Mario Triola for writing an excellent elementary statistics book and for inviting me to prepare this solutions manual.

TABLE OF CONTENTS

STUDENT'S
SOLUTIONS MANUAL

to accompany

ELEMENTARY STATISTICS
SEVENTH EDITION

Chapter 1

Introduction to Statistics

1-2 The Nature of Data

1. continuous, since weight can be any value on a continuum

3. discrete, since the number of subscribers must be an integer

5. discrete, since the number of perfect scores must be an integer

7. continuous, since time can be any value on a continuum

9. ordinal, since the labels give relative position in a hierarchy

11. nominal, since the numbers are used for identification only. Even though SS numbers are assigned chronologically within regions and can be placed in numerical order, there is no meaningful way to compare 208-34-3337 and 517-94-1439. If all the numbers had been assigned chronologically beginning with 000-00-0001, like the order of finishers in a race, then SS numbers would illustrate the ordinal level of measurement.

13. interval, since differences are meaningful but ratios are not

15. nominal, since the numbers are used for identification only. Even though they are assigned alphabetically within regions, zip codes are merely numerical names for post offices. If the numbers had resulted from placing all the post offices in one large list alphabetically (or by city size, or by mail volume, etc.), so that 17356 [Red Lion, PA] was the 17,356th post office in the list, like the order of finishers in a race, then zip codes would illustrate the ordinal level of measurement.

17. ordinal, since the labels give relative position in a hierarchy

19. Years are not data at the ratio level of measurement because the year zero has been arbitrarily assigned so that 1900 does not represent twice as much time as 950 -- consider, for example, the Chinese numerical representations for the given years. Since the time difference between 1900 and 1920 is the same as the time difference between 1920 and 1940, however, years are data at the interval level of measurement.

1-3 Uses and Abuses of Statistics

1. Telephone directories do not contain unlisted numbers, and consumer types electing not be listed would not be represented in the survey. In addition, sometimes numbers appear in directories more than once (e.g., under the husband's name and under the wife's name), and consumer types electing multiple listings would be over-represented in the survey. Randomly generated lists of telephone numbers will include inappropriate (stores, etc.) and non-functioning ones, but those can be ignored whenever they occur. NOTE: From the wording of the problem ("you must conduct a telephone survey") it is assumed that the use of the telephone is non-negotiable. This means that those without telephones for ecomonic, relgous, or other reasons would automatically be denied a chance to participate no matter how the numbers were selected.

3. Studies sponsored by groups with a desire to show a particular result are sometimes biased toward that result. Such bias is not necessarily intentional and may owe to well-intentioned efforts to control the experiment by eliminating subjects and/or situations not conducive to the desired result, or to a subconscious tendency to interpret unclear data in the hoped-for direction. In addition, studies that do not support the desired conclusion will probably be re-done and/or not be reported.

5. The statement that "seventy-nine percent of those who responded..." is correct. Applying that figure to the entire population would not be correct, because those who responded were not necessarily representative of the entire population (or even of the readership of the magazine). Those who subscribe to Glamour are probably more likely to be affluent and female. The subscribers who chose to respond are probably those with strong feelings on the issue.

7. The number that say they've committed a crime is $(.08)(1875) = 150$; the number that say they did so under the influence of alcohol or drugs is $(.62)(150) = 93$.

9. That healthier babies are born to mothers who eat lobsters doesn't mean that eating lobster caused the babies to be healthier. Mothers who eat lobster are probably more affluent than the general population and would tend to eat better, be more knowledgeable about proper pre-natal care, have better health care, etc.

11. Since the poll was sponsored by a shoe polish company, the questions (especially if they were multiple choice and not open ended) may have been designed to encourage and/or suggest responses relating specifically to the appearance of shoes. Similarly, a necktie company might make certain there were questions and/or suggested responses about neck wear but fail to include items about foot wear.

13. Assuming a male can't become a symphony conductor until about age 30, one should eliminate from consideration all males who die before 30. The average age at death for male symphony conductors should then be compared to the average age at death of those males who survived past 30 -- an age that would certainly be higher than 69.5 and possibly very close to 73.4.

15. One possible listing of four major flaws is as follows.
    (1) The question is not objective, but worded to encourage negative answers. Consider the question, "Do you support the development of atomic weapons whose presence could discourage an enemy attack?"
    (2) The sample size of 20 is probably too small.
    (3) The respondents are not necessarily a random sample, but merely those who chose to reply.
    (4) In a sample of size 20, the percentages of "yes" and "no" responses must be multiples of

5 -- 87% and 13% are not mathematically possible.

1-4 Design of Experiments

1. observational study, since specific characteristics are measured but not manipulated

3. experiment, since the effect of an applied treatment is measured

5. convenience, since the sample was simply those choosing to respond

7. stratified, since the set of interest (assumed to be the 4 English classes) was divided into 8 subpopulations (C1 males, C1 females, C2 males, C2 females, C3 males, C3 females, C4 males, C4 females) from which the actual sampling was done

9. random, since each U.S. senator has an equal chance of being selected

11. stratified, since the set of interest was divided into 2 subpopulations from which the actual sampling was done

13. cluster, since the entire set of interest (all hospital leukemia patients) was divided into hospitals, and all the appropriate persons in each selected hospital were interviewed

15. random, since each company has an equal chance of being selected

17. a. Open questions elicit the respondent's true feelings without putting words or ideas into his mind. In addition, open questions might produce responses the pollster failed to consider. Unfortunately open questions sometimes produce responses that are rambling, unintelligible or not relevant.
    b. Closed questions help to focus the respondent and prevent misinterpretation of the question. Sometimes, however, closed questions reflect only the wording and opinions of the pollster and do not allow respondents to express legitimate alternatives.
    c. Closed questions are easier to analyze because the pollster can control the number of possible responses to each question and word responses to establish relationships between questions.

1-5 Statistics with Calculators and Computers

1. $\dfrac{3.44 + 2.67 + 2.09 + 1.87 + 3.11}{5} = \dfrac{13.18}{5} = 2.636$

3. $\sqrt{\dfrac{3(101) - 15^2}{6}} = \sqrt{\dfrac{303 - 225}{6}} = \sqrt{\dfrac{78}{6}} = \sqrt{13} = 3.6056$

5. $\dfrac{1.96^2 \cdot 0.25}{0.03^2} = \dfrac{.9604}{.0009} = 1067.1111$

7. $\dfrac{15!}{9!6!} = \dfrac{15 \cdot 14 \cdot 13 \cdot 12 \cdot 11 \cdot 10 \cdot 9!}{9!6!} = \dfrac{15 \cdot 14 \cdot 13 \cdot 12 \cdot 11 \cdot 10}{6 \cdot 5 \cdot 4 \cdot 3 \cdot 2 \cdot 1} = \dfrac{3603600}{720} = 5005$

9. Methods for loading STATDISK or Minitab, retrieving data sets, and performing procedures vary slightly from institution to institution. See your instructor or appropriate staff person for assistance with your institution's particular computer, network, printer, etc.

Review Exercises

1. a. continuous, since voltage can be any value on a continuum
   b. ratio, since differences between values are consistent and there is a meaningful zero
   c. stratified, since the set of interest (all 9 volt batteries) was divided into subpopulations (by company of manufacture) from which the actual sampling was done
   d. observational study, since specific characteristics are measured but not manipulated
   e. The product in which the battery is to be used may not function properly, may not function at all, or may be damaged by the incorrect voltage.

2. a. ratio, since differences between values are consistent and there is a meaningful zero
   b. ordinal, since the labels give only relative position in a hierarchy
   c. interval, since differences are meaningful but ratios are not
   d. nominal, since the country is used for identification only

3. Since the sample was self-selected and consisted only of those who chose to answer the questionnaire, it was not necessarily representative of the entire set of interest. Those who take such surveys seriously and take the time to respond, for example, may tend to be the more fearful and pessimistic.

4. a. discrete, since the number of adults surveyed must be an integer
   b. continuous, since time can be any value on a continuum
   c. continuous, since weight can be any value on a continuum

5. a. systematic, since products are selected at regular intervals
   b. random, since each car has the same chance of being selected
   c. cluster, since the stocked items were organized into stores and all the items in randomly selected stores were chosen
   d. stratified, since the drivers were divided by gender and age into 8 subpopulations from which the actual sampling was done
   e. convenience, since the sample was composed of those who chose to take a test drive.
      NOTE: If the car maker first selected the particular dealership at random from among all its dealerships, then cluster sampling was used. Assuming the set of interest is all potential customers and not just all who take a test drive, the selections at the local dealership were made by convenience sampling.

6. When giving their own personal data (height, weight, income, age, etc.) people sometimes give round numbers as a subconscious way to avoid invasions of privacy (and they often round in the most favorable direction). When answering for someone else, people give round numbers because they may not know the exact value.

7. If the set of interest is all students, then surveying every 50th student leaving the cafeteria would not produce a representative sample. Missing from the survey would be the responses of students not eating in the cafeteria that day because they had schedule conflicts (e.g., due to labs, sports, jobs, or field trips), don't generally patronize the cafeteria (e.g., commuters), didn't like the menu that day, etc.

8. Even though the word "about" is not used, 8725 is likely an estimate. Because 8725 sounds more precise than 8700 or 8800, the article suggests an accuracy that is probably undeserved. In addition, since the rally was supporting one side of a controversial issue, the estimate may be influenced by the position of the person or the publication making the report.

Cumulative Review Exercises

1. No, the second version is not substantially less confusing. It has merely replaced the double negation "impossible that it never happened" with "certain that it happened." Both questions are confusing because they include two opposing alternatives in such a way that it is unclear which alternative a "yes" response agrees with. It would be much clearer to ask about only one alternative and let a "no" response indicate agreement with the other. Consider, for example, using only the first half of the given versions: "Does it seem possible to you that the Nazi extermination of the Jews never happened?"

2. The information given is the percent of college enrollees aged 25 and over for the years 1972 to 1976. Confusion is created by the unnecessary "upper graph" and by the absence of a meaningful zero, which makes it appear that older enrollment more than doubled from 1973 to 1974 (see section 1-3, exercise #12). The actual values are 1972 - 28.0%, 1973 - 29.2%, 1974 - 32.8%, 1975 - 33.6%, 1976 - 33.0%. One possibility for a more effective graph is given below.

Chapter 2

Describing, Exploring, and Comparing Data

2-2 Summarizing Data

1. Subtracting two consecutive lower class limits indicates that the class width is 6 - 0 = 6. Since there is a gap of 1.0 between the upper class limit of one class and the lower class limit of the next, class boundaries are determined by increasing or decreasing the appropriate class limits by (1.0)/2 = 0.5. The class boundaries and class marks are given in the following table.

absences	class boundaries	class mark	frequency
0 - 5	-0.5 - 5.5	2.5	39
6 - 11	5.5 - 11.5	8.5	41
12 - 17	11.5 - 17.5	14.5	38
18 - 23	17.5 - 23.5	20.5	40
24 - 29	23.5 - 29.5	26.5	42
			200

NOTE: Although they often contain extra decimal points and may involve consideration of how the data were obtained, class boundaries are the key to tabular and pictorial data summaries. Once the class boundaries are obtained, everything else falls into place. In this case, the first class width is readily seen to be 5.5 - (-0.5) = 6.0 and the first class mark is (-0.5 + 5.5)/2 = 2.5. In this manual, class boundaries will typically be calculated first and then be used to determine other values. In addition, since the sum of the frequencies (i.e., the total number of values) is an informative number and used in many subsequent calculations, it will typically be shown as an integral part of each table.

3. Since the gap between classes as presented is 0.1, the appropriate class limits are increased or decreased by (0.1)/2 = .05 to obtain the class boundaries and the following table.

weight (kg)	class boundaries	class mark	frequency
0.0 - 1.9	-0.05 - 1.95	0.95	20
2.0 - 3.9	1.95 - 3.95	2.95	32
4.0 - 5.9	3.95 - 5.95	4.95	49
6.0 - 7.9	5.95 - 7.95	6.95	31
8.0 - 9.9	7.95 - 9.95	8.95	18
			150

The class width is 1.95 - (-0.05) = 2.00 and the first class mark is (-0.05 + 1.95)/2 = 0.95.

5. The relative frequency for each class is found by dividing its frequency by 200, the sum of the frequencies. NOTE: As before, the sum is included as an integral part of the table. For relative frequencies, this should always be 1.000 (i.e., 100%) and serves as a check for the calculations.

absences	relative frequency
0 - 5	.195
6 - 11	.205
12 - 17	.190
18 - 23	.200
24 - 29	.210
	1.000

7. The relative frequency for each class is found by dividing its frequency by 150, the sum of the frequencies. NOTE: In #5, the relative frequencies were expressed as decimals; here they are expressed as percents The choice is arbitrary.

weight (kg)	relative frequency
0.0 - 1.9	13.3%
2.0 - 3.9	21.3%
4.0 - 5.9	32.7%
6.0 - 7.9	20.7%
8.0 - 9.9	12.0%
	100.0%

9. The cumulative frequencies are determined by repeated addition of successive frequencies to obtain the combined number in each class and all previous classes. NOTE: Consistent with the emphasis that has been placed on class boundaries, we choose to use upper class boundaries in the "less than" column. Conceptually, absences (whether for classes in school, days of work, or some other application) occur on a continuum in that a person who misses half the time accumulated 0.5 absences. The integer values reported in this exercise are then treated as the nearest whole number representation of the precise measure of absences. An exact score of 5.7, for example, would be reported as 6 and fall in the second class. The 39 values in the first class, therefore, are better described as being "less than 5.5" (using the upper class boundary) than as being "less than 6." This distinction becomes crucial in the construction of pictorial representations in the next section. In addition, the fact that the final cumulative frequency must equal the total number (i.e, the sum of the frequency column) serves as a check for calculations. The sum of cumulative frequencies, however, has absolutely no meaning and is not included.

absences	cumulative frequency
less than 5.5	39
less than 11.5	80
less than 17.5	118
less than 23.5	158
less than 29.5	200

11. The cumulative frequencies are determined by repeated addition of successive frequencies to obtain the combined number in each class and all previous classes.

weight (kg)	cumulative frequency
less than 1.95	20
less than 3.95	52
less than 5.95	101
less than 7.95	132
less than 9.95	150

13. The data in exercise 1 are approximately evenly distributed across the range of reported values. The data in exercise 2 seem to "bunch up" in the middle of the range of reported values.

15. There is more than one acceptable solution. One possibility is to note that for a range of 4.95 - 0.26 = 4.69 to be covered with 10 classes, there must be at least (4.69)/10 = .469 units per class. Rounding up to a class width of .50 and starting at .25, for example, produces a first class with lower and upper class limits of .25 and .74. NOTE: The second class would then have lower and upper class limits of .75 and 1.24, and the tenth class would have lower and upper class limits of 4.75 and 5.24 that include the maximum value of 4.95.

17. For 11 classes to cover data ranging from a beginning lower class limit of 0 to a maximum value of 514, the class width must be at least $(514 - 0)/11 = 46.7$. A convenient class width would be 50. Since the data is given to the nearest integer, the beginning lower class boundary is $0 - (1)/2 = -0.5$, producing the table at the right. NOTE: Using class limits (0-49, 50-99, etc.) to identify the categories is also acceptable (and "less messy"). Class boundaries, however, promote better understanding of the information and are the appropriate values for the graphic presentations of the next section.

weight (lbs)	frequency
-0.5 - 49.5	6
49.5 - 99.5	10
99.5 - 149.5	10
149.5 - 199.5	7
199.5 - 249.5	8
249.5 - 299.5	2
299.5 - 349.5	4
349.5 - 399.5	3
399.5 - 449.5	3
449.5 - 499.5	0
499.5 - 549.5	1
	54

19. For 7 classes to cover data ranging from a beginning lower class limit of 56 to a maximum value of 104, the class width must be at least $(104 - 56)/7 = 6.86$. The given class width of 8 meets that criterion. Since the data is given to the nearest integer, the beginning lower class boundary is $56 - (1)/2 = 55.5$, producing the table at the right.

time (min)	frequency
55.5 - 63.5	8
63.5 - 71.5	3
71.5 - 79.5	9
79.5 - 87.5	17
87.5 - 95.5	8
95.5 - 103.5	4
103.5 - 111.5	1
	50

21. The total numbers (i.e., sums of the frequency columns) for the men and women were 13,055 and 721 respectively. The relative frequency tables are as follows.

ethanol consumed by men (oz.)	relative frequency
0.0 - 0.9	.019
1.0 - 1.9	.071
2.0 - 2.9	.118
3.0 - 3.9	.171
4.0 - 4.9	.087
5.0 - 9.9	.273
10.0 - 14.9	.142
15.0 or more	.118
	1.000

ethanol consumed by women (oz.)	relative frequency
0.0 - 0.9	.010
1.0 - 1.9	.072
2.0 - 2.9	.173
3.0 - 3.9	.265
4.0 - 4.9	.042
5.0 - 9.9	.279
10.0 - 14.9	.060
15.9 or more	.100
	1.000

The distributions seem very similar except that there are proportionately more women in the 2.0-3.9 range and proportionately less women in the 10.0-14.9 range. NOTE: Due to rounding, the relative frequencies given actually sum to 0.999 for the men and to 1.001 for the women. Discrepancies of 1 or 2 at the last decimal place are probably due to rounding; larger discrepancies should not be so attributed, and such work should be carefully checked.

23. a. classes mutually exclusive? yes
    b. all classes included? yes
    c. same width for all classes? no
    d. convenient class limits? yes
    e. between 5 and 20 classes? no

Two of the five guidelines were not followed. NOTE: This does not mean the table is in error. The guidelines are only suggestions that make most presentations more readable; depending on the context, the given table may be the best way to present the ages.

2-3 Pictures of Data

1. See the figure below. The bars extend from class boundary to class boundary. Each axis is labeled numerically <u>and</u> with the name of the quantity represented. Barring an interval longer than any previously recorded, a minimum stay of 109.5 minutes assures seeing an eruption.

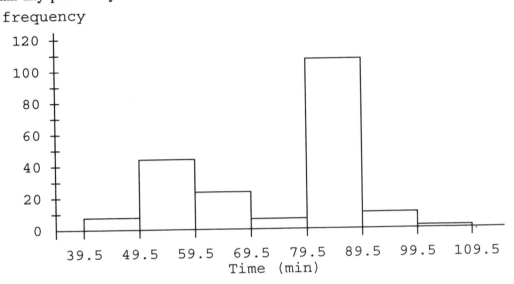

3. See the figure below. The bars extend from class boundary to class boundary. Each axis is labeled numerically <u>and</u> with the name of the quantity represented. Although the posted limit is 30 mph, it appears that the police ticket only those traveling at least 42 mph.

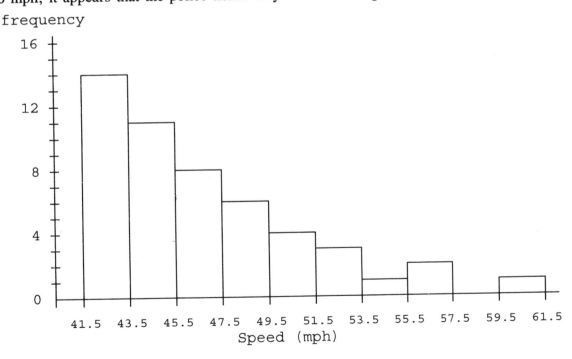

5. The original numbers are listed by the row in which they appear in the stem-and-leaf plot.

```
stem|leaves        original numbers
  57|017            570, 571, 577
  58|13349          581, 583, 583, 584, 589
  59|456678         594, 595, 596, 596, 597, 598
  60|23             602, 603
```

7. The dotplot is constructed using the original scores as follows. Each space represents 1 unit.

```
        .:                .                .
     .:        .:.:     :      .::::   .:.
  --T-----------T----:-----T---------T-----T--
   570        580        590       600       610
```

9. The expanded stem-and-leaf plot below on the left is one possibility. NOTE: The text claims that stem-and-leaf plots "enable us to see the distribution <u>without</u> losing information in the process. Following the suggestion to round the nearest inch not only loses information but also uses subjectivity to round values exactly half way between. Since always rounding such values "up" creates a slight bias, many texts suggest rounding toward the even digit -- so that 33.5 becomes 34, but 36.5 becomes 36. The technique below of using superscripts to indicate the occasional decimals is both mathematically clear and visually uncluttered.

```
stem|leaves                                              stem|leaves
  3|6 7                                                     0|8
  4|0 0 1 3 3⁵                                              1|5
  4|6 6 7 8 8 9                                             2|
  5|0 2 2⁵ 3 3 4                                            3|
  5|7³ 7⁵ 8 9 9 9                                           4|
  6|0 0⁵ 1 1 1⁵ 2 3 3 3 3⁵ 4 4 4                            5|8
  6|5 5 6⁵ 7 7⁵ 8⁵                                          6|00055777889
  7|0 0⁵ 2 2 2 2 3 3⁵                                       7|0223567888
  7|5 6⁵                                                    8|000002356888
                                                            9|027
                                                           10|0
```

11. The stem-and-leaf plot at the right is one possibility.

```
stem|leaves
 5.4|9
 5.5|2333
 5.5|677777888889999
 5.6|00000001222333
 5.6|5666778
 5.7|01123334
 5.7|
 5.8|4
```

13. See the figure below, with bars arranged in order of magnitude. Networking appears to be the most effective job-seeking approach.

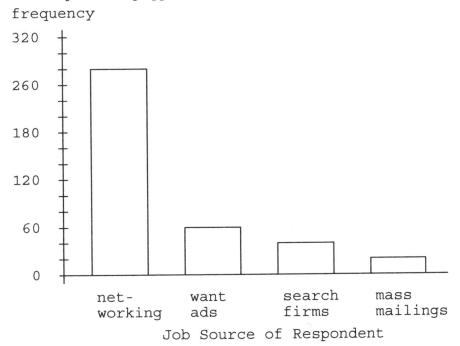

15. See the figure below. The sum of the frequencies is 50; the relative frequencies are 23/50 = 46%, 9/50 = 18%, 12/50 = 24%, and 6/50 = 12%. The corresponding central angles are (.46)360° = 165.6°, (.18)360° = 64.8°, (.24)360° = 86.4°, and (.12)360° = 43.2°.

NOTE: To be complete, the figure needs to be titled with the name of the quantity being measured.

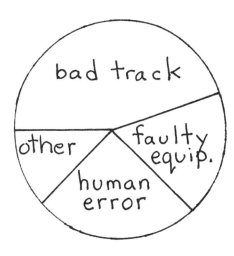

Causes of Train Derailments

17. The scatter diagram is given below. The figure should have a title, and each axis should be labeled both numerically and with the name of the variable. An "x" marks a single occurrence, while numbers indicate multiple occurrences at a point. Cigarettes high in tar also tend to be high in CO. The points cluster about a straight line from (0,0) to (18,18), indicating that the mg of CO tends to be about equal to the mg of tar.

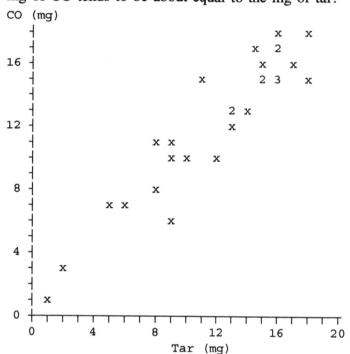

TAR AND CO CONTENT
OF SELECTED 100mm
FILTERED AMERICAN
CIGARETTES

19. a. The frequency distribution and histogram are given below. The bars extend from class boundary to class boundary. While the true midpoints of the classes are 24.5,74.5,..., the labels 25,75,... give a better visual presentation without being misleading.

weight (lbs)	frequency
0 - 49	6
50 - 99	10
100 - 149	10
150 - 199	7
200 - 249	8
250 - 299	2
300 - 349	4
350 - 399	3
400 - 449	3
450 - 499	0
500 - 549	1
	54

b. The distribution of the weights is skewed, with more bears at the lighter end than at the heavier end.

21. There is more than one acceptable solution. To cover a range of 20.58 - 1.65 = 18.93 with 10 classes, there must be at least (18.93)/10 = 1.893 units per class. Rounding up to 2.00 and using 1.00 for the first lower class limit produces first class limits from 1.00 to 2.99.

weight (lbs)	frequency
1.00 - 2.99	4
3.00 - 4.99	4
5.00 - 6.99	12
7.00 - 8.99	9
9.00 - 10.99	12
11.00 - 12.99	10
13.00 - 14.99	5
15.00 - 16.99	3
17.00 - 18.99	1
19.00 - 20.99	2
	62

a. See the figure below. The bars extend from class boundary to class boundary.

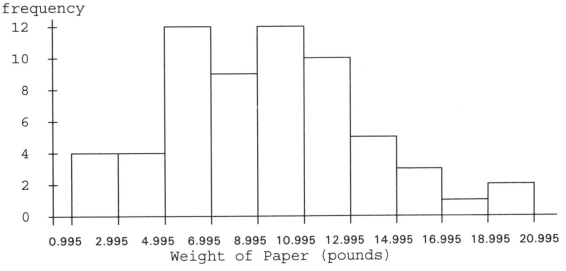

b. The figure shows that the distribution is approximately bell-shaped.

23. According to the figure, 422,000 started and 10,000 returned.
10,000/422,000 = 2.37%

25. The figure indicates the number of men had just dropped to 37,000 on November 9 when the temperature was 16°F (-9°C), and had just dropped to 24,000 on November 14 when the temperature was -6°F (-21°C). The number who died during that time, therefore, was 37,000 - 24,000 = 13,000.

27. The frequency polygon is given below.

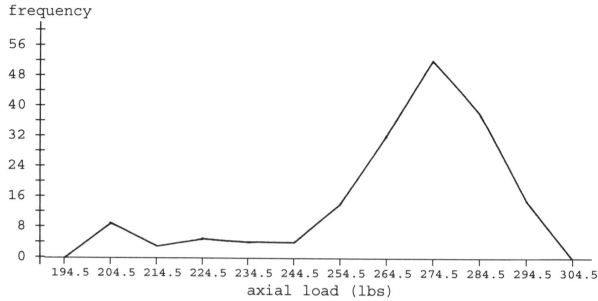

NOTE: The points at which the figure reaches a frequency of 0 are the midpoints of the classes below and and above the lowest and highest classes actually containing data.

29. Doubling the number of classes gives each class about half of its original frequency. If the original vertical scale is maintained, the figure will reach about half of its original height. If the original figure followed the guideline that the height be about three-fourths of the width, the new figure will be "short and fat" and may not give an accurate visual impression.

31. a. The final form of the back-to-back stem-and-leaf plot is given below. NOTE: This is another example of adapting a standard visual form in order to better communicate the data (see exercise #21 above). While such decisions are arbitrary, we choose to display "outward" from the central stem but to keep the actors' ages in increasing order from left to right.

actor's age		actress' age
	2	1466678
122235677899	3	001133444555778
0012233456788	4	111249
13566	5	0
012	6	011
6	7	4
	8	0

b. Female Oscar winners tend to be younger than male Oscar winners. If one assumes that acting ability doesn't peak differently for females and males, the data may reveal a difference in the standards by which females and males are judged.

2-4 Measures of Central Tendency

NOTE: As it is common in mathematics and statistics to use symbols instead of words to represent quantities that are used often and/or that may appear in equations, this manual employs symbols for the measures of central tendency as follows:

mean $= \bar{x}$ mode $= M$
median $= \tilde{x}$ midrange $= m.r.$

Also, this manual follows the author's guideline of presenting means, medians and ranges accurate to one more decimal place than found in the original data. The mode, the only measure which must be one of the original pieces of data, is presented with the same accuracy as the original data.

1. Arranged in order, the 20 scores are: 17 18 18 18 18 19 19 19 19 19
                                           20 20 20 20 20 20 20 20 21 21
    a. $\bar{x} = (\Sigma x)/n = (386)/20 = 19.3$ c. $M = 20$
    b. $\tilde{x} = (19 + 20)/2 = 19.5$ d. $m.r. = (17 + 21)/2 = 19.0$
    No, the steaks do not appear to weigh enough.

NOTE: The median is the middle score <u>when the scores are arranged in order</u>, and the midrange is halfway between the first and last score <u>when the scores are arranged in order</u>. It is therefore usually helpful to begin by placing the scores in order. This will not affect the mean, and it may also aid in identifying the mode. In addition, no measure of central tendency can have a value lower than the smallest score or higher than the largest score -- remembering this helps to protect against gross errors, which most commonly occur when calculating the mean.

3. Arranged in order, the 11 scores are: 4.66 5.21 5.24 5.41 5.53 5.78 6.00 6.40 6.80 6.96 8.23
    a. $\bar{x} = (\Sigma x)/n = (62.22)/11 = 6.020$ c. $M = $ [none]
    b. $\tilde{x} = 5.780$ d. $m.r. = (4.66 + 8.23)/2 = 6.445$

5. Arranged in order, the scores are as follows.
    JV: 6.5 6.6 6.7 6.8 7.1 7.3 7.4 7.7 7.7 7.7
    Pr: 4.2 5.4 5.8 6.2 6.7 7.7 7.7 8.5 9.3 10.0

<u>Jefferson Valley</u>   <u>Providence</u>
$n = 10$ $n = 10$
$\bar{x} = (\Sigma x)/n = (71.5)/10 = 7.15$ $\bar{x} = (\Sigma x)/n = (71.5)/10 = 7.15$
$\tilde{x} = (7.1 + 7.3)/2 = 7.20$ $\tilde{x} = (6.7 + 7.7)/2 = 7.20$
$M = 7.7$ $M = 7.7$
$m.r. = (6.5 + 7.7)/2 = 7.10$ $m.r. = (4.2 + 10.0)/2 = 7.10$

Comparing only measures of central tendency, one might suspect the two sets are identical. The Jefferson Valley times, however, are considerably less variable.

NOTE: This is the reason most banks have gone to the single waiting line. While it doesn't make service faster, it makes service times more equitable by eliminating the "luck of the draw" -- i.e., ending up by pure chance in a fast or slow line and having unusually short or long waits.

7. Arranged in order, the scores are as follows.
    BC: 119 125 126 126 128 128 129 131 131 131 132 138
    AD: 126 126 129 130 131 133 134 136 137 138 139 141

<u>4000 BC skulls</u>   <u>150 AD skulls</u>
$n = 12$ $n = 12$
$\bar{x} = (\Sigma x)/n = (1544)/12 = 128.7$ $\bar{x} = (\Sigma x)/n = (1600)/12 = 133.3$
$\tilde{x} = (128 + 129)/2 = 128.5$ $\tilde{x} = (133 + 134)/2 = 133.5$
$M = 131$ $M = 126$
$m.r. = (119 + 138)/2 = 128.5$ $m.r. = (126 + 141)/2 = 133.5$

Egyptian males of 150 AD had larger skulls than their counterparts of 4000 BC, as verified by 3 of the 4 measures of central tendency.

NOTE: This example illustrates the difficulty with using the mode to measure what is typical. When the observed values are approximately evenly spread out across all the scores and do not "bunch up" around a most popular score, the mode could be any score (no matter how unrepresentative) that just happened to occur twice.

9. Construct a stem-and-leaf plot to arrange the 38 scores in order.

```
96. | 2 6
97. | 0 2 3 4 4 4 4 5 6 7 8
98. | 0 0 0 0 1 2 2 2 2 4 4 6 7 7 8 8 8 8 9 9
99. | 0 0 0 2 4
```

a. $\bar{x} = (\Sigma x)/n = (3728.8)/38 = 98.13$
b. $\tilde{x} = (98.2 + 98.2)/2 = 98.20$
c. M = 97.4, 98.0, 98.2, 98.8 (multi-modal)
d. m.r. = (96.2 + 99.4)/2 = 97.80

11. Construct a stem-and-leaf plot to arrange the 54 scores in order.

```
0 | 26 29 34 40 46 48
0 | 60 62 64 65 76 79 80 86 90 94
1 | 05 14 16 20 25 32 40 40 44 48
1 | 50 50 54 66 66 80 82
2 | 02 02 04 04 12 20 20 36
2 | 62 70
3 | 16 32 44 48
3 | 56 60 65
4 | 16 36 46
4 |
5 | 14
```

a. $\bar{x} = (\Sigma x)/n = (9876)/54 = 182.9$
b. $\tilde{x} = (150 + 150)/2 = 150.0$
c. M = 140, 150, 166, 202, 204, 220
   (multi-modal, see NOTE)
d. m.r. = (26 + 514)/2 = 270

NOTE: There are six weights that occur twice, which is more than any other weight occurs. Strictly speaking, there are six modes and the set of weights is multi-modal. From a practical point of view, the double occurrences have no significance (e.g., would it make any real difference in the study if the two 140's had been 139 and 141?) and most researchers would either say there was no mode or group the data to find a modal class.

13. The x values below are the class midpoints from the given frequency table.

x	f	x·f
44.5	8	356.0
54.5	44	2398.0
64.5	23	1483.5
74.5	6	447.0
84.5	107	9041.5
94.5	11	1039.5
104.5	1	104.5
	200	14870.0

$\bar{x} = (\Sigma x \cdot f)/n$
$= (14870.0)/200$
$= 74.35$

NOTE: The mean time was calculated to be 74.35 minutes. According to the rule given in the text, this value should be rounded to one decimal place. The text describes how many decimal places to present in an answer, but not the actual rounding process. When the figure to be rounded is <u>exactly</u> half-way between two values (i.e., the digit in the position to be discarded is a 5, and there are no further digits because the calculations have "come out even"), there is no universally accepted rounding rule. Some authors say to always round up such a value; others correctly note that always rounding up introduces a consistent bias, and that the value should actually be rounded up half the time and rounded down half the time. And so some authors suggest rounding toward the even value (e.g., .65 becomes .6 and .75 becomes .8), while others simply suggest flipping a coin. In this manual, answers <u>exactly</u> half-way between will be reported without rounding (i.e., stated to one more decimal than usual).

15. The x values below are the class midpoints from the given frequency table.

x	f	x·f
42.5	14	595.0
44.5	11	489.5
46.5	8	372.0
48.5	6	291.0
50.5	4	202.0
52.5	3	157.5
54.5	1	54.5
56.5	2	113.0
58.5	0	0.0
60.5	1	60.5
	50	2335.0

$$\bar{x} = (\Sigma x \cdot f)/n$$
$$= (2335.0)/50$$
$$= 46.7$$

17.

x	w	w·f
60	.20	12.0
84	.20	16.8
90	.20	18.0
88	.40	35.2
	1.00	82.0

$$\bar{x} = (\Sigma w \cdot f)/\Sigma w$$
$$= (82.0)/1.00$$
$$= 82.0$$

19. a. Arranged in order, the 5 scores are:
    108,000 179,000 206,000 236,000 236,000
    $\bar{x} = (\Sigma x)/n = (965,000)/5 = 193,000$
    $\tilde{x} = 206,000$
    M = 236,000
    m.r. = (108,000 + 236,000)/2 = 172,000

    b. Arranged in order, the 5 scores would be:
    108,000 + k   179,000 + k   206,000 + k   236,000 + k   236,000 + k
    $\bar{x} = (\Sigma x)/n = (965,000 + 5k)/5 = 193,000 + k$
    $\tilde{x} = 206,000 + k$
    M = 236,000 + k
    m.r. = [(108,000 + k) + (236,000 + k)]/2 = [344,000 + 2k]/2 = 172,000 + k
    In general, adding (or subtracting) a constant k from each score will add (or subtract) k from each measure of central tendency.

    c. Arranged in order, the 5 scores would be:
    108,000·k   179,000·k   206,000·k   236,000·k   236,000·k
    $\bar{x} = (\Sigma x)/n = (965,000 \cdot k)/5 = 193,000 \cdot k$
    $\tilde{x} = 206,000 \cdot k$
    M = 236,000·k
    m.r. = (108,000·k + 236,000·k)/2 = (344,000·k)/2 = 172,000·k
    In general, dividing (or multiplying) each score by a constant k will divide (or multiply) each measure of central tendency by k.

    d. Arranged in order, the 5 scores would be:
    5.0334   5.2529   5.3139   5.3729   5.3729
    $\overline{(\log x)} = [\Sigma(\log x)]/n = [26.3460]/5 = 5.2692$, but $\log \bar{x} = \log (193,000) = 5.2856$
    In this case, the mean of the log x values does not equal the log of the mean of the x values.

    NOTE: In general, $\log \bar{x}$ does not equal $(\Sigma \log x)/n$. In words, the log of the mean is not the same as the mean of the logs. This is analogous to the more straightforward fact that the square of the average is not the same as the average of the squares and illustrates the principle that the order in which operations are applied can make a difference. Consider the following example, in which the log base 10 is used.

x	log x	x^2
1	0	1
10	1	100
100	2	10000
111	3	10101

$\bar{x} = (\Sigma x)/n = 111/3 = 37$

$(\Sigma \log x)/n = 3/3 = 1$, but $\log \bar{x} = \log 37 = 1.568$

$(\Sigma x^2)/n = 10101/3 = 3367$, but $\bar{x}^2 = 37^2 = 1369$

A similar statement is true for other measures involving mathematical operations (e.g., the median of an even number of scores and the midrange) -- i.e., the midrange of the logs is not the same as the log of the midrange. Measures involving only order or identification (e.g., the median of an odd number of scores and the mode), however, do preserve such direct relationships -- i.e., the mode of the logs is the same as the log of the mode.

21. The geometric mean of the five values is the fifth root of their product.

$$\sqrt[5]{(1.10)(1.08)(1.09)(1.12)(1.07)} = 1.092$$

23. a. The class mark of the last class is $(20.5 + 25.5)/2 = 23$

x	f	x·f
0	5	0
3	96	288
8	57	456
13	25	325
18	11	198
23	6	138
	200	1405

$\bar{x} = (\Sigma x \cdot f)/n$
$= (1405)/200$
$= 7.0$

b. The class mark of the last class is $(20.5 + 30.5)/2 = 25.5$

x	f	x·f
0.0	5	0.0
3.0	96	288.0
8.0	57	456.0
13.0	25	325.0
18.0	11	198.0
25.5	6	153.0
	200	1420.0

$\bar{x} = (\Sigma x \cdot f)/n$
$= (1420)/200$
$= 7.1$

c. The class mark of the last class is $(20.5 + 40.5)/2 = 30.5$

x	f	x·f
0.0	5	0.0
3.0	96	288.0
8.0	57	456.0
13.0	25	325.0
18.0	11	198.0
30.5	6	183.0
	200	1450.0

$\bar{x} = (\Sigma x \cdot f)/n$
$= (1450)/200$
$= 7.25$

In this case, because it involves relatively few scores and there is a reasonable upper limit (e.g, we know no student spends 100 hours a week studying) the interpretation of the open-ended class seems to make little difference and justifies the use of such classes when they make for significantly simpler presentations.

25. a. Arranged in order, the original 54 scores are:

```
 26   29   34   40   46   48   60   62   64   65   76   79   80   86   90   94
105  114  116  120  125  132  140  140  144  148  150  150  154  166  166  180
182  202  202  204  204  212  220  220  236  262  270  316  332  344  348  356
360  365  416  436  446  514
```

$\bar{x} = (\Sigma x)/n = (9876)/54 = 182.9$

b. After trimming the highest and lowest 10% (or 5.4 = 5 scores), the remaining 44 scores are:

```
48   60   62   64   65   76   79   80   86   90   94  105  114  116  120  125
132  140  140  144  148  150  150  154  166  166  180  182  202  202  204  204
212  220  220  236  262  270  316  332  344  348  356  360
```

$\bar{x} = (\Sigma x)/n = (7524)/44 = 171.0$

c. After trimming the highest and lowest 20% (or 10.8 = 11 scores), the remaining 32 scores are:

```
 79   80   86   90   94  105  114  116  120  125  132  140  140  144  148  150
150  154  166  166  180  182  202  202  204  204  212  220  220  236  262  270
```

$\bar{x} = (\Sigma x)/n = (5093)/32 = 159.2$

In this case, the mean gets smaller as more scores are trimmed. In general, means can increase, decrease, or stay the same as more scores are trimmed. The mean decreased here because the higher scores were farther from the original mean than were the lower scores.

2-5 Measures of Variation

NOTE: Although not given in the text, the symbol R will be used for the range throughout this manual. Remember that the range is the difference between the highest and the lowest scores, and not necessarily the difference between the last and the first values as they are listed. Since calculating the range involves only the subtraction of 2 original pieces of data, that measure of variation will be reported with the same accuracy as the original data.

1.

x	x - $\bar{x}$	$(x-\bar{x})^2$	x^2
17	-2.3	5.29	289
18	-1.3	1.69	324
18	-1.3	1.69	324
18	-1.3	1.69	324
18	-1.3	1.69	324
19	-0.3	0.09	361
19	-0.3	0.09	361
19	-0.3	0.09	361
19	-0.3	0.09	361
19	-0.3	0.09	361
20	0.7	0.49	400
20	0.7	0.49	400
20	0.7	0.49	400
20	0.7	0.49	400
20	0.7	0.49	400
20	0.7	0.49	400
20	0.7	0.49	400
20	0.7	0.49	400
21	1.7	2.89	441
21	1.7	2.89	441
386	0	22.20	7472

$\bar{x} = (\Sigma x)/n = 386/20 = 19.3$

$R = 21 - 17 = 4$

by formula 2-4,
$s^2 = \Sigma(x-\bar{x})^2/(n-1)$
$= 22.20/19$
$= 1.16842$
$= 1.2$

by formula 2-6,
$s^2 = [n(\Sigma x^2) - (\Sigma x)^2]/[n(n-1)]$
$= [20(7472) - (386)^2]/[20(19)]$
$= [444]/[380]$
$= 1.16842$
$= 1.2$

$s = \sqrt{1.16842} = 1.1$

NOTE: When finding the square root of the variance to obtain the standard deviation, use all the decimal places of the variance, and not the rounded value reported as the answer. The best way to do this is either to keep the value on the calculator display or to place it in the memory. Do not copy down all the decimal places and then re-enter them to find the square root, as that could introduce round-off and/or copying errors.

When using formula 2-4, constructing a table having the first three columns shown above helps to organize the calculations and makes errors less likely. In addition, verify that $\Sigma(x-\bar{x}) = 0$ before proceeding -- if such is not the case, there is an error and further calculation is fruitless. For

completeness, and as a check, both formulas 2-4 and 2-6 were used above. In general, only formula 2-6 will be used throughout the remainder of this manual for the following reasons:
   (1) When the mean does not "come out even," formula 2-4 involves round-off error and/or many messy decimal calculations.
   (2) The quantities Σx and Σx^2 needed for formula 2-6 can be found directly and conveniently on the calculator from the original data without having to construct a table like the one above.

3. preliminary values: $n = 11$, $\Sigma x = 66.22$, $\Sigma x^2 = 408.9492$
    $R = 8.23 - 4.66 = 3.57$
    $s^2 = [n(\Sigma x^2) - (\Sigma x)^2]/[n(n-1)]$
       $= [11(408.9492) - (66.22)^2]/[11(10)]$
       $= (113.3528)/110 = 1.030$
    $s = 1.015$
NOTE: The quantity $[n(\Sigma x^2) - (\Sigma x)^2]$ cannot be less than zero. A negative value indicates that there is an error and that further calculation is fruitless. In addition, remember to find the value for s by taking the square root of the precise value of s^2 showing on the calculator display before it is rounded to one more decimal place than the original data.

5. Jefferson Valley
    $n = 10$, $\Sigma x = 71.5$, $\Sigma x^2 = 513.27$
    $R = 7.7 - 6.5 = 1.2$
    $s^2 = [n(\Sigma x^2) - (\Sigma x)^2]/[n(n-1)]$
       $= [10(513.27) - (71.5)^2]/[10(9)]$
       $= 20.45/90 = 0.23$
    $s = 0.48$

Providence
    $n = 10$, $\Sigma x = 71.5$, $\Sigma x^2 = 541.09$
    $R = 10.0 - 4.2 = 5.8$
    $s^2 = [n(\Sigma x^2) - (\Sigma x)^2]/[n(n-1)]$
       $= [10(541.09) - (71.5)^2]/[10(9)]$
       $= 298.65/90 = 3.32$
    $s = 1.82$

Exercise #5 of section 2-4 indicated that the mean waiting time was 7.15 minutes at each bank. The Jefferson Valley waiting times, however, are considerably less variable. The range measures the difference between the extremes. The longest and shortest waits at Jefferson Valley differ by a little over 1 minute ($R=1.2$), while the longest and shortest waits at Providence differ by almost 6 minutes ($R=5.8$). The standard deviation measures the typical difference from the mean. A Jefferson Valley customer usually receives service within about ½ minute ($s=0.48$) of 7.15 minutes, while a Providence customer usually receives service within about 2 minutes ($s=1.82$) of the mean.

7. 4000 BC
    $n = 12$, $\Sigma x = 1544$, $\Sigma x^2 = 198898$
    $R = 138 - 119 = 19$
    $s^2 = [n(\Sigma x^2) - (\Sigma x)^2]/[n(n-1)]$
       $= [12(198898) - (1544)^2]/[12(11)]$
       $= 2840/132 = 21.5$
    $s = 4.6$

150 AD
    $n = 12$, $\Sigma x = 1600$, $\Sigma x^2 = 213610$
    $R = 141 - 126 = 15$
    $s^2 = [n(\Sigma x^2) - (\Sigma x)^2]/[n(n-1)]$
       $= [12(213610) - (1600)^2]/[12(11)]$
       $= 3320/132 = 25.2$
    $s = 5.0$

There is approximately the same amount of variability in skull breadths within each of the time periods.

9. preliminary values: $n = 38$, $\Sigma x = 3728.8$, $\Sigma x^2 = 365914.52$
    $s^2 = [n(\Sigma x^2) - (\Sigma x)^2]/[n(n-1)]$
       $= [38(365914.52) - (3728.8)^2]/[38(37)]$
       $= (802.32)/1406 = 0.5706$
    $s = 0.76$

11. preliminary values: $n = 54$, $\Sigma x = 9876$, $\Sigma x^2 = 2592494$
    $s^2 = [n(\Sigma x^2) - (\Sigma x)^2]/[n(n-1)]$
       $= [54(2592494) - (9876)^2]/[54(53)]$
       $= 42459300/2862 = 14835.53$
    $s = 121.8$

13.

x	f	f·x	f·x^2
44.5	8	356.0	15842.00
54.5	44	2398.0	130691.00
64.5	23	1483.5	95685.75
74.5	6	447.0	33301.50
84.5	107	9041.5	764006.75
94.5	11	1039.5	98232.75
104.5	1	104.5	10920.25
	200	14870.0	1148680.00

$s^2 = [n(\Sigma f \cdot x^2) - (\Sigma f \cdot x)^2]/[n(n-1)]$
$\quad = [200(1148680.00) - (1487.0)^2]/[200(199)]$
$\quad = (8619100.00)/29800 = 216.56$
$s = 14.7$

15.

x	f	f·x	f·x^2
42.5	14	595.0	25287.50
44.5	11	489.5	21782.75
46.5	8	372.0	17298.00
48.5	6	291.0	14113.50
50.5	4	202.0	10201.00
52.5	3	157.5	8268.75
54.5	1	54.5	2970.25
56.5	2	113.0	6384.50
58.5	0	0.0	0.00
60.5	1	60.5	3660.25
	50	2334.0	109966.50

$s^2 = [n(\Sigma f \cdot x^2) - (\Sigma f \cdot x)^2]/[n(n-1)]$
$\quad = [50(109966.50) - (2335.0)^2]/[50(49)]$
$\quad = (46100.00)/2450 = 18.82$
$s = 4.3$

17. If $\sigma = 1$ month, the typical battery life is within about one month of the mean; you know approximately when to expect failure and can plan accordingly. If $\sigma = 1$ year, the typical battery life is within about one year of the mean; you may be lucky and get one that lasts more than a year longer than you expect, or you may be unlucky and get one that fails more than a year before you expect. The battery from the lifetimes with the smaller standard deviation is to be preferred. Not only is planning easier, but $\sigma = 1$ month may indicate superior quality control that suggests the product is better in other ways -- e.g., a design with easier installation, recharging, etc.

19. This can be done without measuring the height of anyone in the class! Arrange the students in order by height to find the tallest and shortest students. Use a yardstick to find the range by measuring the distance from the top of the head of the shortest student to the top of the head of the tallest student. Use the estimate s $\approx$ (range)/4.

21. a. The limits 70 and 90 are 1 standard deviation from the mean. The empirical rule for bell-shaped data states that about 68% of the scores should fall within those limits.

b. A distance of 20 is 2 standard deviations from the mean. The empirical rule for bell-shaped data states that about 95% of the scores should fall within those limits.

c. The empirical rule for bell-shaped data states that about 99.7% of the scores should fall within 3 standard deviations of the mean. In this case, that would be within 3(10) = 30 of the mean of 80 -- i.e., from 50 to 110. NOTE: Always remember the units and limitations of the original story problem. If these test scores represent the percentage correct in a written test on typing techniques, then the upper limit is 100 and 50-100 is the correct response. In this case, there is no such stated restriction [the scores could be, for example, the number of words typed per minute] and the correct response is 50-110.

23. NOTE: Because it allows for better appreciation of the concepts involved, formula 2-4 and the following table format are employed for this exercise.

a.

x	x-$\overline{x}$	(x-$\overline{x}$)^2
108,000	-85,000	7,225,000,000
179,000	-14,000	196,000,000
206,000	13,000	169,000,000
236,000	43,000	1,849,000,000
236,000	43,000	1,849,000,000
965,000	0	11,288,000,000

$\overline{x}$ = (Σx)/n
= (965,000)/5
= 193,000

R = 236,000 - 108,000 = 128,000
By formula 2-4, $s^2 = \Sigma(x-\overline{x})^2/(n-1) = (11,288,000,000)/(4) = 2,822,000,000$
    and s = 53122.

NOTE: Assuming the original salaries were accurate to the nearest $1,000, the guideline in the text suggests reporting the standard deviation to the nearest $100 (i.e., to one more place value of accuracy). In this case, however, reporting to the nearest $1 seems more natural.

b.

x	x-$\overline{x}$	(x-$\overline{x}$)^2
108,000 + k	-85,000	7,225,000,000
179,000 + k	-14,000	196,000,000
206,000 + k	13,000	169,000,000
236,000 + k	43,000	1,849,000,000
236,000 + k	43,000	1,849,000,000
965,000 +5k	0	11,288,000,000

$\overline{x}$ = (Σx)/n
= (965,000 + 5k)/5
= 193,000 + k

R = (236,000 + k) - (108,000 + k) = 128,000
By formula 2-4, $s^2 = \Sigma(x-\overline{x})^2/(n-1) = (11,288,000,000)/(4) = 2,822,000,000$
    and s = 53122.
Adding a constant value k to each score does not affect the values of the measures of dispersion. In non-statistical terms, shifting everything by k units does not affect the spread of the scores.

c.

x	x-$\overline{x}$	(x-$\overline{x}$)^2
108,000·k	-85,000·k	7,225,000,000·k^2
179,000·k	-14,000·k	196,000,000·k^2
206,000·k	13,000·k	169,000,000·k^2
236,000·k	43,000·k	1,849,000,000·k^2
236,000·k	43,000·k	1,849,000,000·k^2
965,000·k	0	11,288,000,000·k^2

$\overline{x}$ = (Σx)/n
= (965,000·k)/5
= 193,000·k

R = 236,000·k - 108,000·k = 128,000·k
By formula 2-4, $s^2 = \Sigma(x-\overline{x})^2/(n-1) = (11,288,000,000 \cdot k^2)/(4) = 2,822,000,000 \cdot k^2$
    and s = 53122·k.
Multiplying each score by the value k multiplies both the range and the standard deviation by k. The variance, whose units are the square of the units in the problem, is multiplied by k^2.

d.

y	y-$\overline{y}$	(y-$\overline{y}$)^2
5.0334	-0.2358	.05560164
5.2529	-0.0163	.00026569
5.3139	0.0447	.00199809
5.3729	0.1037	.01075369
5.3729	0.1037	.01075369
26.3460	0	.07937280

$\overline{y}$ = (Σy)/n
= 26.3460/5
= 5.2692

NOTE: Let "y" stand for "log x"

By formula 2-4, $s^2 = \Sigma(y-\overline{y})^2/(n-1) = (.07937280)/(4) = 0.0198432$
    and s = 0.14087.
For the original scores, log s = log 53122 = 4.72527. Since 0.14807 ≠ 4.72527, the standard deviation of the log x values does not equal the log of the standard deviation of the x values.

e. Summarizing the results of this exercise [see also exercise #19 of section 2-4],
    Adding (or subtracting) a constant to each score will add (or subtract) that constant to the mean but will not change the standard deviation.
    Multiplying (or dividing) each score by a constant will multiply (or divide) both the mean and the standard deviation by that constant.

Mathematically, if $y = a \cdot x + b$
  then $\overline{y} = a \cdot \overline{x} + b$
  and $s_y = a \cdot s_x$

Applying this to C and F, the body temperatures in degrees Celsius and degrees Fahrenheit, where $C = 5(F-32)/9 = (5/9) \cdot F - 160/9$
  $\overline{C} = (5/9) \cdot \overline{F} - 32/9 = (5/9)(98.20) - 160/9 = 36.78°C$
  $s_C = (5/9) \cdot s_F = (5/9)(0.62) = 0.34°C$

25. section 1
  $n = 11$, $\Sigma x = 201$, $\Sigma x^2 = 4001$
  $R = 20 - 1 = 19$
  $s^2 = [n(\Sigma x^2) - (\Sigma x)^2]/[n(n-1)]$
    $= [11(4001) - (201)^2]/[11(10)]$
    $= 3610/110 = 32.82$
  $s = 5.7$

      section 2
      $n = 11$, $\Sigma x = 119$, $\Sigma x^2 = 1741$
      $R = 19 - 2 = 17$
      $s^2 = [n(\Sigma x^2) - (\Sigma x)^2]/[n(n-1)]$
        $= [11(1741) - (119)^2]/[11(10)]$
        $= 1990/110 = 45.36$
      $s = 6.7$

The range values give the impression that section 1 had more variability than section 2. The range can be misleading because it is based only on the extreme scores. In this case, the lowest score in section 1 was so distinctly different from the others that to include it in any measure trying to give a summary about the section as a whole would skew the results. For the mean, where the value is only one of 11 used in the calculation, the effect is minimal; for the range, where the value is one of only 2 used in the calculation, the effect is dramatic. The standard deviation values give the impression that section 2 had slightly more variability.
NOTE: In this case, section 2 seems considerably more variable (or diverse), and even the standard deviation by itself fails to accurately distinguish between the sections.

27. Given in section 2-4 are $\overline{x} = 267.1$ and $\tilde{x} = 273.0$. Given in section 2-5 is $s = 22.1$
  $I = 3(\overline{x} - \tilde{x})/s = 3(267.1 - 273.0)/22.1 = -17.7/22.1 = -.801$.
  Since $I < 0$, there is a negative skew; but since $-1 < I < 1$, there is <u>not</u> significant skewness.

2-6 Measures of Position

1. In general, $z = (x - \mu)/\sigma$.
  a. $z_{63} = (63 - 69.0)/2.8 = -2.14$
  b. $z_{85} = (85 - 69.0)/2.8 = 5.71$
  c. $z_{69.72} = (69.72 - 69.0)/2.8 = 0.26$

3. $z = (x - \mu)/\sigma$
  $z_{20.00} = (20.00 - 7.06)/5.32 = 2.43$

5. $z = (x - \mu)/\sigma$
  $z_{70} = (70 - 63.6)/2.5 = 2.56$
  Yes, the height is considered unusual since $2.56 > 2.00$.

7. $z = (x - \mu)/\sigma$
  $z_{5.50} = (5.50 - 5.67)/0.070 = -2.43$
  No, the quarter will not be accepted; it is considered unusual since $-2.43 < -2.00$.

9. In general $z = (x - \bar{x})/s$.
   a. $z_{50} = (60 - 50)/5 = 2.00$
   b. $z_{250} = (250 - 200)/20 = 2.50$
   The score in part b has the better relative position since $2.50 > 2.00$.

11. In general, $z = (x - \bar{x})/s$.
    a. $z_{37} = (37 - 28)/6 = 1.50$
    b. $z_{398} = (398 - 312)/56 = 1.54$
    c. $z_{4.10} = (4.10 - 2.75)/0.92 = 1.47$
    The score in part b has the highest relative position.

13. Let b = # of scores below; let n = total number of scores.
    In general, the percentile of score x is $(b/n) \cdot 100$.
    The percentile of score 254 is $(30/175) \cdot 100 = 17$.

15. Let b = # of scores below x; let n = total number of scores.
    In general, the percentile of score x is $(b/n) \cdot 100$.
    The percentile of score 277 is $(105/175) \cdot 100 = 60$.

17. To find P_{70}, $L = (70/100) \cdot 175 = 122.5$ rounded up to 123.
    Since the 123rd score is 279, $P_{70} = 279$.

19. To find $D_6 = P_{60}$, $L = (60/100) \cdot 175 = 105$ -- a whole number.
    The mean of the 105th and 106th scores, $P_{60} = (276 + 277)/2 = 276.5$.

21. To find $Q_3 = P_{75}$, $L = (75/100) \cdot 175 = 131.25$ rounded up to 132.
    Since the 132nd score is 282, $Q_3 = 282$.

23. To find $D_1 = P_{10}$, $L = (10/100) \cdot 175 = 18.5$ rounded up to 18.
    Since the 18th score is 230, $D_1 = 230$.

NOTE: For exercises 25-36, refer to the ordered cross-numbered chart below. A cross-numbered chart gives position in the list. The column head gives the tens digit and the row lead gives the ones digit, so that **270** is #43 in the ordered list of the 54 weights.

	0	1	2	3	4	5
0		65	120	166	220	365
1	26	76	125	166	236	416
2	29	79	132	180	262	436
3	34	80	140	182	**270**	446
4	40	86	140	202	316	514
5	45	90	144	202	332	
6	48	94	148	204	344	
7	60	105	150	204	348	
8	62	114	150	212	356	
9	64	116	154	220	360	

25. Let b = # of scores below x
          n = total number of scores
    In general, the percentile of score x is $(b/n) \cdot 100$.
    The percentile of score 144 is $(24/54) \cdot 100 = 44$.

27. Let b = # of scores below x
          n = total number of scores
    In general, the percentile of score x is $(b/n) \cdot 100$.
    The percentile of score 316 is $(43/54) \cdot 100 = 80$.

29. To find P_{85}, L = (85/100)·54 = 45.9 rounded up to 46.
Since the 46th score is 344, P_{85} = 344.

31. To find $Q_1 = P_{25}$, L = (25/100)·54 = 13.5 rounded up to 14.
Since the 14th score is 86, Q_1 = 86.

33. To find $D_9 = P_{90}$, L = (90/100)·54 = 48.6 rounded up to 49.
Since the 49th score is 360, D_9 = 360.

35. To find $\tilde{x} = P_{50}$, L = (50/100)·54 = 27 -- a whole number.
The mean of the 27th and 28th scores, P_{50} = (150 + 150)/2 = 150.0.

37. a. The interquartile range is $Q_3 - Q_1$.
    For $Q_3 = P_{75}$, L = (75/100)·175 = 131.25 rounded up to 132.
       Since the 132nd score is 282, Q_3 = 282.
    For $Q_1 = P_{25}$, L = (25/100)·175 = 43.75 rounded up to 44.
       Since the 44th score is 262, Q_1 = 262.
    The interquartile range is 282 - 262 = 20.
   b. The midquartile is $(Q_1 + Q_3)/2$ = (262 + 282)/2 = 272
   c. The 10-90 percentile range is $P_{90} - P_{10}$.
    For P_{90}, L = (90/100)·175 = 157.5 rounded up to 158.
       Since the 158th score is 289, P_{90} = 289.
    For P_{10}, L = (10/100)·106 = 17.5 rounded up to 18.
       Since the 18th score is 230, P_{10} = 230.
    The 10-90 percentile range is 289 - 230 = 59.
   d. Yes, $Q_2 = P_{50}$ by definition. They are always equal.
   e. For $Q_2 = P_{50}$, L = (50/100)·175 = 87.5 rounded up to 88.
       Since the 88th score is 273, Q_2 = 273.
    In this case 273 = $Q_2 \neq (Q_1 + Q_3)/2$ = 272, demonstrating that the median does not necessarily equal the midquartile.

39. Unusual values are those more than two standard deviations from the mean. In this case, that would be any values such that $x < \bar{x} - 2 \cdot s$ or $x > \bar{x} + 2 \cdot s$
$x < 267.1 - 2(22.1)$ or $x > 267.1 + 2(22.1)$
$x < 222.9$ or $x > 311.3$
NOTE: Of the 175 axial loads, 13 are considered unusally low and 0 are considered unusally large.

2-7 Exploratory Data Analysis

NOTE: The exercises in this section may be done much more easily when ordered lists of the values are available. On the next page appear ordered lists for the data of exercises #1-10 (including the even-numbered exercises!). The left-most column gives the ordered ID number for each value.

1. Consider the 29 nicotine values.
    For $Q_1 = P_{25}$, L = (25/100)·29 = 7.25 rounded up to 8.
    For $\bar{x} = Q_2 = P_{50}$, L = (50/100)·29 = 14.5 rounded up to 15.
    For $Q_3 = P_{75}$, L = (75/100)·29 = 21.75 rounded up to 22.
    min = x_1 = 0.1
    $Q_1 = x_8 = 0.8$
    $Q_2 = x_{15} = 1.0$
    $Q_3 = x_{22} = 1.2$
    max = x_{29} = 1.4

0.1   0.8 1.0 1.2 1.4
                                 nicotine (mg)

The following ordered lists are used for exercises #1-10.

	#1 nic	#2 tar	#3 mal	#3 fem	#4 smo	#4 non	#5 mal	#5 fem	#6 rrr	#6 oth	#7 red	#7 yel	#8 qua	#9* 333		#10 pap	#10 pla
01	0.1	1	31	21	52	40	40	58	72	88	.870	.868	5.49	x	cf	1.65	0.15
02	0.2	2	32	24	52	47	47	60	82	90	.872	.876	5.52	0	13	2.41	0.27
03	0.5	5	32	26	60	48	48	60	90	91	.874	.877	5.53	1	29	2.61	0.38
04	0.6	6	32	26	60	54	52	60	92	92	.882	.879	5.53	2	40	2.80	0.58
05	0.7	8	33	26	60	55	52	65	93	93	.888	.879	5.53	3	57	3.36	0.63
06	0.7	8	35	17	60	55	54	65	95	94	.891	.886	5.56	4	71	3.27	0.65
07	0.8	9	36	28	63	58	55	67	96	94	.897	.886	5.57	5	89	3.45	0.72
08	0.8	9	37	30	63	60	55	67	96	94	.898	.892	5.57	6	107	3.69	0.74
09	0.8	9	37	30	66	60	60	67	97	96	.908	.893	5.57	7	125	5.86	0.80
10	0.8	10	38	31	67	60	60	68	98	98	.908	.900	5.57	8	137	5.87	0.85
11	0.9	11	39	31	68	60	60	68	100	99	.908	.096	5.57	9	150	5.88	0.89
12	1.0	12	39	33	69	61	60	69	101	100	.911	.910	5.58			6.05	0.92
13	1.0	13	40	33	71	63	60	70	102	100	.912	.911	5.58			6.16	0.92
14	1.0	13	40	34	72	63	61	72	104	103	.913	.917	5.58			6.33	0.93
15	1.0	13	41	34	73	63	63	72	104	104	.920	.921	5.58			6.38	1.14
16	1.0	14	42	34	75	64	63	73	104	105	.924	.924	5.58			6.44	1.15
17	1.0	14	42	35	78	64	63	75	105	108	.924	.926	5.59			6.67	1.28
18	1.0	15	43	35	80	64	63	76	106	108	.933	.934	5.59			6.83	1.36
19	1.1	15	43	35	82	65	63	77	106	111	.936	.939	5.59			6.96	1.40
20	1.1	15	44	37	83	64	64	78	107	121	.952	.940	5.59			6.98	1.41
21	1.1	16	45	37	88	66	64	78	110	123	.983	.941	5.60			7.57	1.41
22	1.2	16	46	38	90	67	64	78	111	123		.949	5.60			7.72	1.44
23	1.2	16	47	41		67	66	80	114	144		.960	5.60			7.98	1.44
24	1.2	16	48	41		67	66	80	115	168		.968	5.60			8.08	1.45
25	1.2	16	48	41		67	67	80	117	193		.978	5.60			8.26	1.45
26	1.2	16	51	42		68	67	80	119			.989	5.60			8.72	1.48
27	1.3	17	53	44		68	68	80	120				5.60			8.78	1.49
28	1.4	18	55	49		69	69	82	125				5.61			8.82	1.53
29	1.4	18	56	50		69	69	83	129				5.62			8.96	1.53
30			56	60		69	69	85	133				5.62			9.09	1.58
31			60	61		70	70	86	134				5.62			9.19	1.61
32			61	61		70	70	88	139				5.63			9.41	1.68
33			62	74		70	70	88	155				5.63			9.45	1.74
34			76	80		70	70	88	159				5.63			9.46	1.81
35						70	71	90	160				5.65			9.55	2.00
36						71	71	92					5.66			9.64	2.04
37						72	72	97					5.66			9.83	2.10
38						72	72	100					5.66			9.92	2.13
39						72	72						5.67			10.00	2.13
40						72	72						5.67			10.58	2.17
41						72	75						5.68			10.99	2.19
42						75	75						5.70			11.03	2.19
43						75	76						5.71			11.08	2.30
44						76	77						5.71			11.36	2.31
45						76	78						5.72			11.42	2.35
46						77	78						5.73			12.29	2.44
47						77	78						5.73			12.32	2.66
48						78	80						5.73			12.43	2.68
49						78	80						5.74			12.45	2.83
50						78	81						5.84			12.56	2.83
51						78	83									12.73	2.87
52						78	84									13.05	2.88
53						80	90									13.11	2.93
54						80	92									13.31	2.96
55						80										13.61	2.97
56						80										14.33	3.05
57						80										15.09	3.36
58						80										16.08	3.42
59						81										16.39	3.53
60						83										17.65	4.37
61						84										20.12	4.69
62						85										20.58	5.28
63						86											
64						88											
65						88											
66						90											
67						92											
68						92											
69						97											
70						100											

*cumulative frequency

3. Consider the 34 actor and 34 actress values.
   For $Q_1 = P_{25}$, $L = (25/100) \cdot 34 = 8.5$ rounded up to 9.
   For $\tilde{x} = Q_2 = P_{50}$, $L = (50/100) \cdot 34 = 17$ -- an integer, use 17.5.
   For $Q_3 = P_{75}$, $L = (75/100) \cdot 29 = 25.5$ rounded up to 26.

   For the actors
   min $= x_1 = 31$
   $Q_1 = x_9 = 37$
   $Q_2 = x_{17.5} = (42 + 43)/2 = 42.5$
   $Q_3 = x_{26} = 51$
   max $= x_{34} = 76$

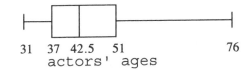

31 37 42.5 51   76
        actors' ages

   For the actresses
   min $= x_1 = 21$
   $Q_1 = x_9 = 30$
   $Q_2 = x_{17.5} = (35 + 35)/2 = 35$
   $Q_3 = x_{26} = 42$
   max $= x_{34} = 80$

21   30 35 42   80
        actress' ages

The ages for the actresses cover a wider range and cluster around a lower value than do the ages of the actors.

NOTE for exercise #5 and all future uses of this data set: The values 8 and 15 are obvious errors that have been eliminated from this and all subsequent analyses. Such obvious errors occur in many real life data sets. Often correct values can be deduced and the data adjusted accordingly. If, for example, all the other values were multiples of four, one could infer that students monitored their pulse for 15 seconds and multiplied by 4 to obtain a per minute rate. In that case the 15 should be 60. The 8 (which could not be a 32) and the non-multiples of 4, however, eliminate that possibility. In this case, eliminate the values 8 and 15. Also be aware that the values in the 40's are highly suspect. It appears that the instructor gathered data by having each student monitor his own pulse rate -- and that there was not careful instruction or a re-take of suspicious values. It also appears that 6 students couldn't find their pulse -- and those places were just left blank. The values in the 40's will be included here and in all subsequent analyses of the pulse data, but the numbers are suspect and should not be taken as accurate renderings of student pulse rates.

5. Consider the 54 values for the males.
   For $Q_1 = P_{25}$, $L = (25/100) \cdot 54 = 13.5$ rounded up to 14.
   For $\tilde{x} = Q_2 = P_{50}$, $L = (50/100) \cdot 54 = 27$ -- an integer, use 27.5.
   For $Q_3 = P_{75}$, $L = (75/100) \cdot 54 = 40.5$ rounded up to 41.

   min $= x_1 = 40$
   $Q_1 = x_{14} = 61$
   $Q_2 = x_{27.5} = (68 + 69)/2 = 68.5$
   $Q_3 = x_{41} = 75$
   max $= x_{54} = 92$

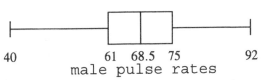

40   61 68.5 75   92
        male pulse rates

   Consider the 38 values for the females.
   For $Q_1 = P_{25}$, $L = (25/100) \cdot 38 = 9.5$ rounded up to 10.
   For $\tilde{x} = Q_2 = P_{50}$, $L = (50/100) \cdot 38 = 19$ -- an integer, use 19.5.
   For $Q_3 = P_{75}$, $L = (75/100) \cdot 38 = 28.5$ rounded up to 29.

   min $= x_1 = 58$
   $Q_1 = x_{10} = 68$
   $Q_2 = x_{19.5} = (77 + 78)/2 = 77.5$
   $Q_3 = x_{29} = 83$
   max $= x_{38} = 100$

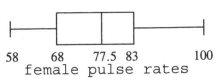

58   68   77.5 83   100
        female pulse rates

The female pulse rates appear to be higher than those for the males.

7. Consider the 21 red M&M's.
   For $Q_1 = P_{25}$, $L = (25/100) \cdot 21 = 5.25$ rounded up to 6.
   For $\tilde{x} = Q_2 = P_{50}$, $L = (50/100) \cdot 21 = 10.5$ rounded up to 11.
   For $Q_3 = P_{75}$, $L = (75/100) \cdot 21 = 15.75$ rounded up to 16.
   min $= x_1 = .870$
   $Q_1 = x_6 = .891$
   $Q_2 = x_{11} = .908$
   $Q_3 = x_{16} = .924$
   max $= x_{21} = .983$

   .870   .891 .908 .924   .983
   weights of red M&M's

   Consider the 26 yellow M&M's.
   For $Q_1 = P_{25}$, $L = (25/100) \cdot 26 = 6.5$ rounded up to 7.
   For $\tilde{x} = Q_2 = P_{50}$, $L = (50/100) \cdot 26 = 13$ -- an integer, use 13.5.
   For $Q_3 = P_{75}$, $L = (75/100) \cdot 26 = 19.5$ rounded up to 20.
   min $= x_1 = .868$
   $Q_1 = x_7 = .886$
   $Q_2 = x_{13.5} = (.911 + .917)/2 = .914$
   $Q_3 = x_{20} = .940$
   max $= x_{26} = .989$

   .868   .886   .914   .940   .989
   weights of yellow M&M's

   The weights of the red M&M's are less variable than the weights of the yellow ones.

9. Consider the 150 Pick Three lottery digits.
   For $Q_1 = P_{25}$, $L = (25/100) \cdot 150 = 37.5$ rounded up to 38.
   For $\tilde{x} = Q_2 = P_{50}$, $L = (50/100) \cdot 150 = 75$ -- an integer, use 75.5.
   For $Q_3 = P_{75}$, $L = (75/100) \cdot 150 = 112.5$ rounded up to 113.
   min $= x_1 = 0$
   $Q_1 = x_{38} = 2$
   $Q_2 = x_{75.5} = 5$
   $Q_3 = x_{113} = 7$
   max $= x_{150} = 9$

   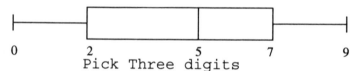

   0   2   5   7   9
   Pick Three digits

   The distribution of the Pick Three digits is approximately uniform.

11. Assuming that cost and all other relevant factors are approximately equal, one should select the brand corresponding to the first boxplot. Its lifetimes are consistently (low variability) long-lasting (high mean).

Review Exercises

1. Subtracting the smallest value from the largest value and dividing by the desired number of classes determines that the class width must be at least $(1128 - 235)/9 = 99.22$. Rounding up to a convenient number, we choose 100. The exercise specifies that the first lower class limit is to be 235, the smallest data value. NOTE: In practice, the first lower class limit and the class width may be any convenient values. Since there are 60 seconds in a minute, for example, it might be reasonable to start at 180 or 210 and use a class width of 90 or 120.

time (seconds)	frequency
235 - 334	4
335 - 434	9
435 - 534	11
535 - 634	9
635 - 734	9
735 - 834	6
835 - 934	8
935 - 1034	2
1035 - 1134	2
	60

2. Obtain the relative frequencies by dividing each frequency by 60, the total number of cases.

time (seconds)	relative frequency
235 - 334	.067
335 - 434	.150
435 - 534	.183
535 - 634	.150
635 - 734	.150
735 - 834	.100
835 - 934	.133
935 - 1034	.033
1035 - 1134	.033
	1.000

3. frequency

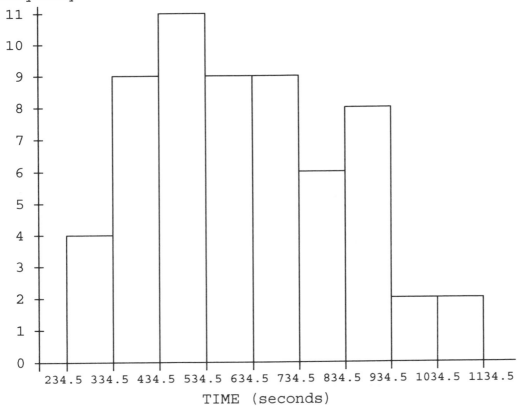

TIME (seconds)

4. The scores are given in order in the final stem-and-leaf plot of exercise #11.
   a. To find $Q_1 = P_{25}$, $L = (25/100) \cdot 60 = 15$ -- a whole number.
      The mean of the 15th and 16th scores, $P_{25} = (447 + 448)/2 = 447.5$.
   b. To find P_{45}, $L = (45/100) \cdot 60 = 27$ -- a whole number.
      The mean of the 27th and 28th scores, $P_{45} = (564 + 587)/2 = 575.5$.
   c. Let b = # of scores below x
        n = total number of scores
      In general, the percentile of score x is $(b/n) \cdot 100$.
      The percentile of score 375 is $(4/60) \cdot 100 = 7$.

5. According to the range rule of thumb, the standard deviation is usually about 1/4 of the range -- i.e., we estimate s to be about $(1/4)(893) = 223.25$.

6.

x	f	f·x	f·x^2
284.5	4	1138.0	323761.00
384.5	9	3460.5	1330562.25
484.5	11	5329.5	2582142.75
584.5	9	5260.5	3074762.25
684.5	9	6160.5	4216862.25
784.5	6	4707.0	3692641.50
884.5	8	7076.0	6258722.00
984.5	2	1969.0	1938480.50
1084.5	2	2169.0	2352280.50
	60	37270.0	25770215.00

$\bar{x} = (\Sigma f \cdot x)/n = 37270.0/60 = 621.2$
$s^2 = [n(\Sigma f \cdot x^2) - (\Sigma f \cdot x)^2]/[n(n-1)]$
$\quad = [60(25770215.00) - (37270.0)^2]/[60(59)]$
$\quad = (157160000)/3540$
$\quad = 44395.5$
$s = 210.7$

7. Use the hundred's digits for the stem and the last two digits -- with spaces between values -- for the leaves.

first pass, by columns

```
 2 | 92 40 35
 3 | 25 37 35 78 63 96 96 45
 4 | 48 43 04 57 47 95 94 20 74 83
 5 | 87 14 06 40 03 64 52
 6 | 26 70 15 88 09 25 76 70 93 66 27
 7 | 56 23 00 94 93 04 78 48
 8 | 61 71 20 53 52 60 62
 9 | 91 29 15
10 | 23 70
11 | 28
```

final form

```
 2 | 35 40 92
 3 | 25 35 37 45 63 78 96 96
 4 | 04 20 43 47 48 57 74 83 94 95
 5 | 03 06 14 40 52 64 87
 6 | 09 15 25 26 27 66 70 70 76 88 93
 7 | 00 04 23 48 56 78 93 94
 8 | 20 52 53 60 61 62 71
 9 | 15 29 91
10 | 23 70
11 | 28
```

8. The scores are given in order in the final stem-and-leaf plot of exercise #7.
For $Q_1 = P_{25}$, $L = (25/100) \cdot 60 = 15$ -- an integer, use 15.5.
For $\tilde{x} = Q_2 = P_{50}$, $L = (50/100) \cdot 60 = 30$ -- an integer, use 30.5.
For $Q_3 = P_{75}$, $L = (75/100) \cdot 60 = 45$ -- an integer, use 45.5.
$min = x_1 = 235$
$Q_1 = x_{15.5} = (447 + 448)/2 = 447.5$
$Q_2 = x_{30.5} = (615 + 625)/2 = 620$
$Q_3 = x_{45.5} = (778 + 793)/2 = 785.5$
$max = x_{60} = 1128$

```
   235         447.5        620       785.5                    1128
                      TIME (seconds)
```

9. The scores arranged in order are: 69 83 85 87 90 121 135 135 159 177 227
preliminary values: n = 11, $\Sigma x = 1368$, $\Sigma x^2 = 193,774$
a. $\bar{x} = (\Sigma x)/n = (1368)/11 = 124.4$
b. $\tilde{x} = 121.0$
c. M = 135
d. m.r. = (69 + 227)/2 = 148.0
e. R = 227 - 69 = 158
f. s = 48.6 (from part g)
g. $s^2 = [n(\Sigma x^2) - (\Sigma x)^2]/[n(n-1)]$
$\quad = [11(193,774) - (1368)^2]/[11(10)]$
$\quad = (260,090)/110 = 2364.5$

10. The scores arranged in order are:
    42 43 46 46 47 48 49 49 50 51 51 51 51 51 52 52 54 54 54 54 55
    55 55 55 56 56 56 57 57 57 57 58 60 61 61 61 62 64 64 65 68 69
    preliminary values: $n = 42$, $\Sigma x = 2304$, $\Sigma x^2 = 128,014$
    a. $\bar{x} = (\Sigma x)/n = (2304)/42 = 54.9$
    b. $\tilde{x} = (55 + 55)/2 = 55.0$
    c. $M = 51$
    d. m.r. $= (42 + 69)/2 = 55.5$
    e. $R = 69 - 42 = 27$
    f. $s = 6.3$ (from part g)
    g. $s^2 = [n(\Sigma x^2) - (\Sigma x)^2]/[n(n-1)]$
       $= [42(128,014) - (2304)^2]/[42(41)]$
       $= (68,172)/1722 = 39.6$
    h. For $Q_1 = P_{25}$, $L = (25/100) \cdot 42 = 10.5$ rounded up to 11. And so $Q_1 = x_{11} = 51$.
    i. For P_{30}, $L = (30/100) \cdot 42 = 12.6$ rounded up to 13. And so $P_{30} = x_{13} = 51$.
    j. For $D_7 = P_{70}$, $L = (70/100) \cdot 42 = 29.4$ rounded up to 30. And so $D_7 = x_{30} = 57$.

11. In general, $z = (x - \mu)/\sigma$. Here, $\mu = 200$ and $\sigma = 40$.
    a. $z_{260} = (260 - 200)/40 = 60/40 = 1.50$. No, this score is not unusually high because it is only 1.50 standard deviations away from the mean. A score is considered unusual if it is more than 2 standard deviations away from the mean (i.e., if $z < -2.00$ or $z > 2.00$).
    b. $z_{185} = (185 - 200)/40 = -15/40 = -0.375$.
    c. About 95%. The empirical rule for bell-shaped distributions says that about 95% of the scores are within 2 standard deviations of the mean, and 120 & 280 are $\mu \pm 2 \cdot \sigma$.
    d. 220. See exercise #19b of section 2-4.
    e. 40. See exercise #23b of section 2-5.

12. Use the class midpoint of 9.5 as the x value for the last class.

x	f	$f \cdot x$	$f \cdot x^2$
4	147	588	2352
5	81	405	2025
6	27	162	972
7	15	105	735
9.5	30	285	2707.5
	300	1545	8791.5

NOTE: $n = \Sigma f = 300$

$\bar{x} = (\Sigma f \cdot x)/n = 1545/300 = 5.15$
$s^2 = [n(\Sigma f \cdot x^2) - (\Sigma f \cdot x)^2]/[n(n-1)]$
$= [300(8791.5) - (1545)^2]/[300(299)]$
$= (250425)/89700 = 2.79$
$s = 1.67$

No, $z_{8.00} = (8.00 - 5.15)/1.67 = 1.71 \leq 2.00$. Since 8 is not more than 2 standard deviations above the mean, it is not considered unusual for an undergraduate to require 8 years to earn a bachelor's degree.

13. The relative frequency for each class is obtained by dividing that class' frequency by
    $n = \Sigma f = 300$.

    ```
    class          relative
    boundaries     frequency
    3.5 -   4.5       .49
    4.5 -   5.5       .27
    5.5 -   6.5       .09
    6.5 -   7.5       .05
    7.5 -  11.5       .10
                     1.00
    ```

    The relative frequency histogram below is constructed from the above relative frequency
    distribution.

    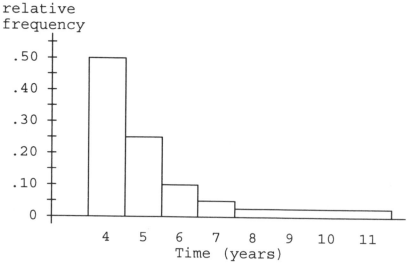

    NOTE: The purpose of a histogram is to show the relative amounts in each class using areas of
    bars -- i.e., the proportion of the total area in each bar should represent the proportion of the
    sample in that class. Since the last class is 4 times as wide as the other classes, its bar must be
    made 1/4 of its normal height in order to preserve proper proportions. The area of the last bar
    should be about equal to the area of the 5.5-6.5 bar and twice the area of the 6.5-7.5 bar; this
    conveys the information that 10% of the data were spread over the years 8,9,10,11 -- assumed
    to be about 2.5% of the data for each of those years.

14. In general $z = (x - \mu)/\sigma$.
    For the first test, $\mu = 72$ and $\sigma = 20$. $z_{57} = (57-72)/20 = -15/20 = -0.75$.
    For the second test, $\mu = 500$ and $\sigma = 80$. $z_{450} = (450 - 500)/80 = -50/80 = -0.625$.
    Assuming that a higher score is better, the score on the second test is better
       because $-0.625 > -0.75$ indicates that it has a higher relative position.

15. Yes, there is a shift in the maximum skull breadth. While both groups have about the same
    variability, the widths are higher for the 150 AD group; the mean for the 150 AD widths is
    higher than the 75th percentile for the 4000 BC widths.

16. Arranging the categories in decreasing order by frequency produces the following figure.

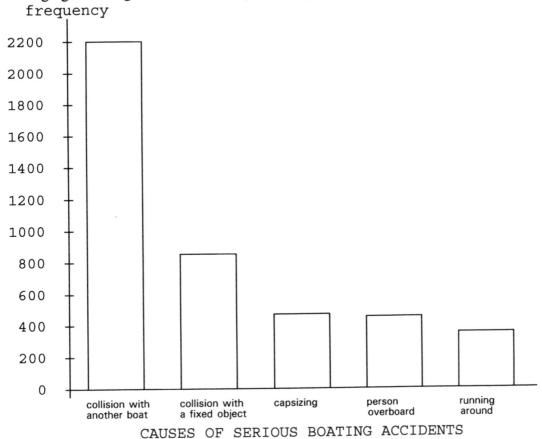

CAUSES OF SERIOUS BOATING ACCIDENTS

Cumulative Review Exercises

1. The scores arranged in order are:
   0.0 0.0 1.0 1.5 1.7 1.8 2.0 2.0 2.1 2.1 2.3 2.4 2.4 2.9 3.3 3.4 3.7 4.4 4.4 4.5
   preliminary values: n = 20, Σx = 47.9, Σx^2 = 146.53
   a. $\bar{x}$ = $(\Sigma x)/n$ = (47.9)/20 = 2.395   M = 0.0, 2.0, 2.1, 2.4, 4.4 (multimodal)
      $\tilde{x}$ = (2.1 + 2.3)/2 = 2.20   m.r. = (0.0 + 4.5)/2 = 2.25
   b. R = 4.5 - 0 = 4.5
      s^2 = $[n(\Sigma x^2) - (\Sigma x)^2]/[n(n-1)]$
         = $[20(146.53) - (47.9)^2]/[20(19)]$
         = (636.19)/380 = 1.67
      s = 1.29
   c. continuous. Time can be any non-negative value on a continuum
   d. ratio. Differences are consistent and there is a meaningful zero; 2.0 hours is twice as much time spent as 1.0 hours.

2. a. mode. The median requires at least ordinal level data, and the mean and the midrange require at least interval level data.
   b. convenience. The group was not selected by any process other than the fact that they happened to be the first names on the list.
   c. cluster. The population was divided into units (election precincts), some of which were selected at random to be examined in their entirety.

3. No, use the state populations as weights to find the weighted mean.

Chapter 3

Probability

3-2 Fundamentals

1. Since $0 \leq P(A) \leq 1$ is always true, the following values less than 0 or greater than 1 cannot be probabilities.
   values less than 0: -0.2
   values greater than 1: 3/2 $\sqrt{2}$

3. Let H = getting a head when a balanced coin is tossed.
   H is one of 2 equally likely different outcomes (head or tail); use Rule 2. *#ways H can occur / #diff. simple events*
   P(H) = 1/2

5. Let R = a plain M&M candy is red.
   R occurred 21 times in a sample of 100; use Rule 1. *#times R occurred / #times experiment was repeated*
   P(R) ≈ 21/100 = .21

7. Let A = an American Airline flight arrives on time.
   A occurred 430 times in a sample of 500; use Rule 1.
   P(A) ≈ 430/500 = .86

   Yes; considering weather and other factors beyond the airline's control, an 86% on time arrival rate is good.

   NOTE: On time arrival rate by itself does not describe an airline's record. By lengthening the projected flight times, the airline could achieve a higher on time arrival rate but would have planes creating problems by arriving "too early" for their unrealistic (and inaccurate) scheduled arrival times.

9. There were 1162 + 2468 = 3630 students surveyed.
   Let C = the selected student cheated.
   C includes 1162 of 3630 equally likely outcomes; use Rule 2.
   P(C) = 1162/3630 = .320

11. There were 225 + 275 = 500 donors surveyed.
    Let O = a person has type O blood.
    A occurred 225 times in a sample of 500; use Rule 1.
    P(O) ≈ 225/500 = .45

13. a. Let B = a person's birthday is October 18.
       B is one of 365 (assumed) equally likely outcomes; use Rule 2.
       P(B) = 1/365 = .00274
    b. Let N = a person's birthday is in November.
       N includes 30 of 365 (assumed) equally likely outcomes; use Rule 2.
       P(N) = 30/365 = .0822

15. There were 132 + 880 = 1012 respondents.
    Let D = a respondent gave the doorstop answer.
    D includes 132 of 1012 equally likely outcomes; use Rule 2.
    P(D) = 132/1012 = .130

17. Let A = a driver in that age bracket has an accident.
    A occurred 136 times in a sample of 400; use Rule 1.
    P(A) ≈ 136/400 = .340
    Yes; since this indicates more than 1 such driver in every 3 is likely to have an accident, this
    should be a concern.
    NOTE: This is properly an estimate that such a person <u>had</u> an accident last year, and not that
    such a person <u>will have</u> an accident next year. To extend the estimate as desired requires the
    additional assumption that relevant factors (speed limits, weather, alcohol laws, the economy,
    etc.) remain the same.

19. There were 70 + 711 = 781 people tested.
    Let D = a user experiences drowsiness.
    D occurred 70 times in a sample of 781; use Rule 1.
    P(D) ≈ 70/781 = .0896
    This factor does not appear worthy of major consideration by Seldane users, except perhaps by
    those who are routinely made drowsy by drugs.
    NOTE: There is no indication that a control group was used. It could be that about 10% of the
    population (especially those with health problems requiring intervention) get drowsy during the
    day even if they take <u>no</u> medication.

21. There were 243 + 85 + 52 + 46 = 426 cases of credit card fraud examined.
    Let C = the case involved a counterfeit card.
    C includes 85 of the 426 equally likely outcomes; use Rule 2.
    P(C) = 85/426 = .200

23. a. The 4 equally likely outcomes are: BB BG GB GG.
    b. P(2 girls) = 1/4 = .25
    c. P(1 of each sex) = 2/4 = .5

25. a. The 8 equally likely outcomes are at the right.   CCC   CNN
        Use Rule 2 and let C = correct   CCN   NCN
                   let N = not correct   CNC   NNC
    b. P(3 correct) = 1/8 = .125   [CCC outcome]   NCC   NNN
    c. P(0 correct) = 1/8 = .125   [NNN outcome]
    d. P(at least 2 correct) = 4/8 = .5 [column 1 outcomes]

27. Let C = guessing the correct answer. Then P(C) = 1/5; P(C̄) = 4/5.
    odds aggainst C = P(C̄)/P(C) = (4/5)/(1/5) = 4/1 or 4:1

29. Let R = getting a 7 in roulette. Then P(R) = 1/38; P(R̄) = 37/38.
    a. odds against R = P(R̄)/P(R) = (37/38)/(1/38) = 37/1 or 37:1
    b. net profit = 72 - 2 = 70
       odds used for payoff = 70/2 = 35/1 or 35:1
    c. The discrepancy between the theoretical odds and the odds used for the payoff gives the
       casino its profit. Without that discrepancy, the casino and the player expect to break even,
       and the casino has no profit from which to pay its employees, expenses, investors, etc.

31. If the odds against A are a:b, then P(A) = b/(a+b).
    Let H = Horse Cents wins the next race.
    If the odds against H are 10:3, then a=10 and b=3 in the statement above.
    P(H) = 3/(3+10) = 3/13 = .231

33. There were 20 managerial responses summarized in the stem-and-leaf plot.
Let M = a manager spends more than 2.0 hours per day on paperwork.
M occurred 12 times in a sample of 20; use Rule 1.
$P(M) \approx 12/20 = .6$

35. Let B = a person's birthday is October 18; use Rule 2.
   a. In one repeating 4 year cycle there are $3 \cdot (365) + 1 \cdot (366) = 1461$ days.
      In one repeating 4 year cycle there are 4 October 18's.
      $P(B) = 4/1461$
   b. In one repeating 400 year cycle there are $303 \cdot (365) + 97 \cdot (366) = 146,097$ days.
      In one repeating 400 year cycle there are 400 October 18's.
      $P(B) = 400/146,097$
   NOTE: Comparing the answers as decimals -- #13a, .00273973
   #35a, .00273785 (a slight decrease from #13a)
   #35b, .00273791 (a slight increase from #35a)

3-3 Addition Rule

1. a. No, it's possible for a person to be a male TV viewer and rarely use the remote.
   b. No, it's possible for a person to be a registered Democrat and oppose all welfare plans.
   c. Yes, if the outcome is an even number then it cannot be a 7.

3. a. $P(\bar{A}) = 1 - P(A)$
      $= 1 - 2/5$
      $= 3/5$
      $= .6$
   b. $P(girl) = P(\overline{boy})$
      $= 1 - P(boy)$
      $= 1 - .513$
      $= .487$

5. a. $P(Club\ or\ Ace) = P(Club) + P(Ace) - P(Club\ and\ Ace)$
      $= 13/52 + 4/52 - 1/52$
      $= 16/52$
      $= .308$
   b. $P(Ace\ or\ Two) = P(Ace) + P(Two) - P(Ace\ and\ Two)$
      $= 4/52 + 4/52 - 0/52$
      $= 8/52$
      $= .154$

7. Refer to the table at the right.
   let S = a person used Seldane
   P = a person used a placebo
   $P(S\ or\ P) = P(S) + P(P) - P(S\ and\ P)$
   $= 781/2072 + 665/2072 - 0/2072$
   $= 1446/2072$
   $= .698$

		HEADACHE?		
		Yes	No	
	Seldane	49	732	781
DRUG	Placebo	49	616	665
	Control	24	602	626
		122	1950	2072

9. Make a chart like the one on the right.
   let A = a person is 18-21
   N = a person does not respond
   There are two correct approaches.

		RESPOND?		
		Yes	No	
AGE	18-21	73	11	84
	22-29	255	20	275
		328	31	359

* Use broad categories and allow for double-counting (i.e., the "formal addition rule").

$$P(A \text{ or } N) = P(A) + P(N) - P(A \text{ and } N)$$
$$= 84/359 + 31/359 - 11/359$$
$$= 104/359$$
$$= .290$$

* Use individual mutually exclusive categories that involve no double-counting (i.e., the "intuitive addition rule"). NOTE: For simplicity in this problem, we use B for "a person is 22-29" and AY for "A and Y."

$$P(A \text{ or } N) = P(AY \text{ or } AN \text{ or } BN)$$
$$= P(AY) + P(AN) + P(BN)$$
$$= 73/359 + 11/359 + 20/359$$
$$= 104/359$$
$$= .290$$

NOTE: In general, using broad categories and allowing for double-counting is a "more powerful" technique that "lets the formula do the work" and requires less analysis by the solver. Except when such detailed analysis is instructive, this manual uses the first approach.

11. Make a chart like the one on the right.
    let M = a person is male
    N = a person says no harassment
    P(N) = 580/1000 = .58

		HARASSMENT?		
		Yes	No	
SEX	male	240	380	620
	female	180	200	380
		420	580	1000

13. Make a chart like the one on the right.
    let D = a person is divorced
    A = a person was always single
    Y = a person smokes
    N = a person does not smoke

		SMOKE?		
		Yes	No	
	married	54	146	200
STATUS	divorced	38	62	100
	always s	11	39	50
		103	247	350

$$P(D \text{ or } Y) = P(D) + P(Y) - P(D \text{ and } Y)$$
$$= 100/350 + 103/350 - 38/350$$
$$= 165/350$$
$$= .471$$

15. The total number of outcomes is 200.
    let H = the interval is at least one hour
    total number of outcomes favorable to H is 23 + 6 + 107 + 11 + 1 = 148
    P(H) = 148/200 = .74

17. Make a chart like the one on the right.

$$P(\bar{O}) = 1 - P(O)$$
$$= 1 - 45/100$$
$$= 55/100$$
$$= .55$$

		Rh FACTOR		
		+	−	
	A	35	5	40
GROUP	B	8	2	10
	AB	4	1	5
	O	39	6	45
		86	14	100

19. Refer to exercise #17.

$$P(B \text{ or } Rh\text{-}) = P(B) + P(Rh\text{-}) - P(B \text{ and } Rh\text{-})$$
$$= 10/100 + 14/100 - 2/100$$
$$= 22/100$$
$$= .22$$

21. Refer to exercise #17.
    P(Rh-) = 14/100 = .14

23. Refer to exercise #17.
    P(AB or Rh-) = P(AB) + P(Rh-) - P(AB and Rh-)
                 = 5/100 + 14/100 - 1/100
                 = 18/100
                 = .18

25. Refer to exercise #17.
    a. P(A or B) = P(A) + P(B) - P(A and B)
            1/3 = P(A) + 1/4 - 1/5
           .333 = P(A) + .250 - .200
           .333 = P(A) + .050
           .283 = P(A)
    b. for mutually exclusive events, P(A or B) = P(A) + P(B) = .4 + .5 = .9
    c. if A and B are not mutually exclusive, P(A and B) > 0
       and so, P(A or B) = P(A) + P(B) - P(A and B)
                         = .4 + .5 - P(A and B)
                         = .9 - P(A and B)
                         = something less than .9
       NOTE: More precisely, whenever A and B are not mutually exclusive it must be true that
                0 < P(A and B) ≤ min[P(A),P(B)]
       here,   0 < P(A and B) ≤ .4
       and so, .5 ≤ P(A or B) < .9

27. If the *exclusive or* is used instead of the *inclusive or*, then the double-counted probability must
    be completely removed (i.e., must be subtracted twice) and the formula becomes
       P(A or B) = P(A) + P(B) - 2·P(A and B)

3-4 Multiplication Rule

1. a. Dependent, attending class does affect the probability of passing a course.
   b. Independent, since sleeping too late does not affect the probability of getting a flat tire.
      NOTE: Sleeping late might cause one to hurry and engage in reckless behavior more likely
      to produce flat tires. This manual ignores such highly speculative hypothetical scenarios.
   c. Dependent, since P(A and B) = P(A)·P(B) = (.4)(.6) = .24 if A and B are independent.

3. let L = a person is left-handed
   P(L) = .10, for each random selection
   P(L_1 and L_2) = P(L_1)·P(L_2)
              = (.1)(.1)
              = .01

5. let A = selecting an ace
   P(A) = 4/52, for the first selection only
   P(A_1 and A_2 and A_3 and A_4) = P(A_1)·P(A_2 | A_1)·P(A_3 | A_1 and A_2)·P(A_4 | A_1 and A_2 and A_3)
              = (4/52)(3/51)(2/50)(1/49)
              = 1/270,725
              = .00000369

7. let F = the alarm clock fails
   P(F) = .02, for each clock
   P(at least one works) = 1 - P(all fail)
   $$= 1 - P(F_1 \text{ and } F_2 \text{ and } F_3)$$
   $$= 1 - P(F_1) \cdot P(F_2) \cdot P(F_3)$$
   $$= 1 - (.02)(.02)(.02)$$
   $$= .999992$$

   NOTE: Rounded to 3 significant digits as usual, the answer is 1.00. In cases when rounding to 3 significant digits produces a probability of 1.00, this manual gives the answer with sufficient significant digits to distinguish the answer from a certainty.

9. let T = the tire named is consistent with all naming the same tire
   P(T) = 4/4 for the first person, but then the others must match him
   P(all name the same tire) = $P(T_1 \text{ and } T_2 \text{ and } T_3 \text{ and } T_4)$
   $$= P(T_1) \cdot P(T_2 \mid T_1) \cdot P(T_3 \mid T_1 \text{ and } T_2) \cdot P(T_4 \mid T_1 \text{ and } T_2 \text{ and } T_3)$$
   $$= (4/4)(1/4)(1/4)(1/4)$$
   $$= 1/64$$
   $$= .0156$$

11. a. let D = a defective motor is selected
    P(D) = 12/30, for the first selection only
    $P(D_1 \text{ and } D_2 \text{ and } D_3) = P(D_1) \cdot P(D_2 \mid D_1) \cdot P(D_3 \mid D_1 \text{ and } D_2)$
    $$= (12/30)(11/29)(10/28)$$
    $$= 1320/24360$$
    $$= .0542$$
    b. let G = a good motor is selcted
    P(G) = 18/30, for the first selection only
    P(at least one defective) = 1 - P(all good)
    $$= 1 - P(G_1 \text{ and } G_2 \text{ and } G_3)$$
    $$= 1 - P(G_1) \cdot P(G_2 \mid G_1) \cdot P(G_3 \mid G_1 \text{ and } G_2)$$
    $$= 1 - (18/30)(17/29)(16/28)$$
    $$= 19464/24360$$
    $$= .799$$

13. let M = an audited taxpayer owes more money
    P(M) = .70, for each selection
    $P(M_1 \text{ and } M_2 \text{ and } \dots \text{ and } M_{12}) = P(M_1) \cdot P(M_2) \cdot \dots \cdot P(M_{12})$
    $$= (.70)^{12}$$
    $$= .0138$$

    Having all 12 really owe money would be an ususual event. It seems that either he is being untruthful or that he overzealously charged the taxpayers more than they really owed.

15. let A = a death is accidental
    P(A) = .0478, for each random selections
    $P(A_1 \text{ and } A_2 \text{ and } A_3 \text{ and } A_4 \text{ and } A_5) = P(A_1) \cdot P(A_2) \cdot P(A_3) \cdot P(A_4) \cdot P(A_5)$
    $$= (.0478)^5$$
    $$= .000000250$$

    NOTE: This is, as the problem requests, the probability of 5 randomly chosen deaths being accidental. The deaths investigated by the Baltimore detective, however, involved special circumstances and were not chosen at random.

17. let N = a checked employee is a not a thief
    $P(N) = 34/36$, for the first selection only
    $P(\text{no thieves}) = P(\text{all not thieves})$
    $\qquad\qquad\qquad = P(N_1 \text{ and } N_2 \text{ and } N_3 \text{ and } N_4)$
    $\qquad\qquad\qquad = P(N_1) \cdot P(N_2 \mid N_1) \cdot P(N_3 \mid N_1 \text{ and } N_2) \cdot P(N_4 \mid N_1 \text{ and } N_2 \text{ and } N_3)$
    $\qquad\qquad\qquad = (34/36)(33/35)(32/34)(31/33)$
    $\qquad\qquad\qquad = 1{,}113{,}024/1{,}413{,}720$
    $\qquad\qquad\qquad = .787$

19. let S = a car is stolen
    $P(S) = .045$, for each car selected at random
    $P(S_1 \text{ and } S_2 \text{ and } S_3 \text{ and } S_4) = P(S_1) \cdot P(S_2) \cdot P(S_3) \cdot P(S_4)$
    $\qquad\qquad\qquad\qquad\qquad\qquad\quad = (.045)^4$
    $\qquad\qquad\qquad\qquad\qquad\qquad\quad = .00000410$

    While the probability of 4 randomly selected cars all being stolen is extremely small, the 4 brothers do not necessarily represent a random selection. Being brothers, even in different parts of the city, they may share characteristics that make them inviting targets -- always leaving keys in car, buying the models that are frequently stolen, etc. The results do suggest, however, that further investigation would be appropriate.

21. let N = an individual tests negative
    $P(N) = .985$, for each person
    $P(\text{group is positive}) = P(\text{at least one person is positive})$
    $\qquad\qquad\qquad\qquad\quad = 1 - P(\text{all are negative})$
    $\qquad\qquad\qquad\qquad\quad = 1 - P(N_1 \text{ and } N_2 \text{ and } N_3 \text{ and } N_4 \text{ and } N_5)$
    $\qquad\qquad\qquad\qquad\quad = 1 - (.985)(.985)(.985)(.985)(.985)$
    $\qquad\qquad\qquad\qquad\quad = 1 - .9272$
    $\qquad\qquad\qquad\qquad\quad = .0728$

    NOTE: This plan is very efficient. Suppose there were 1,000 persons to be tested. Instead of 1,000 tests, use 200 tests of groups of 5. We expect to find about $(.0728)(200) = 15$ positive groups, for which tests will have to be re-done on all 5 individuals. Then the total number of tests required is $15 \times 6 = 90$ (from the positive groups) plus 185 (from the negative groups) -- and 1,000 individuals can be accurately tested with about 275 tests.

23. Refer to the table at the right.
    let S,F,U,H,R,A represent the categories
    a. $P(F \mid R) = 106/505$
    $\qquad\qquad\quad = .210$
    b. $P(F \text{ and } R) = 106/2000$
    $\qquad\qquad\qquad\quad = .053$
    c. $P(R \text{ or } F)$
    $\qquad = P(R) + P(F) - P(R \text{ and } F)$
    $\qquad = 505/2000 + 787/2000 - 106/2000$
    $\qquad = 1186/2000$
    $\qquad = .593$
    d. $P(R_1 \text{ and } R_2) = P(R_1) \cdot P(R_2 \mid R_1)$
    $\qquad\qquad\qquad\quad = (505/2000)(504/1999)$
    $\qquad\qquad\qquad\quad = .637$

		CRIME			
		Hom	Rob	Asl	
	stranger	12	379	727	1118
CRIMINAL	frnd/rel	39	106	642	787
	unknown	18	20	57	95
		69	505	1426	2000

25. Refer to the tables at the right and use the following notation.
    let H = a person is HIV infected
      N = a person is not HIV infected
      + = a person's HIV test is positive
      - = a person's HIV <u>test is negative</u>
      A = a person is from the at-risk population
      G = a person is from the general population
    $P(H \mid A) = 500/5000$
        $= .1$

At-Risk Population
HIV Test Result

		+	−	
STATUS	H	475	25	500
	N	225	4275	4500
		700	4300	5000

General Population
HIV Test Result

		+	−	
STATUS	H	57	3	60
	N	997	18943	19940
		1054	18946	20000

27. Refer to the tables and notation of exercise #25.
    $P(+ \text{ or } H \mid A) = P(+ \mid A) + P(H \mid A) - P(+ \text{ and } H \mid A)$
          $= 700/5000 + 500/5000 - 475/5000$
          $= 725/5000$
          $= .145$

29. Refer to the tables and notation of exercise #25.
    a. For the at-risk population only,
      $P(H \mid +) = 475/700$
         $= .679$
    b. For the general population only,
      $P(H \mid +) = 57/1054$
         $= .0541$
    c. When the test shows positive, a person is over ten times more likely to really have the HIV infection if he is from the at-risk population.

31. a. let D = a birthday is different from any yet selected
      $P(D_1) = 366/366$ NOTE: With nothing to match, it <u>must</u> be different.
      $P(D_2 \mid D_1) = 365/366$
      $P(D_3 \mid D_1 \text{ and } D_2) = 364/366$
      ...
      $P(D_{25} \mid D_1 \text{ and } D_2 \text{ and}...\text{and } D_{24}) = 342/366$
      $P(\text{no match}) = P(D_1 \text{ and } D_2...D_{25}) = (366/366) \cdot (365/366) \cdot (364/366) \cdot \cdot \cdot (342/366)$
               $= .432$

    NOTE: A program to perform this calculation can be constructed using a programming language, or using most spreadsheet or statistical software packages. In BASIC, for example, use

```
10 LET P=1
20   FOR K=1 TO 24
20   LET P=P*(366-K)/366
40   NEXT K
50 PRINT P
60 END
```

    b. $P(\text{at least one match}) = 1 - P(\text{no match}) = 1 - .432 = .568$

33. This is problem can be done by two different methods. In either case,
    let T = getting a 10
      C = getting a club
    * consider the sample space
      The first card could be any of 52 cards; for each first card, there are 51 possible second cards. This makes a total of $52 \cdot 51 = 2652$ equally likely outcomes in the sample space. How many of them are T_1C_2?
      The 10's of hearts, diamonds and spades can be paired with any of the 13 clubs for a total of $3 \cdot 13 = 39$ favorable possibilities. The 10 of clubs can only be paired with any of the remaining 12 members of that suit for a total of 12 favorable possibilities. Since there are

39 + 12 = 51 favorable possibilities among the equally likely outcomes,
$$P(T_1C_2) = 51/2652 = .0192$$

* use the formulas

Let Tc and To represent the 10 of clubs and the 10 of any other suit respectively. Break T_1C_2 into mutually exclusive parts so the probability can be found by adding and without having to consider double-counting.

$$\begin{aligned}
P(T_1C_2) &= P[(Tc_1 \text{ and } C_2) \text{ or } (To_1 \text{ and } C_2)] \\
&= P(Tc_1 \text{ and } C_2) + P(To_1 \text{ and } C_2) \\
&= P(Tc_1) \cdot P(C_2 \mid Tc_1) + P(To_1) \cdot P(C_2 \mid To_1) \\
&= (1/52)(12/51) + (3/52)(13/51) \\
&= 12/2652 + 39/2652 \\
&= 51/2652 \\
&= .0192
\end{aligned}$$

3-5   Probabilities Through Simulations

1. Using even numbers to represent males and odd number to represent females produces the following list of 50 numbers of girls per family of 3 children.

```
0 3 1 0 1 2 1 2 2 3 0 3 2 1 3 2 1 0 1 3 2 0 2 1 1
1 1 3 2 1 1 1 2 3 1 1 2 3 0 2 2 3 1 1 2 3 3 3 2 1
```

Letting x be the number of girls per simulated family produces the following frequency distribution.

x	f
0	6
1	18
2	14
3	12
	50

Letting G be getting a family with 3 girls, we estimate $P(G) \approx 12/50 = .24$.
This is considerably higher than the true value of .125. In this case the simulation did not produce a good estimate.

3. Using 0 to represent a defect and 1-9 to represent good products produces the following list of 50 numbers of defects per batch of 3.

```
3 0 0 0 0 0 0 0 0 0 0 0 0 0 0 1 0 2 1 0 0 0 0 1 1
0 0 0 1 0 0 1 0 0 0 0 0 0 1 0 0 0 0 0 1 0 0 0 0 0
```

Letting x be the number of defects per simulated batch produces the accompanying frequency distribution.

x	f
0	40
1	8
2	1
3	1
	50

Letting D be getting a batch with at least 1 defect, we estimate $P(D) \approx 10/50 = .20$.
This is lower than the true value of .271. In this case the simulation did not produce a good estimate.

5. The distribution at the right summarizes the 150 Pick Three digits. For this exercise, ignore the 0,7,8,9 values (given in parentheses).

x	f
0	(13)
1	16
2	11
3	17
4	14
5	18
6	18
7	(18)
8	(12)
9	(13)
	94

Letting O be getting a 1, we estimate
  $P(O) \approx 16/94 = .170$.
This compares well to the true value of $1/6 = .167$.
In this case the simulation produced a good estimate.

7. Using even digits to represent boys and odd digits to represent girls produces the following list of 50 simulated numbers of girls per 3-child family:

    0 3 1 0 1 2 1 2 2 3 0 3 2 1 3 2 1 0 1 3 2 0 2 1 1
    1 1 3 2 1 1 1 2 3 1 1 2 3 0 2 2 3 1 1 2 3 3 3 2 1

   a. Using the sample mean from the simulation to estimate the true mean number of girls in a 3-child family, $\bar{x} = \Sigma x/n = 82/50 = 1.64$.
      NOTE: The true mean number may be found from probability theory to be 1.5.

   b. Using the sample standard deviation from the simulation to estimate the true standard deviation among the numbers of girls in a 3-child family,
      $s^2 = [n(\Sigma x^2) - (\Sigma x)^2]/[n(n-1)]$
          $= [50(182) - (82)^2]/[50(49)] = 2376/2450 = .970$
      $s = .98$
      NOTE: The true standard deviation among the numbers may be found from probability theory to be .866

3-6 Counting

1. $6! = 6 \cdot 5 \cdot 4 \cdot 3 \cdot 2 \cdot 1 = 720$

3. $100!/97! = (100 \cdot 99 \cdot 98 \cdot 97!)/97! = 100 \cdot 99 \cdot 98 = 970,200$
   NOTE: This technique of "cancelling out" or "reducing" the problem by removing the factors $97! = 97 \cdot 96 \cdot \ldots \cdot 1$ from both the numerator and the denominator is preferred over actually evaluating 100!, actually evaluating 97!, and then dividing those two very large numbers. In general, a smaller factorial in the denominator can be completely divided into a larger factorial in the numerator to leave only the "excess" factors not appearing the in the denominator. This is the technique employed in this manual -- e.g., see #7 below, where the 4! is cancelled from both the numerator and the denominator. In addition, the, the answer to a counting problem (but not a probability problem) must always be a whole number; a fractional number indicates that a mistake has been made.

5. $(10-4)! = (6)! = 6 \cdot 5 \cdot 4 \cdot 3 \cdot 2 \cdot 1 = 720$

7. $_6C_4 = 6!/(2!4!) = (6 \cdot 5)/2! = 30/2 = 15$

9. $_{12}P_9 = 12!/(12-9)! = 12!/3! = 12 \cdot 11 \cdot 10 \cdot 9 \cdot 8 \cdot 7 \cdot 6 \cdot 5 \cdot 4 = 79,833,600$

11. $_{40}C_6 = 40!/(34!6!) = (40 \cdot 39 \cdot 38 \cdot 37 \cdot 36 \cdot 35)/6! = 3,838,380$

13. $_nC_0 = n!/(n!0!) = 1/0! = 1/1 = 1$
    NOTE: In words, this represents the number of ways to choose zero objects from among n objects. There is only 1 way -- to leave them all unselected.

15. $_nP_n = n!/(n-n)! = n!/0! = n!/1 = n!$
    NOTE: In words, this represents the number of ways to arrange n objects from among n objects. Using the general counting rule, this is $n \cdot (n-1) \cdot (n-2) \cdot \ldots \cdot 1 = n!$. This is also what the author calls the "factorial rule."

17. a. $10 \cdot 10 \cdot 10 \cdot 10 = 10,000$
    b. $5 \cdot 10,000 = 50,000$ seconds [or 833.33 minutes]
       $= 13.89$ hours [or 13 hours, 53 minutes, 20 seconds]

19. There are 10 possibilities (0,1,2,...,9) for each of the 9 digits.
    There are $10 \cdot 10 \cdot 10 \cdot 10 \cdot 10 \cdot 10 \cdot 10 \cdot 10 \cdot 10 = 10^9 = 1,000,000,000$ (i.e., one billion) sequences.
    Let Y = selecting your SS number.
    Since any SS is 1 of the equally likely one billion possible sequences,
    $P(Y) = 1/1,000,000,000 = .000000001$ [i.e., 1 in one billion]

21. There are 18 items, 10 of one type and 8 of another type. Use the "permutation rule when some items are identical to others."
    $n!/(n_1!n_2!\ldots n_k!) = 18!/(10!8!) = (18 \cdot 17 \cdot 16 \cdot 15 \cdot 14 \cdot 13 \cdot 12 \cdot 11)/8! = 43,758$

23. There are 7 different letters to arrange. Use the factorial rule with n=7.
    $7! = 5040$.

25. a. $_{40}C_6 = 40!/(34!6!) = (40 \cdot 39 \cdot 38 \cdot 37 \cdot 36 \cdot 35)/6! = 3,838,380$
    b. Let N = selecting the numbers that match the winning ones.
       Since your selection is 1 of the equally likely possible combinations,
       $P(N) = 1/3,838,380 = .000000261$
    c. $P(\bar{N})/P(N) = (3,838,379/3,838,380)/(1/3,838,380) = 3,838,379/1$ or 3,838,379:1

27. The first note is given by *. There are 3 possibilities (R,U,D) for each of the next 15 notes.
    There are $3 \cdot 3 \cdot 3 \cdot 3 \cdot 3 \cdot 3 \cdot 3 \cdot 3 \cdot 3 \cdot 3 \cdot 3 \cdot 3 \cdot 3 \cdot 3 \cdot 3 = 3^{15} = 14,348,907$ possible sequences.
    NOTE: This assumes each song has at least 16 notes.

29. a. $_{50}C_4 = 50!/(46!4!) = (50 \cdot 49 \cdot 48 \cdot 47)/4! = 230,300$
    b. Let M = selecting the agents with the most time in service.
       Since only one of the equally likely possibilities is the one with the most senior agents,
       $P(M) = 1/230,300 = .00000434$

31. $_{15}P_8 = 15!/(15-8)! = 15!/7! = 15 \cdot 14 \cdot 13 \cdot 12 \cdot 11 \cdot 10 \cdot 9 \cdot 8 = 259,459,200$

33. a. There are 10 possibilities (0,1,2,...,9) for each of the 5 digits.
       There are $10 \cdot 10 \cdot 10 \cdot 10 \cdot 10 = 10^5 = 100,000$ sequences.
    b. Let Y = selecting your zip code.
       Since any zip code is 1 of the equally likely 100,000 possible sequences,
       $P(Y) = 1/100,000 = .000001$

35. The number of different groups of 8 that could be hired is
    $_{40}C_8 = 40!/(32!8!) = (40 \cdot 39 \cdot 38 \cdot 37 \cdot 36 \cdot 35 \cdot 34 \cdot 33)/8! = 76,904,685$
    Let Y = selecting the youngest 8.
    Since only one of the equally likely possibilities is the one with the youngest members,
    P(Y) = 1/76,904,685 = .0000000130
    Yes; since choosing the youngest 8 is unlikely to occur in a random selection, it appears that age discrimination is occurring.
    NOTE: The group consisting of the youngest 8, however, is no more or less likely to be chosen at random than any other particular group of 8. It could also be that the decision was based entirely on a certain area (e.g., computer knowledge) in which the most recent graduates (i.e., the youngest ones) were definitely better qualified.

37. Use the factorial rule with n=12 to determine the total number of possible arrangements.
    n! = 479,001,600
    Let I = selecting the houses so that the readings are in increasing order.
    Since only one of the equally likely possibilities is the one with the readings in increasing order,
    P(I) = 1/479,001,600 = .00000000209
    Yes, her concern about the test equipment is justified. Remember, however, that her houses were presumably measured in geographical order and not randomly. If she is working her way toward the center of a radon concentration, such a pattern would be expected. Since radon dissipates in circulating air (home readings are typically taken in basements), an extra measurement in a well ventilated upstairs room may determine whether there is an equipment malfunction.

39. The first digit may be any of 8 -- 2,3,4,5,6,7,8,9.
    The second digit may be any of 2 -- 0,1.
    Regardless of the other digit, the third digit may be any of 9 -- 2,3,4,5,6,7,8,9 and either 0 (if the middle digit is 0) or 1 (if the middle digit is 1).
    According to the fundamental counting rule, the total number of possibilities is
    $8 \cdot 2 \cdot 9 = 144$

41. The numbers of possible names with 1,2,3,...,8 letters must be calculated separately using the fundamental counting rule and then added together to determine the total number of possible names. A chart will help to organize the work.

# of letters		# of possible names	
1	26	= 26 =	26
2	26·36	$= 26 \cdot 36^1 =$	936
3	26·36·36	$= 26 \cdot 36^2 =$	33,696
4	26·36·36·36	$= 26 \cdot 36^3 =$	1,213,056
5	26·36···36	$= 26 \cdot 36^4 =$	43,670,016
6	26·36···36	$= 26 \cdot 36^5 =$	1,572,120,576
7	26·36···36	$= 26 \cdot 36^6 =$	56,596,340,736
8	26·36···36	$= 26 \cdot 36^7 =$	2,037,468,266,496
			2,095,681,645,538

43. a. The calculator factorial key gives $50! = 3.04140932 \times 10^{64}$.
    Using the approximation, K = (50.5)·log(50) + .39908993 - .43429448(50)
       = 85.79798522 + .39908993 - 21.71472400
       = 64.48235115
    And then $n! = 10^K$
       $= 10^{64.48235115}$
       $= 3.036345216 \times 10^{64}$

NOTE: The two answers differ by 5.1×10^{61} (i.e., by 51 followed by 60 zeros -- "zillions and zillions"). While that error may seem large, the numbers being dealt with are very large and that difference is only a $(5.1 \times 10^{61})/(3.04 \times 10^{64}) = 1.7\%$ error.

b. Using the approximation, with n=300,

$K = (n+.5)(\log n) + .39908993 - .43429448n$
$= (300.5)(\log 300) + .39908993 - .43429448(300)$
$= 614.4$
$300! = 10^{614.4}$

Since the number of digits in 10^x is the next whole number above x, the number of digits in 300! is 615.

Review Exercises

1. Refer to the table at the right.
   let Y = a deceased person smoked
   N = a deceased person did not smoke
   C = a person died of cancer
   H = a person died of heart disease
   O = a person died of some other cause

   $P(Y) = 65/1000 = .65$

		SMOKE?		
		Yes	No	
CAUSE OF	Cancer	135	55	190
DEATH	Heart	310	155	465
	Other	205	140	395
		650	350	1000

2. Refer to the table and notation of exercise #1.
   $P(Y \text{ or } H) = P(Y) + P(H) - P(Y \text{ and } H)$
   $= 650/1000 + 465/1000 - 310/1000$
   $= 805/1000$
   $= .805$

3. Refer to the table and notation of exercise #1.
   $P(C_1 \text{ and } C_2) = P(C_1) \cdot P(C_2 \mid C_1)$
   $= (190/1000) \cdot (189/999)$
   $= .0359$

4. Refer to the table and notation of exercise #1.
   $P(N \text{ and } C) = P(N) + P(C) - P(N \text{ and } C)$
   $= 55/1000$
   $= .055$

5. Refer to the table and notation of exercise #1.
   $P(C \text{ or } H) = P(C) + P(H) - P(C \text{ and } H)$
   $= 190/1000 + 465/1000 - 0/1000$
   $= 655/1000$
   $= .655$

6. Refer to the table and notation of exercise #1.
   $P(Y_1 \text{ and } Y_2 \text{ and } Y_3) = P(Y_1) \cdot P(Y_2 \mid Y_1) \cdot P(Y_3 \mid Y_1 \text{ and } Y_2)$
   $= (650/1000) \cdot (649/999) \cdot (648/998)$
   $= .274$

7. Refer to the table and notation of exercise #1.
   $P(Y \mid C) = 135/190$
   $= .711$

8. Refer to the table and notation of exercise #1.
   P(C | Y) = 135/650
            = .208
   No, they are not independent events. Since P(C | Y) ≠ P(C) = 190/1000 = .190, the occurrence of smoking does affect the probability of the occurrence of death from cancer.

9. Let D = an item is defective
   P(D) = .84, for each item
   P(at least one good) = 1 - P(all defective)
   $$= 1 - P(D_1 \text{ and } D_2 \text{ and}...\text{and } D_{12})$$
   $$= 1 - P(D_1) \cdot P(D_2) \cdot ... \cdot P(D_{12})$$
   $$= 1 - (.84)^{12}$$
   $$= .877$$
   No. If it is "very important" to have at least one good item, then more than 12 should be made.

10. Let N = not receiving a ticket when he speeds
    P(N) = .98, for each day
    $$P(N_1 \text{ and } N_2 \text{ and}...\text{and } N_{150}) = P(N_1) \cdot P(N_2) \cdot ... \cdot P(N_{150})$$
    $$= (.98)^{150}$$
    $$= .0483$$
    Since his habitual speeding gives him less than 1 chance in 20 of making it through the year without a ticket, he should begin driving withing the posted speed limit.

11. a. The number of possible committees is $_8C_3$ = 8!/(5!3!) = (8·7·6)/3! = 56.
       Let W = the committee consists of the 3 wealthiest members.
       Since only 1 of the equally likely possibilities is the 3 wealthiest members,
       P(W) = 1/56 = .0179
    b. $_8P_3$ = 8!/(8-3)! = 8!/5! = 8·7·6 = 336

12. Let L = a selected 27 year old male lives for one more year
    P(L) = .9982, for each such person
    $$P(L_1 \text{ and } L_2 \text{ and}...\text{and } L_{12}) = P(L_1) \cdot P(L_2) \cdot ... \cdot P(L_{12})$$
    $$= (.9982)^{12}$$
    $$= .979$$

13. Let E = an even number occurs.
    a. Since there are 18 even numbers among the 38 equally likely outcomes,
       P(E) = 18/38
            = .474
    b. P(Ē)/P(E) = (20/38)/(18/38)
                 = 20/18 or 10:9
    c. With a payoff of 1:1, a person wins $1 profit for every $1 bet. If someone bets $5 and wins, he wins $5 profit.

14. Use the factorial rule with n=5 to determine the number of possible arrangements.
    5! = 120
    Let C = selecting the correct chronological order.
    Since only 1 of the equally likely possibilities is the correct chronological order,
    P(C) = 1/120 = .00833

15. Assume the population is so large that it is not changed by repeated sampling -- i.e., that repeated random selections are independent of each other.
    Let R = a selected voter is a Republican.
    P(R) = .30, for each selection
    $$P(R_1 \text{ and } R_2 \text{ and...and } R_{12}) = P(R_1) \cdot P(R_2) \cdot ... \cdot P(R_{12})$$
    $$= (.30)^{12}$$
    $$= .000000531$$
    No, the pollster's claim that this could easily happen by chance is not correct.

16. There are 16 people, 8 males and 8 females. Use the "permutation rule when ome items are identical to others" to determine the number of possible gender arrangements.
    $n!/(n_1!n_2!...n_k!) = 16!/(8!8!) = (16 \cdot 15 \cdot 14 \cdot 13 \cdot 12 \cdot 11 \cdot 10 \cdot 9)/8! = 12,870$
    Let E = having 8 women in the first group and 8 males in the second
    Since only 1 of the 12,870 equally likely possibilities places all the women in the first group and all the men in the second group,
    P(E) = 1/12,870 = .0000777
    NOTE: If the exercise had asked for the probability of all the women being in one group (not necessarily the first group) and all the men being in the other group, then the probability is 2/12,870.

Cumulative Review Exercises

1. NOTE: Since ages are reported as of one's last birthday, and not to the nearest year, care must be taken when working with ages. If n people each report age 16, for example, the mean reported age is clearly 16.0; since those n people were probably evenly spread out between 16.0 and 16.999 years old, however, their true mean age is about 16.5. The ages is the 0-4 class in this exercise, therefore, include values from 0.000 to 4.999 and the class mark should technically be 2.5. This manual ignores that technicality, treats ages as any other data, and so determines the class marks in the x column below. Those wishing to be more precise should add 0.5 years to the calculated mean in part (a). No correction is necessary for the calculated standard deviation in part (b).

x	f	f·x	f·x^2
2.0	3,843	7,686.0	15,372.00
9.5	4,226	40,147.0	318,396.50
19.5	19,975	389,512.5	7,595,493.75
34.5	27,201	938,434.5	32,375,990.25
54.5	14,733	802,948.5	43,760,693.25
69.5	8,499	590,680.5	41,052,294.75
80.0	16,800	1,344,000.0	107,520,000.00
	95,277	4,113,409.0	232,701,240.50

a. $\bar{x} = (\Sigma x)/n = (4,113,409.0)/95,277 = 43.2$

b. $s^2 = [n \cdot (\Sigma fx^2) - (\Sigma fx)^2]/[n(n-1)]$
    $= [(95,277)(232,701,240.50) - (4,113,409.0)^2]/[(95,277)(95,276)]$
    $= 578.45$
    $s = 24.1$

c. Let x = the age selected.
    $P(x<15 \text{ or } x>64) = P(x<15) + P(x>64) - P(x<15 \text{ and } x>64)$
    $= (3843+4226)/95277 + (8499+16800)/95277 - (0)/95277$
    $= 33368/95277$
    $= .350$
    NOTE: No age can be less than 15 and greater than 64.

d. Let x = the age selected.
P(x < 15 or 5 ≤ x ≤ 44)
= P(x < 15) + P(5 ≤ x ≤ 44) - P(5 ≤ x < 15)
= (3843 + 4226)/95277 + (4226 + 19975 + 27201)/95277 - (4226)/95277
= 55245/95277
= .580

e. Let B = the age selected is between 0 and 4.
P(B) = 3843/95277, for the first selection only
P(B_1 and B_2) = P(B_1) · P(B_2 | B_1)
= (3843/95277) · (3842/95276)
= .00163

2. Let x = the height selected.
The values identified on the boxplot are as follows: x_1 [minimum] = 56.1, P_{25} = 62.2, P_{50} = 63.3, P_{75} = 65.0, x_n [maximum] = 71.1.
NOTE: Assume that the number of heights is so large that (1) repeated sampling does not affect the probabilities for subsequent selections and (2) it can be said that P_a has a% of the heights below it and (100-a)% of the heights above it.

a. P(56.1 < x < 62.2) = P(x_1 < x < P_{25}) = .25 - 0 = .25

b. P(x < 62.2 or x > 63.6) = P(x < 62.2) + P(x > 63.6)
= P(x < P_{25}) + P(x > P_{50})
= .25 + .50
= .75

NOTE: No height is less than 62.2 and greater than 63.6 -- i.e., the addition rule for mutually exclusive events can be used.

c. Let B = selecting a height between 62.2 and 63.6.
P(B) = P(P_{25} < x < P_{50}) = .50 - .25 = .25
P(B_1 and B_2) = P(B_1) · P(B_2 | B_1)
= (.25) · (.25)
= .0625

3. The actual values are 10.5, 10.7,...,13.3.
The summary statistics are: n = 15, Σx = 173.2, $Σx^2$ = 2010.24

a. $\bar{x}$ = (Σx)/n = 173.2/15 = 11.55 cm

b. $\tilde{x}$ = 11.30 cm [from the ordered list]

c. s^2 = [n($Σx^2$) - $(Σx)^2$]/[n)n-1)]
= [15(2010.24) - $(173.2)^2$]/[15(14)]
= 155.36/210 = .7398
s = .86 cm

d. s^2 = .74 cm^2 [from part (c)]

e. ratio, since differences are consistent and there is a meaningful zero

f. Let L = a selected diameter is less than 11.0 cm.
Since 3 of the 15 equally likely possibilities are less than 11.0,
P(L) = 3/15 = .2

g. Let A = a selected diameter is above 11.9 cm.
   Four of the 15 equally likely possibilities are above 11.9.
   P(A) = 4/15, for the first selection only
   P(A_1 and A_2) = P(A_1) · P(A_2 | A_1)
   $\qquad\qquad\quad$ = (4/15) · (3/14)
   $\qquad\qquad\quad$ = .0571

h. Let A = a selected diameter is above 11.9 cm.
   Four of the 15 equally likely possibilities are above 11.9.
   P(A) = 4/15, for the each selection [since selected items are replaced before re-selecting]
   P(A_1 and A_2) = P(A_1) · P(A_2 | A_1)
   $\qquad\qquad\quad$ = (4/15) · (4/15)
   $\qquad\qquad\quad$ = .0711

i. P(x < 11.0 or 10.6 < x < 11.6) = P(x < 11.0) + P(10.6 < x < 11.6) - P(10.6 < x < 11.0)
   $\qquad\qquad\qquad\qquad\qquad\quad$ = 3/15 + 9/15 - 2/15
   $\qquad\qquad\qquad\qquad\qquad\quad$ = 10/15
   $\qquad\qquad\qquad\qquad\qquad\quad$ = .667

Chapter 4

Probability Distributions

4-2 Random Variables

1. continuous, since weight can be any value on a continuum

3. discrete, since the number must be an integer

NOTE: When working with probability distributions and formulas in the exercises that follow, always keep these important facts in mind.
   * If one of the conditions for a probability distribution does not hold, the formulas do not apply and produce numbers that have no meaning.
   * $\Sigma x \cdot P(x)$ gives the mean of the x values and must be a number between the highest and lowest x values.
   * $\Sigma x^2 \cdot P(x)$ gives the mean of the x^2 values and must be a number between the highest and lowest x^2 values.
   * $\Sigma P(x)$ must always equal 1.000.
   * Σx and Σx^2 have no meaning and should not be calculated.
   * The quantity $[\Sigma x^2 \cdot P(x) - \mu^2]$ cannot possibly be negative; if it is, then there is a mistake.
   * Always be careful to use the <u>unrounded</u> mean in the calculation of the variance and to take the square root of the <u>unrounded</u> variance to find the standard deviation.

5. This is a probability distribution since $\Sigma P(x)=1$ is true and $0 \leq P(x) \leq 1$ is true for each x.

x	P(x)	$x \cdot P(x)$	x^2	$x^2 \cdot P(x)$
0	.512	0	0	0
1	.301	.301	1	.301
2	.132	.264	4	.528
3	.055	.165	9	.495
	1.000	.730		1.324

$\mu = \Sigma x \cdot P(x)$
$\quad = .73$, rounded to .7
$\sigma^2 = \Sigma x^2 \cdot P(x) - \mu^2$
$\quad = 1.324 - (.73)^2$
$\quad = .791$, rounded to .8
$\sigma = .9$

7. This is <u>not</u> a probability distribution since $\Sigma P(x) = .475 \neq 1.000$.

9. This is a probability distribution since $\Sigma P(x)=1$ is true and $0 \leq P(x) \leq 1$ is true for each x.

x	P(x)	$x \cdot P(x)$	x^2	$x^2 \cdot P(x)$
0	.36	0	0	0
1	.48	.48	1	.48
2	.16	.32	4	.64
	1.00	.80		1.12

$\mu = \Sigma x \cdot P(x)$
$\quad = .8$
$\sigma^2 = \Sigma x^2 \cdot P(x) - \mu^2$
$\quad = 1.12 - (.80)^2$
$\quad = .48$, rounded to .5
$\sigma = .7$

11. This is a probability distribution since $\Sigma P(x)=1$ is true and $0 \le P(x) \le 1$ is true for each x.

x	P(x)	x·P(x)	x^2	x^2·P(x)	
0	.08	0	0	0	$\mu = \Sigma x \cdot P(x)$
1	.05	.05	1	.05	$= 4.16$, rounded to 4.2
2	.10	.20	4	.40	$\sigma^2 = \Sigma x^2 \cdot P(x) - \mu^2$
3	.13	.39	9	1.17	$= 21.82 - (4.16)^2$
4	.15	.60	16	2.40	$= 4.51$, rounded to 4.5
5	.21	1.05	25	5.25	$\sigma = 2.1$
6	.09	.54	36	3.24	
7	.19	1.33	49	9.31	
	1.00	4.16		21.82	

13.

x	P(x)	x·P(x)	
175	1/38	4.605	$E = \Sigma x \cdot P(x)$
-5	37/38	-4.868	$= \$-0.263$ [i.e., a loss of 26.3¢]
	38/38	-0.263	

Since the expected loss is 26.3¢ for a $5 bet, the expected (or long run) loss is 26.3/5 = 5.26¢ for each $1 a person bets.

15. This problem can be worked in two different ways.

* Considering the $156 she paid up front, the woman loses 156 if she lives and gains 100,000 - 156 = 99844 if she dies.

x	P(x)	x·P(x)	
-156	.9995	-155.92	$E = \Sigma x \cdot P(x)$
99844	.0005	44.92	$= -\$106.00$ (i.e., a loss of $106)
	1.0000	-106.00	

* Ignoring the cost of the insurance, one may calculate the expected value of the policy itself.

x	P(x)	x·P(x)	
0	.9995	0	$E = \Sigma x \cdot P(x)$
100000	.0005	50.00	$= \$50.00$
	1.0000	50.00	

If the policy is worth $50 in expected returns and the woman pays $156 to purchase it, her net expectation is 50 - 156 = -$106.

17. The 8 equally like outcomes in the sample space are given at the right. If x represents the number of girls, counting the numbers of favorable outcomes indicates

outcome	x
BBB	0
BBG	1
BGB	1
GBB	1
GGB	2
GBG	2
BGG	2
GGG	3

$P(x = 0) = 1/8 = .125$
$P(x = 1) = 3/8 = .375$
$P(x = 2) = 3/8 = .375$
$P(x = 3) = 1/8 = .125$

x	P(x)	x·P(x)	x^2	x^2·P(x)	
0	.125	0	0	0	$\mu = \Sigma x \cdot P(x)$
1	.375	.375	1	.375	$= 1.5$
2	.375	.750	4	1.500	$\sigma^2 = \Sigma x^2 \cdot P(x) - \mu^2$
3	.125	.375	9	1.125	$= 3.000 - (1.500)^2$
	1.000	1.500		3.000	$= .75$
					$\sigma = .9$

The range rule of thumb states that $\sigma \approx R/4$, or [equivalently] $R \approx 4\sigma = 4(.9) = 3.6$.

For a symmetric distribution, the estimates are
   minimum $\approx \mu$ - R/2 = 1.5 - 3.6/2 = -0.3 [the true minimum is really 0]
   maximum $\approx \mu$ + R/2 = 1.5 + 3.6/2 = 3.3 [the true maximum is really 3]

19. let D = selecting an item that is defective
       N = selecting an item that is not defective
    P(D) = 2/10, for the first selection only
    P(N) = 8/10, for the first selection only

    let x = the number of defectives selected in 2 trials without replacement
    $P(x=0) = P(N_1$ and $N_2) = P(N_1) \cdot P(N_2 \mid N_1) = (8/10) \cdot (7/9) = 56/90$
    $P(x=2) = P(D_1$ and $D_2) = P(D_1) \cdot P(D_2 \mid D_1) = (2/10) \cdot (1/9) = 2/90$
    Since $P(x=0) + P(x=1) + P(x=2) = 1$,
$$P(x=1) = 1 - P(x=0) - P(x=2)$$
$$= 90/90 - 56/90 - 2/90$$
$$= 32/90$$

x	P(x)	x·P(x)	x²	x²·P(x)	
0	56/90	0	0	0	$\mu = \Sigma x \cdot P(x)$
1	32/90	.356	1	.3556	= .400, rounded to 0.4
2	2/90	.044	4	.0889	$\sigma^2 = \Sigma x^2 \cdot P(x) - \mu^2$
	90/90	.400		.4445	= .4445 - (.400)²
					= .2845, rounded to .3
					σ = 0.5

NOTE: Problems whose probabilities do not "come out even" as decimals can be very sensitive
to rounding errors. Since the P(x) values are used in several subsequent calculations, express
them as exact fractions instead of rounded decimals. It is not, however, necessary to continue
with fractions throughout the entire problem. Use sufficient decimal places (typically one more
than "usual") in the $x \cdot P(x)$ and $x^2 \cdot P(x)$ columns to guard against cumulative rounding errors.

21. In order for P(x) to be a probability distribution, it must be true that
    (1) $\Sigma P(x) = 1$
    (2) $0 \le P(x) \le 1$ for every permissible value of x.
    NOTE: Parts (a) and (d) below use the fact from algebra that
       $1 + r + r^2 + r^3 + r^4 + ... = 1/(1-r)$ for any r such that $-1 < r < 1$
    proof: (1)   let $S = 1 + r + r^2 + r^3 + ...$
            (2)   then $rS = \underline{\quad\quad r + r^2 + r^3 + r^4 + ...}$
       subtracting (1)-(2)   S-rS = 1   [the terms with r's subtract out...]
                          S(1-r) = 1
                              S = 1/(1-r)

    a. P(x) is a probability distribution because for x = 1,2,3...
       (1) $\Sigma P(x) = \Sigma(.5)^x$
               $= (.5) \cdot \Sigma(.5)^{x-1}$
               $= (.5) \cdot [1 + .5 + .5^2 + .5^3 + ...]$
               $= (.5) \cdot [1/(1-.5)]$
               $= (.5) \cdot [1/.5]$
               $= 1$
       (2) $0 \le P(x) \le 1$, since (i) P(x) > 0 for all x and (ii) P(x) = .5 for x = 1
           and (iii) P(x) keeps decreasing as x grows larger

    b. P(x) is not a probability distribution since for x = 1,2,3,...
       (1) $\Sigma P(x) = \Sigma[1/(2x)]$
               = 1/2 + 1/4 + 1/6 + 1/8 + ...
               = .5000 + .2500 + .1666 + .1250 + ...
               = 1.0416 + ...
               > 1

(2) [Condition (2) happens to be satisfied, but that is irrelevant since condition (1) is not satisfied and both conditions must be met to have a probability distribution.]

c. P(x) is a probability distribution because for x = 0,1,2,3
   (1) $\Sigma P(x) = \Sigma 3/[4 \cdot (3-x)! x!]$
       $= 3/[4 \cdot 3! 0!] + 3/[4 \cdot 2! 1!] + 3/[4 \cdot 1! 2!] + 3/[4 \cdot 0! 3!]$
       $= 3/24 + 3/8 + 3/8 + 3/24$
       $= 1$
   (2) $0 \le P(x) \le 1$, since the P(x) values are given above as 1/8, 3/8, 3/8, 1/8 for x = 0,1,2,3

d. P(x) is a probability distribution because for x = 1,2,3...
   (1) $\Sigma P(x) = \Sigma (.4)(.6)^{x-1}$
       $= (.4) \cdot \Sigma (.6)^{x-1}$
       $= (.4) \cdot [1 + .6 + .6^2 + .6^3 + \ldots]$
       $= (.4) \cdot [1/(1-.6)]$
       $= (.4) \cdot [1/.4]$
       $= 1$
   (2) $0 \le P(x) \le 1$, since (i) P(x) > 0 for all x and (ii) P(x) = .4 for x = 1
       and (iii) P(x) keeps decreasing as x grows larger

23. a. For random selections with replacement, P(x) = .1 for each x = 0,1,2,3,4,5,6,7,8,9.

x	P(x)	$x \cdot P(x)$	x^2	$x^2 \cdot P(x)$
0	.1	0	0	0
1	.1	.1	1	.1
2	.1	.2	4	.4
3	.1	.3	9	.9
4	.1	.4	16	1.6
5	.1	.5	25	2.5
6	.1	.6	36	3.6
7	.1	.7	49	4.9
8	.1	.8	64	6.4
9	.1	.9	81	8.1
	1	4.5		28.5

$\mu = \Sigma x \cdot P(x)$
$= 4.5$
$\sigma^2 = \Sigma x^2 \cdot P(x) - \mu^2$
$= 28.5 - (4.5)^2$
$= 8.25$
$\sigma = 2.8723$

b. The z score for each x is $z = (x-\mu)/\sigma = (x-4.5)/2.8723$

z	P(z)	$z \cdot P(z)$	z^2	$z^2 \cdot P(z)$
-1.5667	.1	-.15667	2.45455	.245455
-1.2185	.1	-.12185	1.48485	.148485
-0.8704	.1	-.08704	.75758	.075758
-0.5222	.1	-.05222	.27272	.027272
-0.1741	.1	-.01741	.03030	.003030
0.1741	.1	.01741	.03030	.003030
0.5222	.1	.05222	.27272	.027272
0.8704	.1	.08704	.75758	.075758
1.2185	.1	.12185	1.48485	.148485
1.5667	.1	.15667	2.45455	.245455
	1.0	0		1.000000

$\mu = \Sigma z \cdot P(z)$
$= 0$
$\sigma^2 = \Sigma z^2 \cdot P(z) - \mu^2$
$= 1 - (0)^2$
$= 1$
$\sigma = 1$

NOTE: For any population of z scores,
$\Sigma z = \Sigma(x-\mu)/\sigma = (1/\sigma) \cdot \Sigma(x-\mu) = (1/\sigma) \cdot [\Sigma x - \Sigma \mu] = (1/\sigma) \cdot [n\mu - n\mu] = 0$
$\Sigma z^2 = \Sigma[(x-\mu)/\sigma]^2 = (1/\sigma^2) \cdot \Sigma(x-\mu)^2 = (1/\sigma^2) \cdot n\sigma^2 = n$
$\mu = \Sigma z/n = 0/n = 0$
$\sigma^2 = \Sigma(z-\mu)^2/n = \Sigma z^2/n = n/n = 1$

4-3 Binomial Experiments

NOTE: The four requirements for a binomial experiment are
#1 There are a fixed number of trials.
#2 The trials are independent.
#3 Each trial has two possible named outcomes.
#4 The probabilities remain constant for each trial.

1. No, requirement #3 is not met. There are more than 2 possible outcomes.

3. Yes, all four requirements are met.

5. No, requirement #3 is not met. There are more than 2 possible outcomes.

7. No, requirements #2 and #4 are not met.
Let G represent getting a good item, and let D represent getting a defective one.
Since $P(D_2 \mid D_1) = 4/29 \neq P(D_2 \mid G_1) = 5/29$, requirement #2 is not met.
Since $P(D_2 \mid D_1) = 4/29 \neq P(D_1) = 5/30$, requirement #4 is not met.

9. From table A-1, .243.

11. From table A-1, .075.

NOTE: To use the binomial formula, one must identify 3 quantities: n,x,p. Table A-1, for example, requires only these 3 values to supply a probability. Since what the text calls "q" always equals 1-p, it can be so designated without introducing unnecessary notation [just as no special notation is utilized for the quantity n-x, even though it appears twice in the binomial formula]. This has the additional advantage of ensuring that the probabilities p and 1-p sum to 1.00 and protecting against an error in the separate calculation and/or identification of "q." In addition, reversing the order of (n-x)! and x! in the denominator of the $_nC_x$ coefficient term seems appropriate. That agrees with the $n!/(n_1!n_2!)$ logic of the "permutation rule when some objects are alike" for n_1 objects of one type and n_2 objects of another type, and that places the "x" and "n-x" in the same order in both the denominator of the coefficient term and the exponents. Such a natural ordering also leads to fewer errors. Accordingly, this manual expresses the binomial formula as

$$P(x) = [n!/x!(n-x)!] \cdot p^x \cdot (1-p)^{n-x}$$

13. $P(x) = [n!/x!(n-x)!] \cdot p^x \cdot (1-p)^{n-x}$
$\quad P(x=2) = [3!/2!1!] \cdot (1/4)^2 \cdot (3/4)^1$
$\qquad\qquad = [3] \cdot (1/16) \cdot (3/4)$
$\qquad\qquad = 9/64$
$\qquad\qquad = .141$

15. $P(x) = [n!/x!(n-x)!] \cdot p^x \cdot (1-p)^{n-x}$
$\quad P(x=4) = [10!/4!6!] \cdot (.35)^4 \cdot (.65)^6$
$\qquad\qquad = [210] \cdot (.0473) \cdot (.0754)$
$\qquad\qquad = .238$

IMPORTANT NOTE: The intermediate values of 210, .0473 and .0754 are given to help those with incorrect answers to identify the portion of the problem in which the mistake was made. This practice will be followed in all problems (i.e., not just binomial problems) throughout the manual. In practice, all calculations can be done in one step on the calculator. You may choose to (or be asked to) write down such intermediate values for your own (or the instructor's) benefit, but never round off in the middle of a problem. Do not write the values down on paper and then re-enter them in the calculator -- use the memory to let the calculator remember with complete accuracy any intermediate values that will be used in subsequent calculations. In addition, always make certain that the quantity [n!/x!(n-x)!] is a whole number and that the final answer is between 0 and 1.

17. Let x = the # of teenagers out of 8 that use a video game player. Refer to the table reproduced at the right.
NOTE: Due to rounding, the P(x) values given in the table sum to 1.0001. Depending on how the table is used, it is possible to obtain answers differing by .0001 from the ones given in this manual.

$P(x=6) = .3087$

x	P(x)
0	.0000
1	.0002
2	.0025
3	.0165
4	.0689
5	.1844
6	.3087
7	.2953
8	.1236
	1.0000

19. Refer to exercise #17.
$$P(x<7) = 1 - P(x \geq 7)$$
$$= 1 - [P(x=7) + P(x=8)]$$
$$= 1 - [.2953 + .1236]$$
$$= 1 - .4189$$
$$= .5811$$

21. let x = the number of newborns that are girls
NOTE: Since x represents the number of girls (i.e., being a girl is considered a "success"), then p must represent the probability of a newborn being a girl. Beginning each problem with a statement identifying what x represents helps avoid errors caused by inconsistent identifications.

a. n = 10, p = .5, x = 4. Use Table A-1.
   $P(x=4) = .205$

b. n = 10, p = .5, x ≥ 4. Use Table A-1.
   $$P(x \geq 4) = 1 - P(x<4)$$
   $$= 1 - [P(x=0) + P(x=1) + P(x=2) + P(x=3)]$$
   $$= 1 - [.001 + .010 + .044 + .117]$$
   $$= 1 - .172$$
   $$= .828$$

c. n = 20, p = .5, x = 8. Use the binomial formula.
   $$P(x) = [n!/x!(n-x)!] \cdot p^x \cdot (1-p)^{n-x}$$
   $$P(x=8) = [20!/8!12!] \cdot (.5)^8 \cdot (.5)^{12}$$
   $$= [125970] \cdot (.00391) \cdot (.00024)$$
   $$= .120$$

23. let x = the number of M&M's that are red
n = 15, p = .20, x = 3. Use Table A-1.
$P(x=3) = .250$

25. let x = the number of correct answers by guessing
n = 10, p = .20, x ≥ 6. Use Table A-1.
$$P(x \geq 6) = P(x=6) + P(x=7) + P(x=8) + P(x=9) + P(x=10)$$
$$= .006 + .001 + 0^+ + 0^+ + 0^+$$
$$= .007$$

No, the probability is not high enough to make random guessing a reasonable substitute for studying.

27. let x = the number of homes burglarized at least once
$n = 15$, $p = .05$, $x = 4$. Use Table A-1.
$P(x \geq 4) = 1 - P(x < 4)$
$= 1 - [P(x=0) + P(x=1) + P(x=2) + P(x=3)]$
$= 1 - [.463 + .366 + .135 + .031]$
$= 1 - .995$
$= .005$

Either the community is very unlucky or (more likely) there is something about its homes that make them different from the typical U.S. household.

NOTE: The exercise asked for the probability of getting "4 or more burglaries in a community of 15 homes." According to the given information, the answer given is actually the probability that "4 or more homes in a community of 15 will experience at least 1 burglary." Those events are not the same and illustrate the importance of careful wording.

29. let x = the number of defective items selected
$n = 20$, $p = .10$, $x \leq 1$. Use the binomial formula.
$P(x) = [n!/x!(n-x)!] \cdot p^x \cdot (1-p)^{n-x}$
$P(x \leq 1) = P(x=0) + P(x=1)$
$= [20!/0!20!] \cdot (.10)^0 \cdot (.90)^{20} + [20!/1!19!] \cdot (.10)^1 \cdot (.90)^{19}$
$= [1] \cdot (1) \cdot (.1216) + [20] \cdot (.10) \cdot (.1351)$
$= .1216 + .2702$
$= .392$

No, these results are not evidence that the new measures are effective. Even if the measures had no effect, we expect results this good or better about 40% of the time. NOTE: Even 0 defectives in a sample of 20 would not be evidence that the new measures are effective, since we expect that result 12% of the time under the old system. A sample larger than 20 is necessary to provide evidence that the new measures are effective.

31. let x = the number of consumers that recognize Coke
$n = 15$, $p = .95$, $x \leq 10$. Use Table A-1.
$P(x \leq 10) = 1 - P(x > 10)$
$= 1 - [P(x=11) + P(x=12) + P(x=13) + P(x=14) + P(x=15)]$
$= 1 - [.005 + .031 + .135 + .366 + .463]$
$= 1 - [1.000]$
≈ 0

No, it is very unlikely that the pollster was dealing with a valid random sample of all consumers. NOTE: Because of rounding in Table A-1, more accuracy is needed to obtain a more meaningful answer. Application of the binomial formula yields the following.
$P(x \leq 10) = 1 - P(x > 10)$
$= 1 - [P(x=11) + P(x=12) + P(x=13) + P(x=14) + P(x=15)]$
$= 1 - [.00485 + .03073 + .13475 + .36576 + .46329]$
$= 1 - [.99938]$
$= .00062$

33. let x = the number of components tested to find 1st defect
$p = .2$, $x = 7$. Use the geometric formula.
$P(x) = p \cdot (1-p)^{x-1}$
$P(x=7) = (.2) \cdot (.8)^6$
$= (.2) \cdot (.2621)$
$= .0524$

35. Extending the pattern to cover 6 types of outcomes, where $\Sigma x = n$ and $\Sigma p = 1$,

$P(x_1,x_2,x_3,x_4,x_5,x_6) = [n!/(x_1!x_2!x_3!x_4!x_5!x_6!)] \cdot p_1^{x_1} \cdot p_2^{x_2} \cdot p_3^{x_3} \cdot p_4^{x_4} \cdot p_5^{x_5} \cdot p_6^{x_6}$

$n = 20$
$p_1 = p_2 = p_3 = p_4 = p_5 = p_6 = 1/6$
$x_1 = 5,\ x_2 = 4,\ x_3 = 3,\ x_4 = 2,\ x_5 = 3,\ x_6 = 3$
Use the multinomial formula.

$$\begin{aligned}
P(x_1,x_2,x_3,x_4,x_5,x_6) &= [n!/(x_1!x_2!x_3!x_4!x_5!x_6!)] \cdot p_1^{x_1} \cdot p_2^{x_2} \cdot p_3^{x_3} \cdot p_4^{x_4} \cdot p_5^{x_5} \cdot p_6^{x_6} \\
&= [20!/(5!4!3!2!3!3!)] \cdot (1/6)^5 \cdot (1/6)^4 \cdot (1/6)^3 \cdot (1/6)^2 \cdot (1/6)^3 \cdot (1/6)^3 \\
&= [20!/(5!4!3!2!3!3!)] \cdot (1/6)^{20} \\
&= [1.955 \cdot 10^{12}] \cdot (2.735 \cdot 10^{-16}) \\
&= .000535
\end{aligned}$$

4.4 Mean, Variance and Standard Deviation for the Binomial Distribution

1. $\mu = n \cdot p = (64) \cdot (.5) = 32.0$
   $\sigma^2 = n \cdot p \cdot (1-p) = (64) \cdot (.5) \cdot (.5) = 16.0$
   $\sigma = 4.0$

3. $\mu = n \cdot p = (1068) \cdot (.25) = 267.0$
   $\sigma^2 = n \cdot p \cdot (1-p) = (1068) \cdot (.25) \cdot (.75) = 200.25$
   $\sigma = 14.2$

5. let x = the number of correct answers per student; $n = 25$, $p = .5$
   $\mu = n \cdot p = (25) \cdot (.5) = 12.5$
   $\sigma^2 = n \cdot p \cdot (1-p) = (25) \cdot (.5) \cdot (.5) = 6.25$
   $\sigma = 2.5$

7. let x = the number of times 7 appears; $n = 500$, $p = 1/38$
   $\mu = n \cdot p = (500) \cdot (1/38) = 13.2$
   $\sigma^2 = n \cdot p \cdot (1-p) = (500) \cdot (1/38) \cdot (37/38) = 12.8$
   $\sigma = 3.6$

9. let x = the number of consumers recognizing the Coke Brand name; $n = 1200$, $p = .95$
   a. $\mu = n \cdot p = (1200) \cdot (.95) = 1140.0$
      $\sigma^2 = n \cdot p \cdot (1-p) = (1200) \cdot (.95) \cdot (.05) = 57.0$
      $\sigma = 7.5$
   b. Unusual values are those outside $\mu \pm 2 \cdot \sigma$
      $1140.0 \pm 2 \cdot (7.5)$
      1140.0 ± 15.0
      1125.0 to 1155.0
   Yes, since 1170 is not within the above limits it would be considered an unusual result.

11. let x = the number of letter e's per page; $n = 2600$, $p = .130$
    a. $\mu = n \cdot p = (2600) \cdot (.130) = 338.0$
       $\sigma^2 = n \cdot p \cdot (1-p) = (2600) \cdot (.130) \cdot (.870) = 294.1$
       $\sigma = 17.1$
    b. Unusual values are those outside $\mu \pm 2 \cdot \sigma$
       $338.0 \pm 2 \cdot (17.1)$
       338.0 ± 34.2
       303.8 to 372.2
    No, since 307 is within the above limits it would not be considered an unusual result.

13. let x = the number of TV's tuned to *NFL Monday Night Football*; n = 4000, p = .30
    a. μ = n·p = (4000)·(.30) = 1200.0
      σ^2 = n·p·(1-p) = (4000)·(.30)·(.70) = 840.0
      σ = 29.0
    b. Unusual values are those outside $\mu \pm 2 \cdot \sigma$
                        1200.0 ± 2·(29.0)
                        1200.0 ± 58.0
                        1142.0 to 1258.0
    Yes, since 1272 is not within the above limits it would be considered an unusual result. The high number is likely caused by either the teams playing (e.g., popular teams or teams with popular individuals, etc) or the importance of the game (e.g., play-off berth at stake).

15. let x = the number of deaths attributed to myocardial infarctions; n = 5000, p = .149
    a. μ = n·p = (5000)·(.149) = 745.0
      σ^2 = n·p·(1-p) = (5000)·(.149)·(.851) = 634.0
      σ = 25.2
    b. Unusual values are those outside $\mu \pm 2 \cdot \sigma$
                        745.0 ± 2·(25.2)
                        745.0 ± 50.4
                        694.6 to 795.4
    Yes, since 896 is not within the above limits it would be considered an unusual result. There is cause for concern and/or further investigation. The high number may be attributed to a correctable cause (e.g., pollution) or an uncorrectable one (e.g., ethnicity).

17. let x = the number of computer requiring warranty repairs within 1 month; n = 279, p =.16
    a. μ = n·p = (279)·(.16) = 44.6
      σ^2 = n·p·(1-p) = (279)·(.16)·(.84) = 37.5
      σ = 6.1
    b. Unusual values are those outside $\mu \pm 2 \cdot \sigma$ NOTE: Carrying one more decimal
                        44.6 ± 2·(6.1)   gives 44.64 ± 2·(6.12)
                        44.6 ± 12.2   and   32.4 to 56.9
                        32.4 to 56.8
    And so 32 or fewer would be unusually low; 57 or more would be unusually high.

4-5 The Poisson Distribution

1. P(x) = $\mu^x \cdot e^{-\mu}/x!$ with μ = 2 and x = 3
  P(x=3) = $(2)^3 \cdot e^{-2}/3!$
         = (8)·(.135)/6
         = .180

3. P(x) = $\mu^x \cdot e^{-\mu}/x!$ with μ = .845 and x = 2
  P(x=2) = $(.845)^2 \cdot e^{-.845}/2!$
         = (.714)·(.430)/2
         = .153

5. let x = the number of births per day
  P(x) = $\mu^x \cdot e^{-\mu}/x!$ with μ = 2.25
  a. P(x=0) = $(2.25)^0 \cdot e^{-2.25}/0!$ = (1)·(.105)/1 = .105
  b. P(x=1) = $(2.25)^1 \cdot e^{-2.25}/1!$ = (2.25)·(.105)/1 = .237
  c. P(x=4) = $(2.25)^4 \cdot e^{-2.25}/4!$ = (25.629)·(.105)/24 = .113

7. let x = the number of accidents per week that require medical attention
   $P(x) = \mu^x \cdot e^{-\mu}/x!$ with $\mu = .2$
   a. $P(x=0) = (.2)^0 \cdot e^{-.2}/0! = (1) \cdot (.819)/1 = .819$
   b. $P(x=1) = (.2)^1 \cdot e^{-.2}/1! = (.2) \cdot (.819)/1 = .164$
   c. $P(x=2) = (.2)^2 \cdot e^{-.2}/2! = (.04) \cdot (.819)/2 = .0164$

9. let x = the number of defects per 500 feet of tape
   $P(x) = \mu^x \cdot e^{-\mu}/x!$ with $\mu = 2$
   $P(x>1) = 1 - P(x \le 1)$
   $\quad\quad\quad = 1 - [P(x=0) + P(x=1)]$
   $\quad\quad\quad = 1 - [(2)^0 \cdot e^{-2}/0! + (2)^1 \cdot e^{-2}/1!]$
   $\quad\quad\quad = 1 - [(1) \cdot (.135)/1 + (2) \cdot (.135)/1]$
   $\quad\quad\quad = 1 - [.1353 + .2707]$
   $\quad\quad\quad = 1 - .4060$
   $\quad\quad\quad = .594$

11. let x = the number of horse-kick deaths per corps per year
    $P(x) = \mu^x \cdot e^{-\mu}/x!$ with $\mu = 196/280 = .7$
    a. $P(x=0) = (.7)^0 \cdot e^{-.7}/0! = (1) \cdot (.497)/1 = .497$
    b. $P(x=1) = (.7)^1 \cdot e^{-.7}/1! = (.7) \cdot (.497)/1 = .348$
    c. $P(x=2) = (.7)^2 \cdot e^{-.7}/2! = (.49) \cdot (.497)/2 = .122$
    d. $P(x=3) = (.7)^3 \cdot e^{-.7}/3! = (.343) \cdot (.497)/6 = .0284$
    e. $P(x=4) = (.7)^4 \cdot e^{-.7}/4! = (.240) \cdot (.497)/24 = .00497$

    The following table compares the actual relative frequencies to the Poisson probabilities.

x	f	r.f.	P(x)	
0	144	.5143	.4966	
1	91	.3250	.3476	note: r.f. = f/Σf
2	32	.1143	.1217	
3	11	.0393	.0284	
4	2	.0071	.0050	
5 or more	.	.	.0007	(by subtraction)
	280	1.0000	1.0000	

    The agreement between the observed relative frequencies and the probabilities predicted by the Poisson formula is very good.

13. let x = the number of successes observed
    binomial experiment with n = 15, p = .01
    a. from Table A-1
       $P(x=1) = .130$
    b. from the Poisson formula with $\mu = n \cdot p = (15) \cdot (.01) = .15$
       $P(x) = \mu^x \cdot e^{-\mu}/x!$
       $P(x=1) = (.15)^1 \cdot e^{-.15}/1!$
       $\quad\quad\quad = (.15) \cdot (.861)/1$
       $\quad\quad\quad = .129$

    The approximate Poisson value is very close to the true binomial value, even though n = 15 < 100 does not meet the conditions recommended for using the Poisson to approximate the binomial.

Review Exercises

1. a. A random variable is a characteristic that assumes a single value (usually a numerical value), determined by chance, for each outcome of an experiment.
   b. A probability distribution is a statement of the probabilities associated with each value of a random variable. To be valid it must be true for each value x in the distribution that $0 \leq P(x) \leq 1$ and that $\Sigma P(x) = 1$.
   c. Yes, the given table is a valid probability distribution because it meets the definition and conditions in part (b) above.

The table below summarizes the calculations for parts (d) and (e).

x	P(x)	x·P(x)	x^2	x^2·P(x)
0	.0004	0	0	0
1	.0094	.0094	1	.0094
2	.0870	.1740	4	.3480
3	.3562	1.0686	9	3.2058
4	.5470	2.1880	16	8.7520
	1.0000	3.4400		12.3152

d. $\mu = \Sigma x \cdot P(x) = 3.4400$
   $= 3.4$ (rounded)
e. $\sigma^2 = \Sigma x^2 \cdot P(x) - \mu^2$
   $= 12.3152 - (3.4400)^2$
   $= .4816$
   $\sigma = .7$

2. let x = the number of small cars that are dark green
   binomial problem with n = 50, p = .15
   a. $E(x) = \mu = n \cdot p = (50) \cdot (.15) = 7.5$
   b. $\mu = n \cdot p = (50) \cdot (.15) = 7.5$
   c. $\sigma^2 = n \cdot p \cdot (1-p) = (50) \cdot (.15) \cdot (.85) = 6.375$
      $\sigma = 2.5$
   d. Unusual values are those outside $\mu \pm 2 \cdot \sigma$
      $7.5 \pm 2 \cdot (2.5)$
      7.5 ± 5.0
      2.5 to 12.5
      Yes, since 15 is not within these limits it would be considered an unusual result.
   e. $P(x) = [n!/x!(n-x)!] \cdot p^x \cdot (1-p)^{n-x}$
      $P(x=9) = [50!/9!41!] \cdot (.15)^9 \cdot (.85)^{41}$
      $= [2.51 \cdot 10^9] \cdot (3.84 \cdot 10^{-8}) \cdot (1.28 \cdot 10^{-3})$
      $= .123$

3. let x = the number of students owning videocassette recorders
   binomial problem with n = 10, p = .30. Use Table A-1.
   a. $P(x=5) = .103$
   b. $P(x \geq 5) = P(x=5) + P(x=6) + P(x=7) + P(x=8) + P(x=9) + P(x=10)$
      $= .103 + .037 + .009 + .001 + .000^+ + .000^+$
      $= .150$
   c. $\mu = n \cdot p = (10) \cdot (.30) = 3.0$
      $\sigma^2 = n \cdot p \cdot (1-p) = (10) \cdot (.30) \cdot (.70) = 2.1$
      $\sigma = 1.4$

4. let x = the number of workers fired for inability to get along with others
binomial problem with n = 5, p = .17. Use the binomial formula.
$P(x) = [n!/x!(n-x)!] \cdot p^x \cdot (1-p)^{n-x}$
   a. $P(x=0) = [5!/0!5!] \cdot (.17)^0 \cdot (.83)^5 = (1) \cdot (1) \cdot (.394) = .394$
   b. $P(x=4) = [5!/4!1!] \cdot (.17)^4 \cdot (.83)^1 = (5) \cdot (.000835) \cdot (.83) = .00347$
   c. $P(x=5) = [5!/5!0!] \cdot (.17)^5 \cdot (.83)^0 = (1) \cdot (.000142) \cdot (1) = .000142$
   d. $P(x=3) = [5!/3!2!] \cdot (.17)^3 \cdot (.83)^2 = (10) \cdot (.00491) \cdot (.689) = .0338$
      $P(x \geq 3) = P(x=3) + P(x=4) + P(x=5)$
      $= .03385 + .00347 + .00014$
      $= .0375$

5. let x = the number of workers fired for inability to get along with others
binomial problem with n = 5, p = .17
   a. $\mu = n \cdot p = (5) \cdot (.17) = .85$ [use 2-decimal accuracy since the mean is less than 1]
   b. $\sigma^2 = n \cdot p \cdot (1-p) = (5) \cdot (.17) \cdot (.83) = .7055$
      $\sigma = .84$
   c. Unusual values are those outside $\mu \pm 2 \cdot \sigma$
      $.85 \pm 2 \cdot (.84)$
      $.85 \pm 1.68$
      $-.83$ to 2.53
      0 to 2.53, since x cannot be negative
   Yes, since 4 is not within these limits it would be considered an unusual result.

6. let x = the number of breakdowns per day
Poisson problem with $\mu = 84/365 = .230$
   a. $\mu = 84/365 = .230$
   b. $P(x) = \mu^x \cdot e^{-\mu}/x!$
      $P(x=2) = (.230)^2 \cdot e^{-.230}/2!$
      $= (.0530) \cdot (.794)/2$
      $= .0210$

Cumulative Review Exercises

1. a. The table at the right was constructed
   using $\Sigma f = 89$ and r.f. $= f/\Sigma f$.
   b. Yes, use the r.f. column for the P(x)
   values and note that $0 \leq P(x) \leq 1$ and
   $\Sigma P(x) = 1$.
   c. $P(x \geq 5) = P(x=5) + P(x=6) + P(x=7)$
      $= .247 + .225 + .382$
      $= .854$
   d. let L = a series goes the 7 game limit
      $P(L) = 34/89$, only for the first series selected
      $P(L_1 \text{ and } L_2) = P(L_1) \cdot P(L_2 \mid L_1)$
      $= (34/89) \cdot (33/88)$
      $= .143$

x	r.f.
4	.146
5	.247
6	.225
7	.382
	1.000

The table below summarizes the calculations for parts (e) and (f).

x	P(x)	x·P(x)	x^2	x^2·P(x)
4	.146	.584	16	2.336
5	.247	1.235	25	6.175
6	.225	1.350	36	8.100
7	.382	2.674	49	18.718
	1.000	5.843		35.329

e. $\mu = \Sigma x \cdot P(x) = 5.843$
$= 5.8$ (rounded)
f. $\sigma^2 = \Sigma x^2 \cdot P(x) - \mu^2$
$= 35.329 - (5.843)^2$
$= 1.188$
$\sigma = 1.1$

g. $E(x) = \mu = 5.843$, rounded to 5.8
$E(\# \text{ of hot dogs}) = (30,000) \cdot (5.843) = 175,290$
NOTE: This is the long run average. Because x must be an integer, 5.8 games is not a possibility of any one series. If planning for one particular series, a more practical expectation for the number of hot dogs is $(30,000) \cdot (6) = 180,000$.

2. a. The summary statistics are $n = 20$, $\Sigma x = 193$, $\Sigma x^2 = 2027$
$\bar{x} = (\Sigma x)/n = 193/20 = 9.65$
$s^2 = [n \cdot (\Sigma x^2) - (\Sigma x)^2]/[n \cdot (n-1)]$
$= [20 \cdot (2027) - (193)^2]/[20 \cdot (19)]$
$= 3291/380$
$= 8.66$
$s = 2.9$

b. Of the 20 sample results, 8 were 12's. Use the relative frequency to estimate the probability.
estimated $P(x=12) = 8/20 = .400$
correct $P(x=12) = 1/36 = .0278$
These results do not agree. It appears that the dice are not balanced.

c. binomial problem with $n = 20$, $p = 1/36$. Use the binomial formula.
$P(x) = [n!/x!(n-x)!] \cdot p^x \cdot (1-p)^{n-x}$
$P(x=0) = [20!/0!20!] \cdot (1/36)^0 \cdot (35/36)^{20} = (1) \cdot (1) \cdot (.569) = .569$
$P(x \geq 1) = 1 - P(x < 1)$
$= 1 - P(x=0)$
$= 1 - .569$
$= .431$

d. The results obtained in court would be difficult to dispute. Here are some possible ideas and where they lead.

* The probability of at least one 12 is reasonable -- .431, from part (c). Maybe the probability of eight 12's isn't as low as one would guess.
binomial problem with $n = 20$, $p = 1/36$. Use the binomial formula.
$P(x) = [n!/x!(n-x)!] \cdot p^x \cdot (1-p)^{n-x}$
$P(x=8) = [20!/8!12!] \cdot (1/36)^8 \cdot (35/36)^{12}$
$= (125.970) \cdot (3.54 \cdot 10^{-13}) \cdot (.713)$
$= .0000000318$
Oops! That's one calculation that the defense definitely shouldn't show the court.

* Perhaps the criterion for what constitutes and unusual event would be helpful.
For a balanced pair of dice, the court has agreed that $\mu = 7.0$ and $\sigma = 2.4$.
Unusual values are those outside $\mu \pm 2 \cdot \sigma$
$7.0 \pm 2 \cdot (2.4)$
7.0 ± 4.8
2.2 to 11.8
The sample average in part (a) was $\bar{x} = 9.65$. Since this is well within these limits, does that mean the result from the accused person's dice should not be considered unusual?

Unfortunately, not. The above limits are for <u>one</u> x value, not for the <u>mean</u> of 20 x values. Rolling the dice <u>once</u> and getting a 9 or 10 would not be considered unusual. The statement says nothing about rolling a pair of dice 20 times and getting a <u>mean</u> of 9 or 10. [The Central Limit Theorem in the following chapter does address this concern, but the results would be very damaging to the accused.]

* The only hope appears to be to complain in non-statistical language that the sample is too small with a diversionary speech such as the following: "We all know that these dice could be rolled millions and millions of times. Using a sample of size 20 to reach a conclusion about all the millions and millions of possible results within these dice is just as foolish as allowing the opinions of a haphazard sample of size 20 to speak for the millions and millions of law-abiding citizens in this great country."

Chapter 5

Normal Probability Distributions

5-2 The Standard Normal Distribution

1. The height of the rectangle is .2. Probability corresponds to area, and the area of a rectangle is (width)·(height).
   $$P(x > 2) = \text{(width)}\cdot\text{(height)}$$
   $$= (5\text{-}2)\cdot(.2)$$
   $$= (3)\cdot(.2)$$
   $$= .6$$

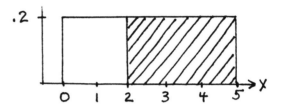

3. The height of the rectangle is .2. Probability corresponds to area, and the area of a rectangle is (width)·(height).
   $$P(2 < x < 4) = \text{(width)}\cdot\text{(height)}$$
   $$= (4\text{-}2)\cdot(.2)$$
   $$= (2)\cdot(.2)$$
   $$= .4$$

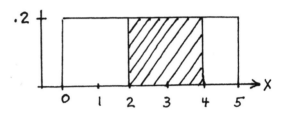

NOTE: The sketch is the key to exercises 5-28. It tells whether to add two Table A-2 probabilities, to subtract two Table A-2 probabilities, to subtract a Table A-2 probability from .5000, to add a Table A-2 probability to .5000, etc. It also often provides a check against gross errors by indicating at a glance whether the final probability is less than or greater than .5000. Remember that the symmetry of the normal curve implies two important facts:
   * There is always .5000 above and below the middle (i.e., at $z = 0$).
   * $P(\text{-}a < z < 0) = P(0 < z < a)$ for all values of "a."

5. $P(0 < z < 3.00)$
   $= .4987$

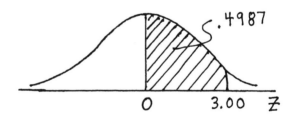

7. $P(\text{-}2.33 < z < 0)$
   $= P(0 < z < 2.33)$
   $= .4901$

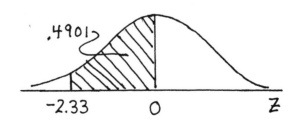

9. $P(z > 2.58)$
   $= P(z > 0) - P(0 < z < 2.58)$
   $= .5000 - .4951$
   $= .0049$

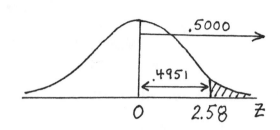

11. $P(z < -2.09)$
    $= P(z < 0) - P(-2.09 < z < 0)$
    $= .5000 - .4817$
    $= .0183$

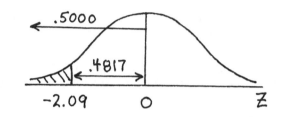

13. $P(1.34 < z < 2.67)$
    $= P(0 < z < 2.67) - P(0 < z < 1.34)$
    $= .4962 - .4099$
    $= .0863$

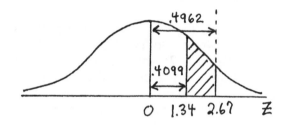

15. $P(-2.22 < z < -1.11)$
    $= P(-2.22 < z < 0) - P(-1.11 < z < 0)$
    $= .4868 - .3665$
    $= .1203$

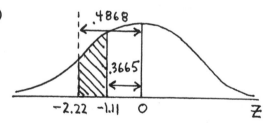

17. $P(z < 0.08)$
    $= P(z < 0) + P(0 < z < 0.08)$
    $= .5000 + .0319$
    $= .5319$

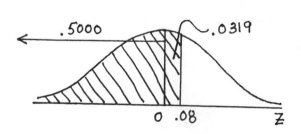

19. P(z > -2.29)
    = P(-2.29 < z < 0) + P(z > 0)
    = .4890 + .5000
    = .9890

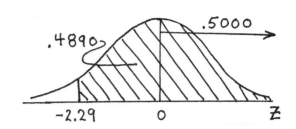

21. P(-1.99 < z < 2.01)
    = P(-1.99 < z < 0) + P(0 < z < 2.01)
    = .4767 + .4778
    = .9545

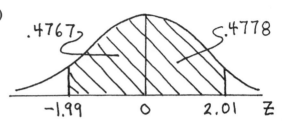

23. P(-1.00 < z < 4.00)
    = P(-1.00 < z < 0) + P(0 < z < 4.00)
    = .3413 + .4999
    = .8412

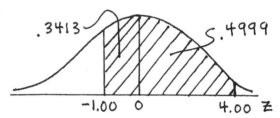

25. P(z > 2.33)
    = P(z > 0) - P(0 < z < 2.33)
    = .5000 - .4901
    = .0099

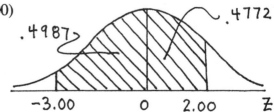

27. P(-3.00 < z < 2.00)
    = P(-3.00 < z < 0) + P(0 < z < 2.00)
    = .4987 + .4772
    = .9759

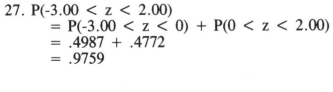

NOTE: The sketch is the key to exercises 29-36. It tells what probability is between 0 and the z score of interest (i.e. the area A to look up when reading Table A-2 "backwards." It also provides a check against gross errors by indicating at a glance whether a z score is above or below 0. Remember that the symmetry of the normal curve implies two important facts:
    * There is always .5000 above and below the middle (i.e., at z = 0).
    * P(-a < z < 0) = P(0 < z < a) for all values of "a."

29. For P_{90}, A = .4000.
    The closest entry is A = .3997, for which z = 1.28 [positive, since it is to the right of the middle, where z = 0].

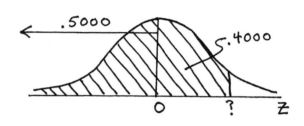

31. For Q_1, A = .2500.
    The closest entry is A = .2486, for which z = -.67 [negative, since it is to the left of the middle, where z = 0].

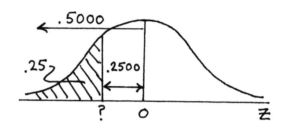

33. For the highest 4%, A = .4600.
    The closest entry is A = .4599, for which z = 1.75 [positive, since it is to the right of the middle, where z = 0].

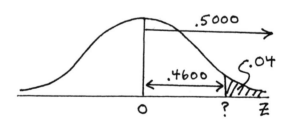

35. For the lowest 2%, A = .4800.
    The closest entry is A = .4798, for which z = -2.05 [negative, since it is to the left of the middle, where z = 0].

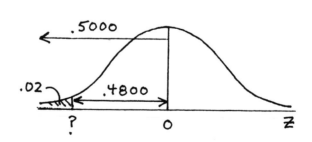

37. a. P(0 < z < a) = .3212
       A = .3212
       a = .92

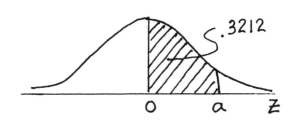

   b. P(-b < z < b) = .3182
      P(0 < z < b) = .3182/2 = .1591
      A = .1591
      b = .41

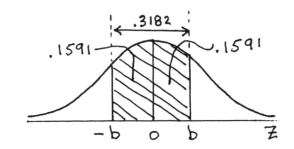

   c. P(z > c) = .2358
      P(0 < z < c) = .5000 - .2358 = .2642
      A = .2642
      c = .72

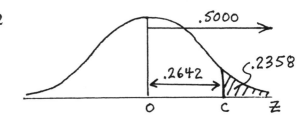

   d. P(z > d) = .7517
      P(d < z < 0) = .7517 - .5000 = .2517
      A = .2517
      d = -.68 [negative, since it is
            to the left of the middle,
            where z = 0]

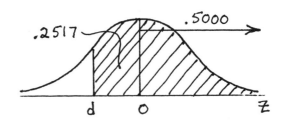

   e. P(z < e) = .4090
      P(e < z < 0) = .5000 - .4090 = .0910
      A = .0910
      e = -.23 [negative since it is
            to the left of the middle,
            where z = 0]

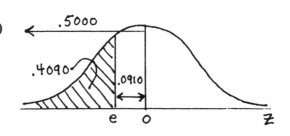

39.

x	$[2.7^{-(x^2)/2}]/2.5$	y
-4	$(2.7)^{-8}/2.5$	.00014
-3	$(2.7)^{-4.5}/2.5$	.00458
-2	$(2.7)^{-2}/2.5$	.05487
-1	$(2.7)^{-.5}/2.5$	.24343
0	$(2.7)^{-0}/2.5$	.40000
1	$(2.7)^{-.5}/2.5$	.24343
2	$(2.7)^{-2}/2.5$	.05487
3	$(2.7)^{-4.5}/2.5$	.00458
4	$(2.7)^{-8}/2.5$	.00014

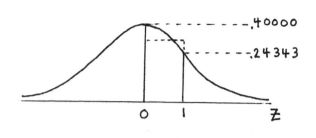

Approximate the area between z=0 and z=1 using a rectangle with width 1.0 and height (.40000 + .24343)/2 = .3217, the average of the heights of the curve at z=0 and z=1. The approximate area of (1.0)·(.3217) = .3217 compares well with the true area from Table A-2 of .3413.

5-3 Nonstandard Normal Distributions:
Finding Probabilities

NOTE: In each nonstandard normal distribution, x scores are converted to z scores using the formula $z = (x-\mu)/\sigma$ and rounded to two decimal places. As in the previous section, drawing and labeling the sketch is the key to successful completion of the exercises.

1. $\mu = 63.6$
   $\sigma = 2.5$
   $P(63.6 < x < 65.0)$
   $= P(0 < z < .56)$
   $= .2123$

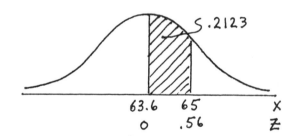

3. $\mu = 63.6$
   $\sigma = 2.5$
   $P(x > 58.1)$
   $= P(z > -2.20)$
   $= .4861 + .5000$
   $= .9861$

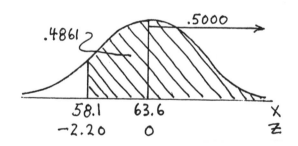

5. $\mu = 63.6$
   $\sigma = 2.5$
   $P(65.5 < x < 68.0)$
   $= P(.76 < z < 1.76)$
   $= .4608 - .2764$
   $= .1844$

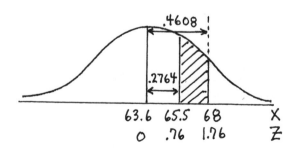

7. $\mu = 8.2$
$\quad \sigma = 1.1$
$\quad P(x < 7.0)$
$\qquad = P(z < -1.09)$
$\qquad = .5000 - .3621$
$\qquad = .1379$

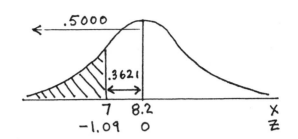

9. $\mu = 9.4$
$\quad \sigma = 4.2$
$\quad P(5.0 < x < 8.0)$
$\qquad = P(-1.05 < z < -.33)$
$\qquad = .3531 - .1293$
$\qquad = .2238$

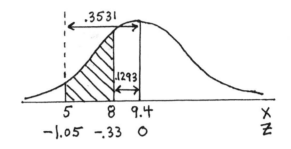

11. $\mu = 268$
$\quad \sigma = 15$
$\quad P(x \geq 308)$
$\qquad = P(z \geq 2.67)$
$\qquad = .5000 - .4962$
$\qquad = .0038$

Such a pregnancy is rare, but not impossible. Out of every one million pregnancies, we expect 3,800 to last 308 days or longer.

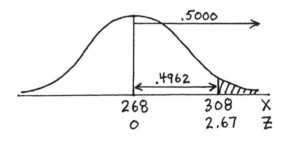

13. $\mu = 11.4$
$\quad \sigma = 1.8$
$\quad P(x \geq 10)$
$\qquad = P(z \geq -.78)$
$\qquad = .2823 + .5000$
$\qquad = .7823$

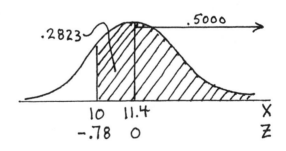

15. $\mu = 100$
$\quad \sigma = 15$

$\quad$ a. $P(x > 131.5) = P(z > 2.10)$
$\qquad\qquad\qquad\quad = .5000 - .4821$
$\qquad\qquad\qquad\quad = .0179$

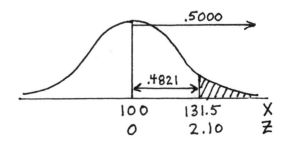

$\quad$ b. $(.0179) \cdot (75,000) = 1,342.5$
$\qquad$ about 1,350 people eligible

17. $\mu = 178.1$
 $\sigma = 40.7$
 $P(200 < x < 250)$
  $= P(.54 < z < 1.77)$
  $= .4616 - .2054$
  $= .2562$

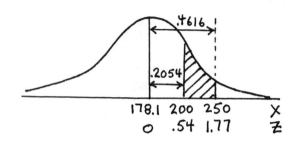

19. $\mu = 69.0$
 $\sigma = 2.8$
 $P(64 < x < 78)$
  $= P(-1.79 < z < 3.21)$
  $= .4633 + .4999$
  $= .9632$

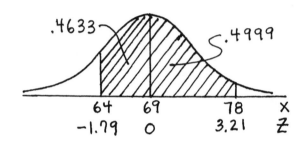

21. preliminary values: $n = 100$, $\Sigma x = 91.470$, $\Sigma x^2 = 83.802512$, $x_1 = 0.838$, $x_{100} = 1.033$

 a. $n = 100$: use about 8 classes
  class width: $(x_n - x_1)/8 = (1.033 - 0.838)/8$
   $= .0244$
   $= .025$ (rounded up)
  convenient first class midpoint: .850
  histogram: given at the right
    approximately normal

 b. $\bar{x} = \Sigma x/n = 91.470/100 = .9147$
  $s^2 = [n(\Sigma x^2) - (\Sigma x)^2]/[n(n-1)]$
   $= [100(83.802512) - (91.470)^2]/[100(99)]$
   $= (13.4903)/9900$
   $= .001363$
  $s = .03691$

 c. $\mu \approx .9147$
  $\sigma \approx .03691$
  $P(x > 1.000) = P(z > 2.31)$
    $= .5000 - .4896$
    $= .0104$
 NOTE: This compares reasonably well
 with the observed r.f. $= 3/100 = .03$.

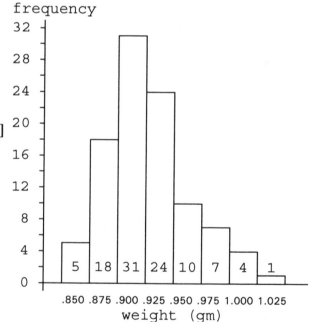

23. preliminary values: $n = 41$, $\Sigma x = 1331.4$, $\Sigma x^2 = 44489.46$, $x_1 = 21.6$, $x_{41} = 42.9$

  a. $n = 41$: use about 7 classes
    class width: $(x_n - x_1)/7 = (42.9 - 21.6)/7$
    $\qquad\qquad\qquad = 3.04$
    $\qquad\qquad\qquad = 3.5$ (rounded up)
    convenient first class midpoint: 22.0
    histogram: given at the right
    $\qquad\qquad$ approximately normal

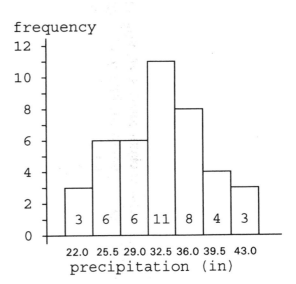

  b. $\bar{x} = \Sigma x/n = 1331.4/41 = 32.47$
    $s^2 = [n(\Sigma x^2) - (\Sigma x)^2]/[n(n-1)]$
    $\quad = [41(44489.46) - (1331.4)^2]/[41(40)]$
    $\quad = (51441.9)/1640$
    $\quad = 31.3673$
    $s = 5.601$

  c. $\mu \approx 32.47$
    $\sigma \approx 5.601$
    $P(x < 40.0) = P(z < 1.34)$
    $\qquad\qquad = .5000 + .4099$
    $\qquad\qquad = .9099$

  NOTE: This compares very well
  with the observed r.f. $= 37/41 = .9024$

5-4 Nonstandard Normal Distributions: Finding Scores

NOTE: The relationship between standardized and raw scores is given by $z = (x-\mu)/\sigma$. That formula may be solved, as needed, for any of the quantities on the right to yield

$\quad x = \mu + z\sigma \qquad\qquad \sigma = (x-\mu)/z \qquad\qquad \mu = x - z\sigma$

As in the previous section, drawing and labeling the sketch is the key to successful completion of the exercises. Remember that the z score for the mean is zero, and that z scores to the left of the mean must be negative.

1. $\mu = 63.6$
  $\sigma = 2.5$
  For P_{85}, $A = .3500$ [.3508] and $z = 1.04$.
  $x = \mu + z\sigma$
  $\quad = 63.6 + (1.04)(2.5)$
  $\quad = 63.6 + 2.6$
  $\quad = 66.2$ inches

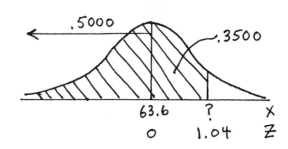

3. $\mu = 63.6$
  $\sigma = 2.5$
  For P_{15}, $A = .3500$ [.3508] and $z = -1.04$.
  $x = \mu + z\sigma$
  $\quad = 63.6 + (-1.04)(2.5)$
  $\quad = 63.6 - 2.6$
  $\quad = 61.0$ inches

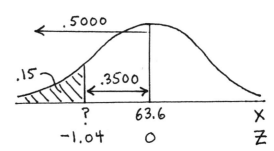

5. $\mu = 8.2$
   $\sigma = 1.1$
   For the top 20%, A = .3000 [.2995] and z = .84.
   $x = \mu + z\sigma$
   $ = 8.2 + (.84)(1.1)$
   $ = 8.2 + 0.9$
   $ = 9.1$ years

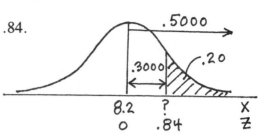

7. $\mu = 9.4$
   $\sigma = 4.2$
   For the bottom 33%, A = .1700 and z = -.44.
   $x = \mu + z\sigma$
   $ = 9.4 + (-.44)(4.2)$
   $ = 9.4 - 1.8$
   $ = 7.6$ lbs

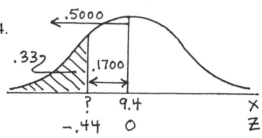

9. $\mu = 268$
   $\sigma = 15$
   For the lowest 4%, A = .4600 [.4599] and z = -1.75.
   $x = \mu + z\sigma$
   $ = 268 + (-1.75)(15)$
   $ = 268 - 26$
   $ = 242$ days

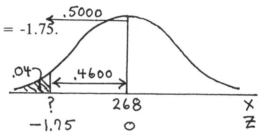

11. $\mu = 100$
    $\sigma = 15$
    For the top 1%, A = .4900 [.4901] and z = 2.33.
    $x = \mu + z\sigma$
    $ = 100 + (2.33)(15)$
    $ = 100 + 35$
    $ = 135$

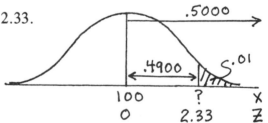

13. $\mu = 178.1$
    $\sigma = 40.7$

    a. $P(x < 200) = P(z < .54)$
    $\phantom{a. P(x < 200)} = .5000 + .2054$
    $\phantom{a. P(x < 200)} = .7054$

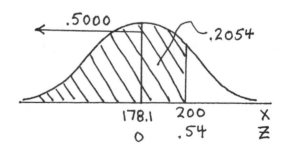

b. For the top 7%, A = .4300 [.4306] and z = 1.48.
   x = μ + zσ
      = 178.1 + (1.48)(40.7)
      = 178.1 + 60.2
      = 238.3 mg/100 mL

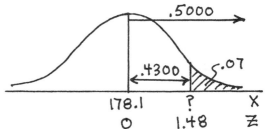

15. μ = 63.6
    σ = 2.5

   a. P(58 < x < 73) = P(-2.24 < z < 3.76)
                    = .4875 + .4999
                    = .9874

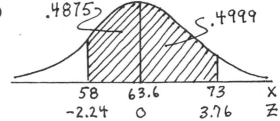

   b. For the shortest 1%, A = .4900 [.4901] and z = -2.33.
      x = μ + zσ
         = 63.6 + (-2.33)(2.5) = 63.6 - 5.8 = 57.8 inches
      For the tallest 1%, A = .4900 [.4901] and z = 2.33.
      x = μ + zσ
         = 63.6 + (2.33)(2.5) = 63.6 + 5.8 = 69.4 inches

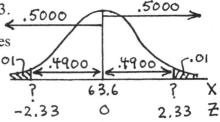

17. Because of the symmetry, either Q_1 or Q_2 gives the same answer.
    μ = 70
    σ = ?

    For Q_1 = P_{25},   OR For Q_3 = P_{75},
    A_L = .2500 [.2486] A_U = .2500 [.2486]
    z_L = -.67 z_U = .67
    z = (x-μ)/σ   z = (x-μ)/σ
    σ = (x-μ)/z   σ = (x-μ)/z
      = (62-70)/(-.67)   = (78-70)/(.67)
      = (-8)/(-.67)   = (8)/(.67)
      = 11.9   = 11.9

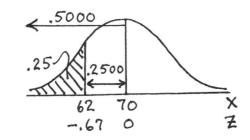

19. μ = 475
    σ = ?
    For the top 17%,
    A = .3300 [.3289]
    z = .95
    z = (x-μ)/σ
    σ = (x-μ)/z
      = (600-475)/.95
      = 125/.95
      = 131.6

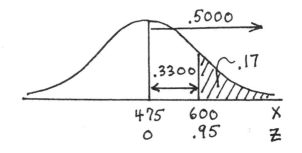

For P_{99},
A = .4900 [.4901]
z = 2.33
$x = \mu + z\sigma$
  = 475 + (2.33)(131.6)
  = 475 + 306.6
  = 782

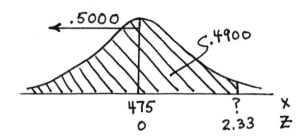

5-5 The Central Limit Theorem

NOTE: When using individual scores (i.e., making a statement about one x score from the original distribution), convert x to z using the mean and standard deviation of the x's and $z = (x-\mu)/\sigma$.

When using a sample of n scores (i.e., making a statement about $\bar{x}$), convert $\bar{x}$ to z using the mean and standard deviation of the $\bar{x}$'s and $z = (\bar{x}-\mu_{\bar{x}})/\sigma_{\bar{x}}$.

IMPORTANT NOTE: After calculating $\sigma_{\bar{x}}$, <u>STORE IT</u> in the calculator to recall it with total accuracy whenever it is needed in subsequent calculations. <u>DO NOT</u> write it down on paper rounded off (even to several decimal places) and then re-enter it in the calculator whenever it is needed. This avoids both round-off errors and recopying errors.

1. a. normal distribution
   $\mu = 63.6$
   $\sigma = 2.5$
   $P(63.6 < x < 64.6) = P(0 < z < .40)$
   $\qquad\qquad\qquad\qquad = .1554$

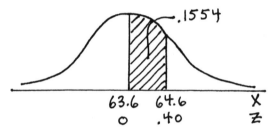

   b. normal distribution, since the original distribution is so
   $\mu_{\bar{x}} = \mu = 63.6$
   $\sigma_{\bar{x}} = \sigma/\sqrt{n} = 2.5/\sqrt{36} = .417$
   $P(63.6 < \bar{x} < 64.6) = P(0 < z < 2.40)$
   $\qquad\qquad\qquad\qquad = .4918$

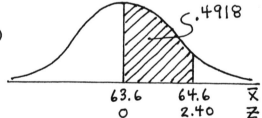

3. a. normal distribution
   $\mu = 63.6$
   $\sigma = 2.5$
   $P(x > 64.0) = P(z > .16)$
   $\qquad\qquad = .5000 - .0636$
   $\qquad\qquad = .4364$

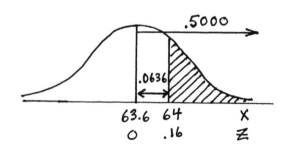

b. normal distribution, since the original distribution is so

$\mu_{\bar{x}} = \mu = 63.6$

$\sigma_{\bar{x}} = \sigma/\sqrt{n} = 2.5/\sqrt{50} = .354$

$P(\bar{x} > 64.0) = P(z > 1.13)$
$= .5000 - .3708$
$= .1292$

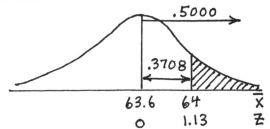

5. normal distribution, since the original distribution is so

$\mu_{\bar{x}} = \mu = 8.2$

$\sigma_{\bar{x}} = \sigma/\sqrt{n} = 1.1/\sqrt{40} = .174$

$P(\bar{x} < 8.0) = P(z < -1.15)$
$= .5000 - .3749$
$= .1251$

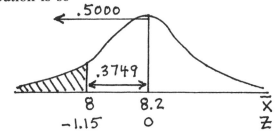

7. normal distribution, since the original distribution is so

$\mu_{\bar{x}} = \mu = 11.4$

$\sigma_{\bar{x}} = \sigma/\sqrt{n} = 1.8/\sqrt{33} = .313$

$P(11.0 < \bar{x} < 12.0) = P(-1.28 < z < 1.91)$
$= .3997 + .4719$
$= .8716$

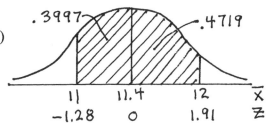

9. a. normal distribution

$\mu = 114.8$

$\sigma = 13.1$

$P(x > 120) = P(z > .40)$
$= .5000 - .1554$
$= .3446$

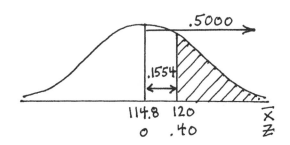

b. normal distribution, since the original distribution is so

$\mu_{\bar{x}} = \mu = 114.8$

$\sigma_{\bar{x}} = \sigma/\sqrt{n} = 13.1/\sqrt{12} = 3.78$

$P(\bar{x} > 120) = P(z > 1.38)$
$= .5000 - .4162$
$= .0838$

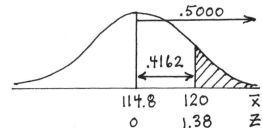

c. If the original x values are normally distributed, then the $\bar{x}$'s from a random sample on size n will also be normally distributed regardless of the sample size.

11. normal distribution, by the Central Limit Theorem
[since 31/630 < .05, the finite population correction factor is not needed]
$\mu_{\bar{x}} = \mu = 13.0$
$\sigma_{\bar{x}} = \sigma/\sqrt{n} = 7.9/\sqrt{35} = 1.34$
$P(\bar{x} > 15.0) = P(z > 1.50)$
$\qquad = .5000 - .4332$
$\qquad = .0668$

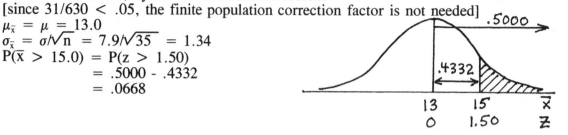

13. normal distribution, by the Central Limit Theorem
$\mu_{\bar{x}} = \mu = .500$
$\sigma_{\bar{x}} = \sigma/\sqrt{n} = .289/\sqrt{45} = .043$
$P(\bar{x} < .565) = P(z < 1.51)$
$\qquad = .5000 + .4345$
$\qquad = .9345$

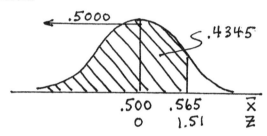

15. normal distribution, since the original distribution is so
$\mu_{\bar{x}} = \mu = 27.44$
$\sigma_{\bar{x}} = \sigma/\sqrt{n} = 12.46/\sqrt{4872} = .179$
$P(\bar{x} > 27.88) = P(z > 2.46)$
$\qquad = .5000 - .4931$
$\qquad = .0069$

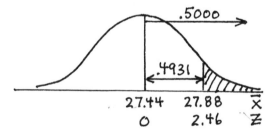

The system is currently acceptable and can expect to be overloaded only (.0069)(52) = .36 weeks a year -- or about once every three years.

17. a. normal distribution
$\mu = 268$
$\sigma = 15$
$P(x < 260) = P(z < -.53)$
$\qquad = .5000 - .2019$
$\qquad = .2981$

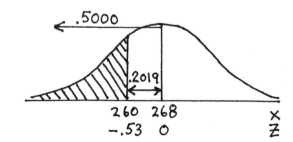

b. normal distribution, since the original distribution is so

$\mu_{\bar{x}} = \mu = 268$

$\sigma_{\bar{x}} = \sigma/\sqrt{n} = 15/\sqrt{25} = 3.00$

$P(\bar{x} < 260) = P(z < -2.67)$

$= .5000 - .4968$

$= .0038$

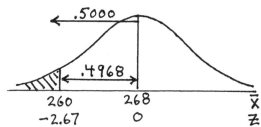

c. Yes; there is reason for concern. Because the diet seems to shorten the length of pregnancy, it may have an adverse affect on fetal development.

19. $\mu = .9147$

$\sigma = .0369$

a. This cannot be answered without knowing the shape of the distribution.

NOTE: If the weights of the M&M's are normally distributed, then Table A-2 can be used and $P(x > .9085) = P(z > -.17) = .0675 + .5000 = .5675$.

b. normal distribution, by the Central Limit Theorem

$\mu_{\bar{x}} = \mu = .9147$

$\sigma_{\bar{x}} = \sigma/\sqrt{n} = .0369/\sqrt{1498} = .000953$

$P(\bar{x} > .9085) = P(z > -6.50)$

$= .4999 + .5000$

$= .9999$

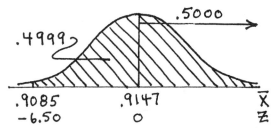

c. Yes; the company is providing the amount claimed on the label with almost absolute certainty.

21. normal distribution, since the original distribution is so

[since 32/500 > .05, the finite population correction factor must be used]

$\mu_{\bar{x}} = \mu = 173$

$\sigma_{\bar{x}} = [\sigma/\sqrt{n}] \cdot \sqrt{(N-n)/(N-1)}$

$= [30/\sqrt{32}] \cdot \sqrt{(500-32)/(500-1)}$

$= [30/\sqrt{32}] \cdot \sqrt{(468)/(499)}$

$= 5.136$

$P(\bar{x} > 186) = P(z > 2.53)$

$= .5000 - .4943$

$= .0057$

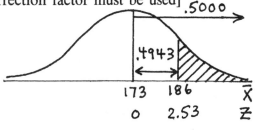

23. normal distribution, since the original distribution appears so

[since 15/350 < .05, the finite population correction factor is not needed]

NOTE: From the boxplot we determine,

$\bar{x} = 50$

$s = 22.4$

In a normal distribution, $P(0 < z < .67) = .2500$.

The boxplot indicates that $Q_2 = 50$ and $Q_3 = 65$, and so $P(50 < x < 65) = .2500$.

Solving $.67 = z_{65} = (x-\bar{x})/s$ for s: $s = (x-\bar{x})/z$

$= (65-50)/.67 = 15/.67 = 22.4$

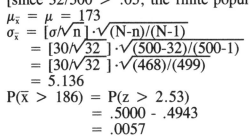

Using $\bar{x}$ and s to estimate μ and σ
$\mu_{\bar{x}} = \mu = 50$
$\sigma_{\bar{x}} = \sigma/\sqrt{n} = 22.4/\sqrt{15} = 5.78$
$P(\bar{x} > 55) = P(z > .86)$
$\qquad\qquad = .5000 - .3051$
$\qquad\qquad = .1949$

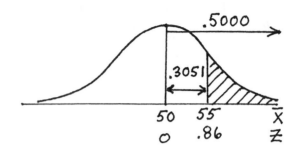

5-6 Normal Distribution as Approximation to Binomial Distribution

NOTE: As in the previous sections, P(E) represents the probability of an event E; this manual uses $P_c(E)$ to represent the probability of an event E with the continuity correction applied.

1. the area to the right of 35.5
   in symbols, $P(x > 35) = P_c(x > 35.5)$

3. the area to the left of 41.5
   in symbols, $P(x < 42) = P_c(x < 41.5)$

5. the area to the left of 72.5
   in symbols, $P(x \leq 72) = P_c(x < 72.5)$

7. the area from 124.5 to 150.5
   in symbols, $P(125 \leq x \leq 150) = P_c(124.5 < x < 150.5)$

IMPORTANT NOTE: As in the previous sections, store σ in the calculator so that it may be recalled with complete accuracy whenever it is needed in subsequent calculations.

9. binomial: n = 14 and p = .50
   a. from Table A-1, P(x = 8) = .183
   b. normal approximation appropriate since
      $np = 14(.50) = 7 \geq 5$
      $n(1-p) = 14(.50) = 7 \geq 5$
      $\mu = np = 14(.50) = 7$
      $\sigma = \sqrt{np(1-p)} = \sqrt{14(.50)(.50)} = 1.871$
      $P(x = 8) = P_c(7.5 < x < 8.5)$
      $\qquad\qquad = P(.27 < z < .80)$
      $\qquad\qquad = .2881 - .1064$
      $\qquad\qquad = .1817$

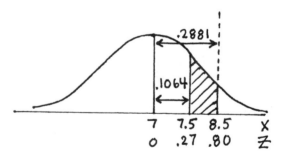

11. binomial: n = 15 and p = .80
    a. from Table A-1, $P(x \geq 8) = P(x = 8) + P(x = 9) +...+ P(x = 15)$
       $\qquad\qquad\qquad\qquad = .014 + .043 +...+ .035$
       $\qquad\qquad\qquad\qquad = .996$
    b. normal approximation <u>not</u> appropriate since
       $np = 15(.80) = 12 \geq 5$
       but $n(1-p) = 15(.20) = 3 < 5$

13. binomial: n = 100 and p = .50
    normal approximation appropriate since
    $\quad np = 100(.50) = 50 \geq 5$
    $\quad n(1-p) = 100(.50) = 50 \geq 5$
    $\mu = np = 100(.50) = 50$
    $\sigma = \sqrt{np(1-p)} = \sqrt{100(.50)(.50)} = 5.000$
    $P(x \geq 55) = P_c(x > 54.5)$
    $\qquad\qquad = P(z > .90)$
    $\qquad\qquad = .5000 - .3159$
    $\qquad\qquad = .1841$

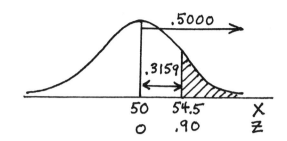

15. binomial: n = 50 and p = .50
    normal approximation appropriate since
    $\quad np = 50(.50) = 25 \geq 5$
    $\quad n(1-p) = 50(.50) = 25 \geq 5$
    $\mu = np = 50(.50) = 25$
    $\sigma = \sqrt{np(1-p)} = \sqrt{50(.50)(.50)} = 3.536$
    $P(x \geq 30) = P_c(x > 29.5)$
    $\qquad\qquad = P(z > 1.27)$
    $\qquad\qquad = .5000 - .3980$
    $\qquad\qquad = .1020$

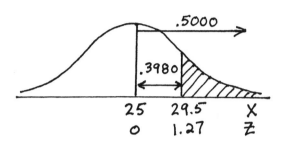

17. binomial: n = 100 and p = .80
    normal approximation appropriate since
    $\quad np = 100(.80) = 80 \geq 5$
    $\quad n(1-p) = 100(.20) = 20 \geq 5$
    $\mu = np = 100(.80) = 80$
    $\sigma = \sqrt{np(1-p)} = \sqrt{100(.80)(.20)} = 4.000$
    $P(x = 85) = P_c(84.5 < x < 85.5)$
    $\qquad\qquad = P(1.125 < z < 1.375)$
    $\qquad\qquad = .4162 - .3686$
    $\qquad\qquad = .0476$

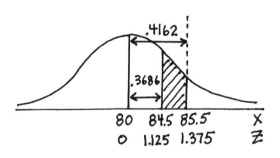

NOTE: Answers may vary due to rounding the z scores. The above answer was obtained by rounding values exactly half way between to the even digit and using z = 1.12, 1.38. Rounding both values up yields .4162 - .3708 = .0454, rounding both values down yields .4147 - .3686 = .0461. Other more complex (and not recommended) options include interpolating from Table A-2 and averaging two final answers. By the binomial formula, the correct answer is .0481.

19. binomial: n = 100 and p = .10
    normal approximation appropriate since
    $\quad np = 100(.10) = 10 \geq 5$
    $\quad n(1-p) = 100(.90) = 90 \geq 5$
    $\mu = np = 100(.10) = 10$
    $\sigma = \sqrt{np(1-p)} = \sqrt{100(.10)(.90)} = 3.000$
    $P(x \leq 5) = P_c(x < 5.5)$
    $\qquad\qquad = P(z < -1.50)$
    $\qquad\qquad = .5000 - .4332$
    $\qquad\qquad = .0668$

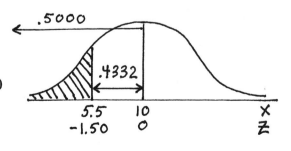

No, it is not very unusual to get 5 or fewer blue M&M's in a group of 100. Such an occurrence can be expected about 7% of the time.

21. binomial: n = 400 and p = .16
    normal approximation appropriate since
       np = 400(.16) = 64 ≥ 5
       n(1-p) = 400(.84) = 336 ≥ 5
    μ = np = 400(.16) = 64
    σ = $\sqrt{np(1-p)}$ = $\sqrt{400(.16)(.84)}$ = 7.332
    P(x ≤ 38) = P_c(x < 38.5)
                = P(z < -3.48)
                = .5000 - .4999
                = .0001

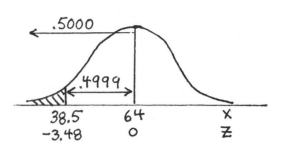

No; it does not seem plausible that 16% of the state's arrestees are women.

23. binomial: n = 150 and p = .038
    normal approximation appropriate since
       np = 150(.038) = 5.7 ≥ 5
       n(1-p) = 150(.962) = 144.3 ≥ 5
    μ = np = 150(.038) = 5.7
    σ = $\sqrt{np(1-p)}$ = $\sqrt{150(.038)(.962)}$ = 2.342
    P(x ≥ 10) = P_c(x > 9.5)
                = P(z > 1.62)
                = .5000 - .4474
                = .0526

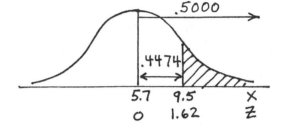

No, it is not very unusual to get 10 or more positive results in a group of 150. Such an occurrence can be expected about 5% of the time.

25. binomial: n = 300 and p = .184
    normal approximation appropriate since
       np = 300(.184) = 55.2 ≥ 5
       n(1-p) = 300(.816) = 244.80 ≥ 5
    μ = np = 300(.184) = 55.2
    σ = $\sqrt{np(1-p)}$ = $\sqrt{300(.184)(.816)}$ = 6.711
    P(x ≥ 72) = P_c(x > 71.5)
                = P(z > 2.43)
                = .5000 - .4925
                = .0075

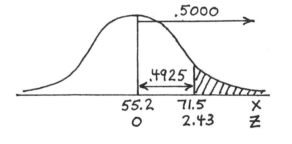

Under the conditions of the problem, getting 72 or more smokers by chance alone is very unlikely. It appears either that the 18.4% figure is incorrect or that there is something wrong with the sample.

27. binomial: n = 400 and p = .45
    normal approximation appropriate since
       np = 400(.45) = 180 ≥ 5
       n(1-p) = 400(.55) = 220 ≥ 5
    μ = np = 400(.45) = 180
    σ = $\sqrt{np(1-p)}$ = $\sqrt{400(.45)(.55)}$ = 9.950
    P(x ≥ 177) = P_c(x > 176.5)
                 = P(z > -.35)
                 = .1368 + .5000
                 = .6368

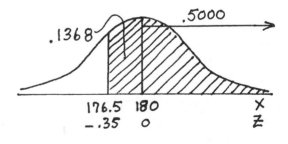

29. normal: $\mu = 8.2$ and $\sigma = 1.1$

$$P(x > 10.0) = P(z > 1.64)$$
$$= .5000 - .4495$$
$$= .0505$$

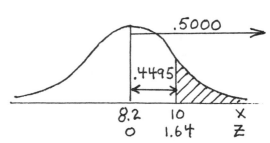

binomial: $n = 250$ and $p = .0505$
normal approximation appropriate since
   $np = 250(.0505) = 12.625 \geq 5$
   $n(1-p) = 250(.9495) = 237.375 \geq 5$
$\mu = np = 250(.0505) = 12.625$
$\sigma = \sqrt{np(1-p)} = \sqrt{250(.0505)(.9495)} = 3.462$

$$P(x \geq 15) = P_c(x > 14.5)$$
$$= P(z > .54)$$
$$= .5000 - .2054$$
$$= .2946$$

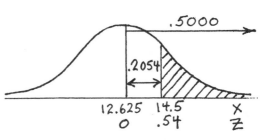

31. The normal approximation to the binomial applies in all parts of this exercise.

a. $\mu = np = 20(.5) = 10$
   $\sigma = \sqrt{np(1-p)} = \sqrt{20(.5)(.5)} = 2.236$

without the correction
$$P(x \geq 11) = P(z > .45)$$
$$= .5000 - .1736$$
$$= .3264$$

with the correction
$$P(x \geq 11) = P_c(x > 10.5)$$
$$= P(z > .22)$$
$$= .5000 - .0871$$
$$= .4129$$

The difference is $.4129 - .3264 = .0865$. Assuming the value obtained using the correction is closer to the correct value, the percent error is $.0865/.4129 = 20.9\%$.

b. $\mu = np = 40(.5) = 20$
   $\sigma = \sqrt{np(1-p)} = \sqrt{40(.5)(.5)} = 3.162$

without the correction
$$P(x \geq 22) = P(z > .63)$$
$$= .5000 - .2357$$
$$= .2643$$

with the correction
$$P(x \geq 22) = P_c(x > 21.5)$$
$$= P(z > .47)$$
$$= .5000 - .1808$$
$$= .3192$$

The difference is $.3192 - .2643 = .0549$. Assuming the value obtained using the correction is closer to the correct value, the percent error is $.0549/.3192 = 17.2\%$.

c. $\mu = np = 400(.5) = 200$
   $\sigma = \sqrt{np(1-p)} = \sqrt{400(.5)(.5)} = 10.000$

without the correction
$$P(x \geq 220) = P(z > 2.00)$$
$$= .5000 - .4772$$
$$= .0228$$

with the correction
$$P(x \geq 220) = P_c(x > 219.5)$$
$$= P(z > 1.95)$$
$$= .5000 - .4744$$
$$= .0256$$

The difference is $.0256 - .0228 = .0028$. Assuming the value obtained using the correction is closer to the correct value, the percent error is $.0028/.0256 = 10.9\%$.

As the sample size increases, the difference between the continuity-corrected and uncorrected normal approximation to the binomial grows smaller -- both in absolute size and in relative size.

Review Exercises

1. normal: $\mu = 69.0$ and $\sigma = 2.8$
   For the shortest 5%, A = .4500 and z = -1.645
   $x = \mu + z\sigma$
   $\quad = 69.0 + (-1.645)(2.8)$
   $\quad = 69.0 - 4.6$
   $\quad = 64.4$

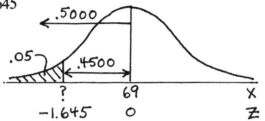

   For the tallest 5%, A = .4500 and z = 1.645
   $x = \mu + z\sigma$
   $\quad = 69.0 + (1.645)(2.8)$
   $\quad = 69.0 + 4.6$
   $\quad = 73.6$

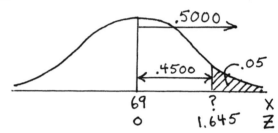

   Stock suits for men with heights between 64.4 and 73.6 inches.

2. normal: $\mu = 69.0$ and $\sigma = 2.8$
   $P(x \geq 74)$
   $\quad = P(z > 1.79)$
   $\quad = .5000 - .4633$
   $\quad = .0367$

3. normal: $\mu = 69.0$ and $\sigma = 2.8$
   $P(62.0 < x < 76.0)$
   $\quad = P(-2.50 < z < 2.50)$
   $\quad = .4938 + .4938$
   $\quad = .9876$

4. normal, since the original distribution is so
   $\mu_{\bar{x}} = \mu = 69.0$
   $\sigma_{\bar{x}} = \sigma/\sqrt{n} = 2.8/\sqrt{45} = .417$
   $P(70.0 < \bar{x} < 71.0)$
   $\quad = P(2.40 < z < 4.79)$
   $\quad = .4999 - .4918$
   $\quad = .0081$

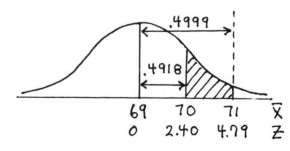

5. binomial: n = 500 and p = .26
   normal approximation appropriate since
   $\quad np = 500(.26) = 130 \geq 5$
   $\quad n(1-p) = 500(.74) = 370 \geq 5$
   $\mu = np = 500(.26) = 130$
   $\sigma = \sqrt{np(1-p)} = \sqrt{500(.26)(.74)} = 9.808$
   $P(125 \leq x \leq 150) = P_c(124.5 < x < 150.5)$
   $\qquad\qquad\qquad\quad = P(-.56 < z < 2.09)$
   $\qquad\qquad\qquad\quad = .2123 + .4817$
   $\qquad\qquad\qquad\quad = .6940$

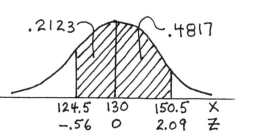

6. normal distribution with $\mu = 0$ and $\sigma = 1$
   [Because this is a standard normal, the variable is already expressed in terms of z scores.]

   a. $P(0 < z < 1.25)$
      $\quad = .3944$

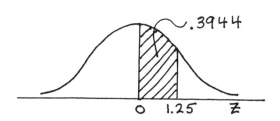

   b. $P(z > .50)$
      $\quad = .5000 - .1915$
      $\quad = .3085$

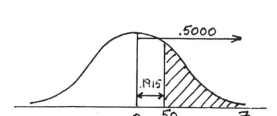

   c. $P(z > -1.08)$
      $\quad = .3599 + .5000$
      $\quad = .8599$

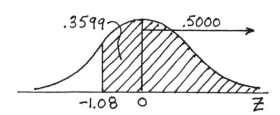

   d. $P(-.50 < z < 1.50)$
      $\quad = .1915 + .4332$
      $\quad = .6247$

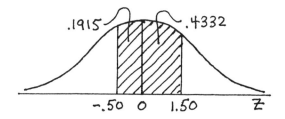

   e. $P(-1.00 < z < -.25)$
      $\quad = .3413 - .0987$
      $\quad = .2426$

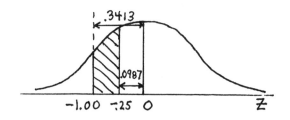

7. normal: $\mu = 8.0$ and $\sigma = 2.6$
   $P(6.0 < x < 7.0)$
   $= P(-.77 < z < -.38)$
   $= .2794 - .1480$
   $= .1314$

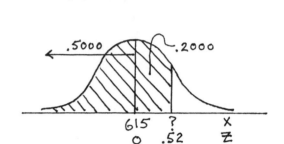

   $(.1314)(600) = 78.8$, or about 79 people

8. normal: $\mu = 615$ and $\sigma = 107$
   For P_{70}, $A = .2000$ [.1985] and $z = .52$
   $x = \mu + z\sigma$
   $= 615 + (.52)(107)$
   $= 615 + 55.6$
   $= 671$

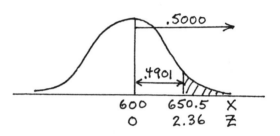

9. binomial: $n = 2500$ and $p = .24$
   normal approximation appropriate since
      $np = 2500(.24) = 600 \geq 5$
      $n(1-p) = 2500(.76) = 1900 \geq 5$
   $\mu = np = 2500(.24) = 600$
   $\sigma = \sqrt{np(1-p)} = \sqrt{2500(.24)(.76)} = 21.354$
   $P(x > 650) = P_c(x > 650.5)$
   $\qquad\qquad = P(z > 2.36)$
   $\qquad\qquad = .5000 - .4901$
   $\qquad\qquad = .0091$

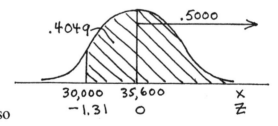

10. a. normal: $\mu = 35,600$ and $\sigma = 4275$
    $P(x > 30,000)$
    $\qquad = P(z > -1.31)$
    $\qquad = .4049 + .5000$
    $\qquad = .9049$

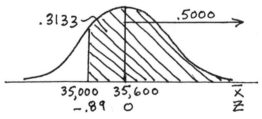

   b. normal, since the original distribution is so
    $\mu_{\bar{x}} = \mu = 35,600$
    $\sigma_{\bar{x}} = \sigma/\sqrt{n} = 4275/\sqrt{40} = 675.9$
    $P(\bar{x} > 35,000)$
    $\qquad = P(z > -.89)$
    $\qquad = .3133 + .5000$
    $\qquad = .8133$

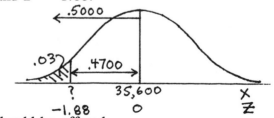

   c. For the lowest 3%, $A = .4700$ [.4699] and $z = -1.88$.
    $x = \mu + z\sigma$
    $\qquad = 35,600 + (-1.88)(4275)$
    $\qquad = 35,600 - 8037$
    $\qquad = 27,563$

   A guarantee of 27,000 or 27,500 miles should be offered.

Cumulative Review Exercises

1. a. Let L = a person is left-handed
   P(L) = .10, for each random selection
   P(L₁ and L₂ and L₃) = $P(L_1) \cdot P(L_2) \cdot P(L_3)$
                        = $(.10) \cdot (.10) \cdot (.10)$
                        = .001

   b. Let N = a person is not left-handed
   P(N) = .90, for each random selection
   P(at least one left-hander) = 1 - P(no left-handers)
                        = 1 - P(N₁ and N₂ and N₃)
                        = $1 - P(N_1) \cdot P(N_2) \cdot P(N_3)$
                        = 1 - (.90) · (.90) · (.90)
                        = 1 - .729
                        = .271

   c. binomial: n = 3 and p = P(left-hander) = .10
   normal approximation <u>not</u> appropriate since
      np = 3(.10) = 0.3 < 5
      [and n(1-p) = 3(.90) = 2.7 < 5, also]

   d. binomial: n = 50 and p = .10
   μ = np = 50(.10) = 5

   e. binomial: n = 50 and p = .10
   $\sigma = \sqrt{np(1-p)} = \sqrt{50(.10)(.90)}$ = 2.121

   f. An unusual score is one that is more than 2 standard deviations from the mean.
   Since z = (x-μ)/σ, z₈ = (8-5)/2.212 = 1.41.
   Since 8 is 1.41 standard deviations from the mean, it is <u>not</u> an unusual score.

2. summary values: n = 40, Σx = 8018.4, Σx^2 = 1,607,371.66

   a. $\bar{x}$ = (Σx)/n = (8018.4)/40 = 200.46 ms

   b. $\tilde{x}$ = (200.4 + 200.5)/2 = 200.45 ms

   c. M = 200.5 ms

   d. $s^2 = [n(\Sigma x^2) - (\Sigma x)^2]/[n(n-1)]$
      = [40(1,607,371.66) - (8018.4)²]/[40(39)]
      = 127.84/1560
      = .081949
   s = .29 ms

   e. z = (x-$\bar{x}$)/s
      z₂₀₀.₅ = (200.5 - 200.46)/.2863 = .14

   f. r.f. of scores greater than 201.0 is 3/40 = .075

   g. normal: μ = 200.46 and σ = .2863
      P(x > 201.0) = P(z > 1.89)
                   = .5000 - .4706
                   = .0294

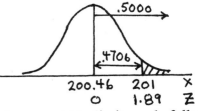

   h. Yes, since none of the 40 sample results were outside the acceptable limits and, following
   the methods of part (g),
   P(unacceptable result)
      = P(x < 197.0 or x > 202.0)
      = P(x < 197.0) + P(x > 202.0)
      = P(z < -12.09) + P(z > 5.38)
      = (.5000 - .4999) + (.5000 - .4999)
      = .0001 + .0001
      = .0002

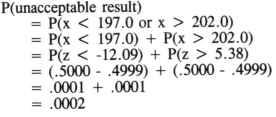
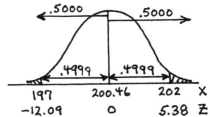

Chapter 6

Estimates and Sample Sizes

6-2 Estimating a Population Mean: Large Samples

1. $\alpha = .01$, $\alpha/2 = .005$
   $A = .5000 - .005 = .4950$
   $z_{.005} = 2.575$

3. $\alpha = .02$, $\alpha/2 = .01$
   $A = .5000 - .01 = .4900$ [.4901]
   $z_{.01} = 2.33$

5. $n > 30$, use z (with s for σ)
   $\alpha = .05$, $\alpha/2 = .025$, $z_{.025} = 1.96$
   a. $E = z_{.025} \cdot \sigma/\sqrt{n}$
      $= 1.96 \cdot (2.4)/\sqrt{50}$
      $= 0.7$
   b. $\bar{x} \pm E$
      63.4 ± 0.7
      $62.7 < \mu < 64.1$ (inches)

7. $n > 30$, use z (with s for σ)
   $\alpha = .10$, $\alpha/2 = .05$, $z_{.05} = 1.645$
   a. $E = z_{.05} \cdot \sigma/\sqrt{n}$
      $= 1.645 \cdot (14.2)/\sqrt{150}$
      $= 1.9$
   b. $\bar{x} \pm E$
      77.6 ± 1.9
      $75.7 < \mu < 79.5$

9. $n > 30$, use z (with s for σ)
   $\bar{x} \pm z_{.025} \cdot \sigma/\sqrt{n}$
   $134.5 \pm 1.96 \cdot 3.48/\sqrt{35}$
   134.5 ± 1.2
   $133.3 < \mu < 135.7$ (mm)

11. $n > 30$, use z (with s for σ)
    $\bar{x} \pm z_{.005} \cdot \sigma/\sqrt{n}$
    $5.15 \pm 2.575 \cdot 1.68/\sqrt{4400}$
    $5.15 \pm .07$
    $5.08 < \mu < 5.22$ (years)

13. $E = 1.5$ and $\alpha = .02$
    $n = [z_{.01} \cdot \sigma/E]^2$
    $= [2.33 \cdot 15/1.5]^2$
    $= 542.89$ rounded up to 543

15. n > 30, use z (with s for σ)
   $\bar{x} \pm z_{.01} \cdot \sigma/\sqrt{n}$
   $5.622 \pm 2.33 \cdot 0.68/\sqrt{50}$
   5.622 ± 0.022
   $5.600 < \mu < 5.644$ (grams)
   While this confidence interval does not include 5.670, it does not contradict the claim that quarters are produced at that weight. The claim is about new quarters, but the sample was taken from general circulation and had experienced wear.

17. E = 2.0 and α = .05
   $n = [z_{.025} \cdot \sigma/E]^2$
   $= [1.96 \cdot 21.2/2.0]^2$
   $= 431.64$ rounded up to 432

19. n > 30, use z (with s for σ)
   $\bar{x} \pm z_{.05} \cdot \sigma/\sqrt{n}$
   $191.7 \pm 1.645 \cdot 41.0/\sqrt{1525}$
   191.7 ± 1.7
   $190.0 < \mu < 193.4$ (mg/100 mL)
   No; since we are 90% confident that the true mean is less than 200, we are 90% confident that the doctor is not correct.

21. E = .25 and α = .04
   $n = [z_{.02} \cdot \sigma/E]^2$
   $= [2.05 \cdot 1.87/.25]^2$
   $= 235.13$ rounded up to 236

23. E = .25 and α = .02
   $\sigma \approx R/4 = (15-3)/4 = 3.00$
   $n = [z_{.01} \cdot \sigma/E]^2$
   $= [2.33 \cdot 3.00/.25]^2$
   $= 781.76$ rounded up to 782

25. preliminary values: n = 62, Σx = 584.54, Σx^2 = 6570.8216
   $\bar{x} = (\Sigma x)/n = (584.54)/62 = 9.4281$
   $s^2 = [n(\Sigma x^2) - (\Sigma x)^2]/[n(n-1)] = [62(6570.8216) - (584.54)^2]/[62(61)] = 17.3728$
   $s = 4.1681$
   n > 30, use z (with s for σ)
   $\bar{x} \pm z_{.025} \cdot \sigma/\sqrt{n}$
   $9.4281 \pm 1.96 \cdot 4.1681/\sqrt{62}$
   9.4281 ± 1.0375
   $8.391 < \mu < 10.466$ (lbs)
   NOTE: Here and in similar problems, $\bar{x}$ and s are shown with one more decimal accuracy than usual, but the values were stored and recalled with full accuracy for use in subsequent calculations.

27. preliminary values: n = 33, Σx = 30.123, Σx^2 = 27.546792
   $\bar{x} = (\Sigma x)/n = (30.123)/33 = .91282$
   $s^2 = [n(\Sigma x^2) - (\Sigma x)^2]/[n(n-1)] = [33(27.546792) - (30.123)^2]/[33(32)] = .001562$
   $s = .03952$
   n > 30, use z (with s for σ)

$\bar{x} \pm z_{.015} \cdot \sigma/\sqrt{n}$
.91282 $\pm$ 2.17 $\cdot$ .03952$/\sqrt{33}$
.91282 $\pm$ .01493
.8979 $< \mu <$ .9277 (grams)

No; for the blue M&M's n=5 is not a large enough sample to use z scores from Table A-2 while using s to estimate σ.

29. E = 1.5
α = .02
N = 200
$n = [N\sigma^2(z_{.01})^2]/[(N-1)E^2 + \sigma^2(z_{.01})^2]$
$= [(200)(15)^2(2.33)^2]/[(199)(1.5)^2 + (15)^2(2.33)^2]$
$= [244300.5]/[1669.2525]$
= 146.35 rounded up to 147

31. preliminary values: n = 30, Σx = 19962.2, Σx^2 = 13,284,510.99
$\bar{x} = (\Sigma x)/n = (19962.2)/30 = 665.41$
$s^2 = [n(\Sigma x^2) - (\Sigma x)^2]/[n(n-1)] = [30(13,284,510.99) - (19962.2)^2]/[30(29)] = 52.760$
s = 7.264
n = 810 > 30, use z (with s for σ)
$E = z_{.025} \cdot \sigma/\sqrt{n}$
$= 1.96 \cdot (7.264)/\sqrt{810}$
= .50 milligrams

6-3 Estimating a Population Mean: Small Samples

IMPORTANT NOTE: This manual uses the following conventions.
(1) The designation "df" stands for "degrees of freedom."
(2) Since the t value depends on both the degrees of freedom and the probability lying beyond it, double subscripts are used to identify points on t distributions. The t distribution with 15 degrees of freedom and .025 beyond it, for example, is designated $t_{15,.025}$ = 2.132.
(3) When df$\geq$30 the difference between the t and z distributions is negligible and Table A-3 uses one final row of z values to cover all such cases. Consequently, the z scores for certain "popular" α and $\alpha/2$ values may be found by reading Table A-3 "frontwards" instead of reading Table A-2 "backwards." This is not only easier but also more accurate, since Table A-3 includes one more decimal place. The manual uses this technique from this point on.

1. n = 10; df = 9
α = .01; $\alpha/2$ = .005
$t_{9,.005}$ = 3.250

3. n = 21; df = 20
α = .02; $\alpha/2$ = .01
$t_{20,.01}$ = 2.528

5. n $\leq$ 30 and σ unknown, use t
a. $E = t_{9,.025} \cdot s/\sqrt{n}$
$= 2.262 \cdot 2.4/\sqrt{10}$
= 1.7
b. $\bar{x} \pm E$
63.4 $\pm$ 1.7
61.7 $< \mu <$ 65.1 (inches)

7. $n \leq 30$ and σ unknown, use t
   a. $E = t_{15,.05} \cdot s/\sqrt{n}$
      $= 1.753 \cdot 14.2/\sqrt{16}$
      $= 6.2$
   b. $\bar{x} \pm E$
      77.6 ± 6.2
      $71.4 < \mu < 83.8$

9. $n \leq 30$ and σ unknown, use t
   $\bar{x} \pm t_{14,.005} \cdot s/\sqrt{n}$
   $1786 \pm 2.977 \cdot 937/\sqrt{15}$
   1786 ± 720
   $1066 < \mu < 2506$ (dollars)

11. $n \leq 30$ and σ unknown, use t
    $\bar{x} \pm t_{19,.025} \cdot s/\sqrt{n}$
    $2.40 \pm 2.093 \cdot 1.30/\sqrt{20}$
    2.40 ± 0.61
    $1.79 < \mu < 3.01$ (hours)

13. $n > 30$, use z (with s for σ)
    $\bar{x} \pm z_{.025} \cdot \sigma/\sqrt{n}$
    $3.94 \pm 1.960 \cdot .75/\sqrt{231}$
    3.94 ± 0.10
    $3.84 < \mu < 4.04$

15. $n \leq 30$ and σ unknown, use t
    $\bar{x} \pm t_{24,.01} \cdot s/\sqrt{n}$
    $39,271 \pm 2.492 \cdot 18,933/\sqrt{25}$
    $39,271 \pm 9436$
    $29,835 < \mu < 48,707$ (dollars)

17. preliminary values: $n = 16$, $\Sigma x = 127.4$, $\Sigma x^2 = 1052.72$
    $\bar{x} = (\Sigma x)/n = (127.4)/16 = 7.96$
    $s^2 = [n(\Sigma x^2) - (\Sigma x)^2]/[n(n-1)] = [16(1052.72) - (127.4)^2]/[16(15)] = 2.553$
    $s = 1.598$
    $n \leq 30$ and σ unknown, use t
    $\bar{x} \pm t_{15,.025} \cdot s/\sqrt{n}$
    $7.96 \pm 2.132 \cdot 1.598/\sqrt{16}$
    7.96 ± 0.85
    $7.11 < \mu < 8.81$

18. preliminary values: $n = 21$, $\Sigma x = 19.104$, $\Sigma x^2 = 17.394278$
    $\bar{x} = (\Sigma x)/n = (19.104)/21 = .9097$
    $s^2 = [n(\Sigma x^2) - (\Sigma x)^2]/[n(n-1)] = [21(17.394278) - (19.104)^2]/[21(20)] = .0007548$
    $s = .02747$
    $n \leq 30$ and σ unknown, use t
    $\bar{x} \pm t_{20,.025} \cdot s/\sqrt{n}$
    $.9097 \pm 2.086 \cdot .02747/\sqrt{21}$
    $.9097 \pm .0125$
    $.8972 < \mu < .9222$ (grams)
    NOTE: This compares very closely with the $.8979 < \mu < .9277$ confidence interval of exercise #27 in section 6-2 found using the 33 brown M&M's and the large sample methods of the previous section.

19. preliminary values: n = 100, Σx = 91.470, Σx^2 = 83.802512
$\bar{x}$ = $(\Sigma x)/n$ = (91.470)/100 = .9147
s^2 = $[n(\Sigma x^2) - (\Sigma x)^2]/[n(n-1)]$ = $[100(83.802512) - (91.470)^2]/[100)(99)]$ = .001363
s = .03691
n > 30, use z (with s for σ)
$\bar{x} \pm z_{.025} \cdot \sigma/\sqrt{n}$
.9147 $\pm$ 1.960 $\cdot$ .03691/$\sqrt{100}$
.9147 $\pm$ .0072
.9075 < μ < .9219 (grams)
This confidence interval is narrower (.9219-.9075 = .0144) than the one in exercise #18 (.9222-.8972 = .0250). There was actually more variability among the individual candies with all the colors mixed together than with just the red alone (.03691 vs. .02747), but the larger sample size allows a more precise statement.

21. For any α, the z value is smaller than the corresponding t value (although the difference decreases as n increases). This creates a smaller E and a narrower confidence interval than one is entitled to -- i.e., it does not take into consideration the extra uncertainty created by using the sample s instead of the true population σ.

6-4 Estimating a Population Proportion

IMPORTANT NOTE: When calculating confidence intervals using the formula
$\hat{p} \pm E$
$\hat{p} \pm z_{\alpha/2}\sqrt{\hat{p}\hat{q}/n}$
do not round off in the middle of the problem. This may be accomplished conveniently on most calculators having a memory as follows.
  (1) Calculate $\hat{p}$ = x/n and STORE the value
  (2) Calculate E as 1 - RECALL = * RECALL = $\div$ n = $\sqrt{}$ * $z_{\alpha/2}$ =
  (3) With the value of E showing on the display, the upper confidence limit is calculated by + RECALL.
  (4) With the value of the upper confidence limit showing on the display, the lower confidence limit is calculated by - RECALL $\pm$ + RECALL
You must become familiar with your own calculator. [Do your homework using the same type of calculator you will be using for the exams.] The above procedure works on most calculators; make certain you understand why it works and verify whether it works on your calculator. If it does not seem to work on your calculator, or if your calculator has more than one memory so that you can STORE both $\hat{p}$ and E at the same time, ask your instructor for assistance.

NOTE: It should be true that $0 \le \hat{p} \le 1$ and that E $\le$ .5 [usually, much less than .5]. If such is not the case, an error has been made.

1. $\hat{p}$ = x/n = 200/800 = .25
   α = .05
   E = $z_{.025}\sqrt{\hat{p}\hat{q}/n}$
     = 1.960$\sqrt{(.25)(.75)/800}$
     = .0300

3. $\hat{p}$ = x/n = 2576/4275 = .603
   α = .02
   E = $z_{.01}\sqrt{\hat{p}\hat{q}/n}$
     = 2.327$\sqrt{(.603)(.397)/4275}$
     = .0174

5. $\hat{p} = x/n = 600/800 = .750$
   $\alpha = .05$
   $\hat{p} \pm z_{.025}\sqrt{\hat{p}\hat{q}/n}$
   $.750 \pm 1.960\sqrt{(.750)(.250)/800}$
   $.750 \pm .030$
   $.720 < p < .780$

7. $\hat{p} = x/n = 992/2475 = .401$
   $\alpha = .10$
   $\hat{p} \pm z_{.05}\sqrt{\hat{p}\hat{q}/n}$
   $.401 \pm 1.645\sqrt{(.401)(.599)/2475}$
   $.401 \pm .016$
   $.385 < p < .417$

9. $\hat{p}$ unknown, use $\hat{p} = .5$
   $n = [(z_{.025})^2\hat{p}\hat{q}]/E^2$
   $\ \ = [(1.960)^2(.5)(.5)]/(.02)^2$
   $\ \ = 2401$

11. $\hat{p} = .20$
    $n = [(z_{.005})^2\hat{p}\hat{q}]/E^2$
    $\ \ = [(2.575)^2(.20)(.80)]/(.04)^2$
    $\ \ = 663.06$ rounded up to 664

13. a. $\hat{p} = x/n = 544/800 = .640 = 64.0\%$
    b. $\hat{p} \pm z_{.05}\sqrt{\hat{p}\hat{q}/n}$
    $.640 \pm 1.645\sqrt{(.640)(.360)/800}$
    $.640 \pm .027$
    $.613 < p < .667$
    $61.3\% < p < 66.7\%$

15. a. $\hat{p} = .27$
    $n = [(z_{.025})^2\hat{p}\hat{q}]/E^2$
    $\ \ = [(1.960)^2(.27)(.73)]/(.04)^2$
    $\ \ = 473.24$ rounded up to 474
    b. $\hat{p}$ unknown, use $\hat{p} = .5$
    $n = [(z_{.05})^2\hat{p}\hat{q}]/E^2$
    $\ \ = [(1.960)^2(.5)(.5)]/(.04)^2$
    $\ \ = 600.25$ rounded up to 601

17. $\hat{p} = x/n = 1054/1220 = .864$
    $\alpha = .02$
    $\hat{p} \pm z_{.01}\sqrt{\hat{p}\hat{q}/n}$
    $.864 \pm 2.327\sqrt{(.864)(.156)/1220}$
    $.864 \pm .023$
    $.841 < p < .887$
    $84.1\% < p < 88.7\%$

19. $\hat{p} = x/n = 56/400 = .140$
$\alpha = .01$
$\hat{p} \pm z_{.005}\sqrt{\hat{p}\hat{q}/n}$
$.140 \pm 2.575\sqrt{(.140)(.860)/400}$
$.140 \pm .045$
$.095 < p < .185$
$9.5\% < p < 18.5\%$

Yes; we are 99% confident that the rate is higher for \$100,000+ Greybar returns. Either the firm is under scrutiny or it handles returns (e.g., particular occupations) under scrutiny.

21. $\hat{p}$ unknown, use $\hat{p} = .5$
$n = [(z_{.015})^2\hat{p}\hat{q}]/E^2$
$= [(2.17)^2(.5)(.5)]/(.02)^2$
$= 2943.06$ rounded up to 2944

23. $\hat{p} = x/n = .84$
$\alpha = .01$
$\hat{p} \pm z_{.005}\sqrt{\hat{p}\hat{q}/n}$
$.84 \pm 2.575\sqrt{(.84)(.16)/6503}$
$.84 \pm .01$
$.83 < p < .85$
$83\% < p < 85\%$

Yes; we are 99% confident that $p > .80$ (i.e., that the question is easy).

NOTE: Since $\hat{p}$ was given with only 2 decimal accuracy and the actual value of x was not given, the final answer is limited to 2 decimal accuracy.

25. a. $\hat{p} = x/n = 20/1234 = .0162$
$\hat{p} \pm z_{.025}\sqrt{\hat{p}\hat{q}/n}$
$.0162 \pm 1.960\sqrt{(.0162)(.9838)/1234}$
$.0162 \pm .0070$
$.0092 < p < .0233$
b. $\hat{p} = x/n = 20/1234 = .0162$
$n = [(z_{.005})^2\hat{p}\hat{q}]/E^2$
$= [(2.575)^2(.0162)(.9838)]/(.005)^2$
$= 4228.95$ rounded up to 4229

27. Of the 60 movies, 35 were rated R.
$\hat{p} = x/n = 35/60 = .583$
$\hat{p} \pm z_{.025}\sqrt{\hat{p}\hat{q}/n}$
$.583 \pm 1.960\sqrt{(.583)(.417)/60}$
$.583 \pm .125$
$.459 < p < .708$
$45.9\% < p < 70.8\%$

Yes; since .58 is within the interval, the result is consistent with the claim.

29. $\hat{p}$ unknown, use $\hat{p} = .5$
$n = [(z_{.015})^2\hat{p}\hat{q}N]/[(z_{.015})^2\hat{p}\hat{q} + (N-1)E^2]$
$= [(2.17)^2(.5)(.5)(5000)]/[(2.17)^2(.5)(.5) + (4999)(.02)^2]$
$= [5886.125]/[3.176825]$
$= 1852.83$ rounded up to 1853

31. $E = .002$, $\hat{p} = .08$ and $n = 47000$

Solve $E = z_{\alpha/2}\sqrt{\hat{p}\hat{q}/n}$ for $z_{\alpha/2}$ to get

$$z_{\alpha/2} = E/\sqrt{\hat{p}\hat{q}/n}$$
$$= .002/\sqrt{(.08)(.92)/47000}$$
$$= .002/.00125$$
$$= 1.60$$

Since $P(-1.60 < z < 1.60) = .4452 + .4452$
$$= .8904,$$
the level of confidence is .8904, or about 89%.

33. $\hat{p} = x/n = 630/750 = .840$

$\hat{p} - z_{.05}\sqrt{\hat{p}\hat{q}/n}$

$.840 - 1.645\sqrt{(.840)(.160)/750}$

$.840 - .022$

$.818 < p$ [i.e., we are 95% confident that $p > 81.8\%$]

6-5 Estimating a Population Variance

1. $\chi^2_L = \chi^2_{25,.975} = 13.120$
$\chi^2_R = \chi^2_{25,.025} = 40.646$

3. $\chi^2_L = \chi^2_{59,.95} = 43.188$
$\chi^2_R = \chi^2_{59,.05} = 79.082$
NOTE: Use df = 60, the closest entry.

5. $(n-1)s^2/\chi^2_{9,.025} < \sigma^2 < (n-1)s^2/\chi^2_{9,.975}$
$(9)(2.4)^2/19.023 < \sigma^2 < (9)(2.4)^2/2.700$
$2.73 < \sigma^2 < 19.20$
$1.7 < \sigma < 4.4$ (inches)

7. $(n-1)s^2/\chi^2_{15,.05} < \sigma^2 < (n-1)s^2/\chi^2_{15,.95}$
$(15)(14.2)^2/24.996 < \sigma^2 < (15)(14.2)^2/7.261$
$121.00 < \sigma^2 < 416.55$
$11.0 < \sigma < 20.4$

9. From the upper right section of Table 6-2, $n = 20$.

11. From the lower left section of Table 6-2, $n = 171$.

NOTE: When raw scores are available, $\bar{x}$ and s should be calculated as the primary descriptive statistics -- but use the unrounded value of s^2 in the confidence interval formula. In addition, always make certain that the confidence interval for σ includes the calculated value of s.

13. summary information

n = 18	$\bar{x}$ = 3787.0
Σx = 68,166	s^2 = 3066.82
Σx^2 = 258,196,778	s = 55.4

$(n-1)s^2/\chi^2_{17,.005} < \sigma^2 < (n-1)s^2/\chi^2_{17,.995}$

$(17)(3066.82)/35.718 < \sigma^2 < (17)(3066.82)/5.697$

$1459.65 < \sigma^2 < 9151.47$

$38.2 < \sigma < 95.7$ (mL)

No; the interval indicates 99% confidence that $\sigma > 30$.

15. $(n-1)s^2/\chi^2_{34,.025} < \sigma^2 < (n-1)s^2/\chi^2_{34,.975}$
$(34)(3.5)^2/46.979 < \sigma^2 < (34)(3.5)^2/16.791$
$8.87 < \sigma^2 < 24.80$
$3.0 < \sigma < 5.0$ (mm)

NOTE: The χ^2 values used are those for 30 df, which is the closest entry from Table A-4. Since df = n-1 is a part of the confidence interval formula, and since the expected value of the χ^2 statistic is n-1, some instructors and text books recommend using in the formula a value for n-1 that agrees with the df of the tabled χ^2

17. summary information
n = 10 $\bar{x} = 7.15$
$\Sigma x = 71.5$ $s^2 = .2272$
$\Sigma x^2 = 513.27$ $s = .48$

$(n-1)s^2/\chi^2_{9,.025} < \sigma^2 < (n-1)s^2/\chi^2_{9,.975}$
$(9)(.2272)/19.023 < \sigma^2 < (9)(.2272)/2.700$
$.1075 < \sigma^2 < .7573$
$.33 < \sigma < .87$ (minutes)

19. a. $s \approx (42.9 - 21.6)/4 = 5.325$

   b. summary information
   n = 41 $\bar{x} = 32.47$
   $\Sigma x = 1331.4$ $s^2 = 31.3670$
   $\Sigma x^2 = 44,489.46$ $s = 5.60$

   $(n-1)s^2/\chi^2_{40,.005} < \sigma^2 < (n-1)s^2/\chi^2_{40,.995}$
   $(40)(31.3670)/66.766 < \sigma^2 < (40)(31.3670)/20.707$
   $18.792 < \sigma^2 < 60.592$
   $4.33 < \sigma < 7.78$ (inches)

   c. Yes, $4.33 < 5.325 < 7.78$.

21. a. The given interval $2.8 < \sigma < 6.0$
   $7.84 < \sigma^2 < 36.00$
   and the usual calculations $(n-1)s^2/\chi^2_{19,\alpha/2} < \sigma^2 < (n-1)s^2/\chi^2_{19,1-\alpha/2}$
   $(19)(3.8)^2/\chi^2_{19,\alpha/2} < \sigma^2 < (19)(3.8)^2/\chi^2_{19,1-\alpha/2}$
   $274.36/\chi^2_{19,\alpha/2} < \sigma^2 < 274.36/\chi^2_{19,1-\alpha/2}$
   imply that $7.84 = 274.37/\chi^2_{19,\alpha/2}$ and $36.00 = 274.36/\chi^2_{19,1-\alpha/2}$
   $\chi^2_{19,\alpha/2} = 274.36/7.84$ $\chi^2_{19,1-\alpha/2} = 274.36/36.00$
   $= 34.99$ $= 7.62$

   The closest entries in Table A-4 are $\chi^2_{19,\alpha/2} = 34.805$ and $\chi^2_{19,1-\alpha/2} = 7.633$
   which imply $\alpha/2 = .01$ $1 - \alpha/2 = .99$
                                        $\alpha = .02$ $\alpha/2 = .01$
                                                                                     $\alpha = .02$
   The level of confidence is therefore is $1-\alpha = 98\%$.

   b. $(n-1)s^2/\chi^2_{11,.025} < \sigma^2 < (n-1)s^2/\chi^2_{11,.975}$
   using the lower endpoint   OR   using the upper endpoint
   $(11)s^2/21.920 = (19.1)^2$ $(11)s^2/3.816 = (45.8)^2$
   $s^2 = 726.97$ $s^2 = 727.69$
   $s = 27.0$ $s = 27.0$

Review Exercises

1. distribution approximately normal: $n = 54$, $\bar{x} = 12.95$, $s = 2.14$

   a. $n > 30$, use z (with s for σ)
   $\bar{x} \pm z_{.01} \cdot \sigma/\sqrt{n}$
   $12.95 \pm 2.327 \cdot 2.14/\sqrt{54}$
   $12.95 \pm .68$
   $12.27 < \mu < 13.63$ (inches)

   b. $\quad (n-1)s^2/\chi^2_{53,.01} < \sigma^2 < (n-1)s^2/\chi^2_{53,.99}$
   $(53)(2.14)^2/76.154 < \sigma^2 < (53)(2.14)^2/29.707$
   $3.19 < \sigma^2 < 8.17$
   $1.79 < \sigma < 2.86$ (inches)

   NOTE: The χ^2 values used are those for 50 df.
   See the NOTE for exercise #15, section 6-5.

   c. $n = [z_{.01} \cdot \sigma/E]^2 = [2.327 \cdot 2.14/.25] = 396.77$ rounded up to 397

2. $\hat{p}$ unknown, use $\hat{p} = .5$
   $n = [(z_{.015})^2\hat{p}\hat{q}]/E^2 = [(2.17)^2(.5)(.5)]/(.02)^2 = 2943.06$ rounded up to 2944

3. $n \leq 30$ and σ unknown, use t
   $\bar{x} \pm t_{24,.025} \cdot s/\sqrt{n}$
   $7.01 \pm 2.064 \cdot 3.74/\sqrt{25}$
   7.01 ± 1.54
   $5.47 < \mu < 8.55$ (years)

4. $\quad (n-1)s^2/\chi^2_{24,.025} < \sigma^2 < (n-1)s^2/\chi^2_{24,.975}$
   $(24)(3.74)^2/39.364 < \sigma^2 < (24)(3.74)^2/12.401$
   $8.528 < \sigma^2 < 27.071$
   $2.92 < \sigma < 5.20$ (years)

5. $\hat{p} = x/n = x/1475 = .320$
   $\hat{p} \pm z_{.025}\sqrt{\hat{p}\hat{q}/n}$
   $.320 \pm 1.960\sqrt{(.320)(.680)/1475}$
   $.320 \pm .024$
   $.296 < p < .344$
   The margin of error that should be reported with the 32.0% is 2.4%.

6. $\hat{p} = .93$
   $n = [(z_{.01})^2\hat{p}\hat{q}]/E^2 = [(2.327)^2(.93)(.07)]/(.04)^2 = 220.3$ rounded up to 221

7. $n = 16 \qquad \Sigma x = 1162.6 \qquad \Sigma x^2 = 84577.34$
   $\bar{x} = 72.66$ [do not round, store all the digits in the calculator]
   $s = 2.581$ [do not round, store all the digits in the calculator]
   $n \leq 30$ and σ unknown, use t
   $\bar{x} \pm t_{15,.025} \cdot \sigma/\sqrt{n}$
   $72.66 \pm 2.132 \cdot 2.581/\sqrt{16}$
   72.66 ± 1.38
   $71.29 < \mu < 74.04$ (cm)

8. $n > 30$, use z (with s for σ)
$\bar{x} \pm z_{.005} \cdot \sigma/\sqrt{n}$
$40.7 \pm 2.575 \cdot 10.2/\sqrt{40}$
40.7 ± 4.2
$36.5 < \mu < 44.9$

9. $E = 4$ and $\alpha = .05$
$n = [z_{.025} \cdot \sigma/E]^2 = [1.960 \cdot 41.0/4]^2 = 403.61$ rounded up to 404

10. $\hat{p} = .24$
$\hat{p} \pm z_{.005}\sqrt{\hat{p}\hat{q}/n}$
$.24 \pm 2.575\sqrt{(.24)(.76)/1998}$
$.24 \pm .02$
$.22 < p < .26$
$22\% < p < 26\%$
NOTE: Since $\hat{p}$ was given with only two decimal accuracy and the actual value of x was not given, the final answer is limited to two decimal accuracy. Any x from 470 to 489 rounds to 24% with varying digits for the next decimal point.

Cumulative Review Exercises

1. Begin by making a stem-and-leaf plot and calculating summary statistics.

66.	8	
67.		$n = 24$
68.	34	
69.	0224699	$\Sigma x = 1683.1$
70.	002245688	
71.	078	$\Sigma x^2 = 118,076.37$
72.	68	

a. $\bar{x} = (\Sigma x)/n = (1683.1)/24 = 70.13$ in
b. $\tilde{x} = (70.0 + 70.2)/2 = 70.10$ in
c. $M = 69.2, 69.9, 70.0, 70.2, 70.8$ in (multi-modal)
d. m.r. $= (66.8 + 72.8)/2 = 69.80$ in
e. $R = 72.8 - 66.8 = 6.0$ in
f. $s^2 = [n(\Sigma x^2) - (\Sigma x)^2]/[n(n-1)]$
$= [24(118,076.37) - (1683.1)^2]/[24(23)]$
$= 1.82$ in^2
g. $s = 1.35$ in
h. for $Q_1 = P_{25}$, $L = (25/100)(24) = 6$ -- a whole number
$Q_1 = (x_6 + x_7)/2 = (69.2 + 69.4)/2 = 69.3$ in
i. for $Q_2 = P_{50}$, $L = (50/100)(24) = 12$ -- a whole number
$Q_2 = (x_{12} + x_{13})/2 = (70.0 + 70.2)/2 = 70.1$ in
j. for $Q_3 = P_{75}$, $L = (75/100)(24) = 18$ -- a whole number
$Q_3 = (x_{18} + x_{19})/2 = (70.8 + 70.8)/2 = 70.8$ in
k. ratio, since differences are consistent and there is a meaningful zero
l.

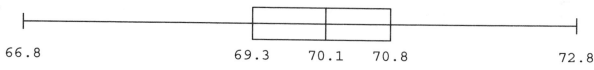

66.8   69.3   70.1   70.8   72.8

m. Answers will vary. Here is one possibility.
Choose to use about 5 classes with at least $(72.8 - 68.8)/5 = 6.0/5 = 1.25$ inches per class.
Choose 1.5 inches per class.
Since $5(1.5) = 7.5$ and $7.5 - 6.0 = 1.5$, begin about $1.5/2 = .75$ below the lowest score.
Since $66.8 - .75 = 66.05$, round to 66.25.
The resulting frequency table and histogram are as follows.

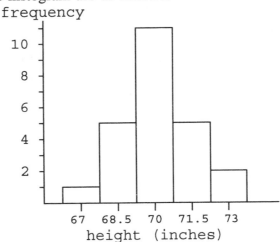

height	f
66.25 - 67.75	1
67.75 - 69.25	5
69.25 - 70.75	11
70.75 - 72.25	5
72.25 - 73.75	2
	24

The heights appear to be approximately normally distributed.

n. $n \leq 30$ and σ unknown, use t
$\bar{x} \pm t_{23,.005} \cdot s/\sqrt{n}$
$70.13 \pm 2.807 \cdot 1.35/\sqrt{24}$
70.13 ± 0.77
$69.36 < \mu < 70.90$ (inches)

o. $(n-1)s^2/\chi^2_{23,.005} < \sigma^2 < (n-1)s^2/\chi^2_{23,.995}$
$(23)(1.82)^2/44.181 < \sigma^2 < (23)(1.82)^2/9.260$
$.9499 < \sigma^2 < 4.5324$
$0.97 < \sigma < 2.13$ (inches)

p. $n = [z_{.005} \cdot \sigma/E]^2$
$= [2.575 \cdot 1.35/.2]^2$
$= 302.48$ rounded up to 303

q. No; the values from the general population do not match those for the recruits.
Since 69.7 is between 69.36 and 70.90, the mean heights are about the same for each group.
Since 2.8 is not between 0.97 and 2.13, however, there is significantly more variability in
height in the general population than among the recruits.

2. binomial: $n = 200$ and $p = .25$
$np = 200(.25) = 50 \geq 5$
$n(1-p) = 200(.75) = 150 \geq 5$
use the normal approximation with
$\mu = np = 200(.25) = 50$
$\sigma = \sqrt{n(p)(1-p)} = \sqrt{200(.25)(.75)} = 6.214$

a. $P(x \geq 65)$
$= P_c(x > 64.5)$
$= P(z > 2.37)$
$= .5000 - .4911$
$= .0089$

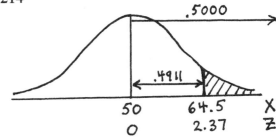

b. $\hat{p} = 65/200 = .325$
$\hat{p} \pm z_{.025}\sqrt{\hat{p}\hat{q}/n}$
$.325 \pm 1.960\sqrt{(.325)(.675)/200}$
$.325 \pm .065$
$.260 < p < .390$

c. No; the expert's value of .25 does not seem correct for two reasons.
   From part (a): if the expert is correct, then the probability of getting the sample obtained is very small -- less than 1%.
   From part (b): according to the sample obtained, there is 95% confidence that the true value is between .26 and .39 -- that is, there is 95% confidence that the value is not .25.

Chapter 7

Hypothesis Testing

7-2 Fundamentals of Hypothesis testing

1. a. $\mu \neq 650$
   b. $H_o:\mu = 650$
   c. $H_1:\mu \neq 650$
   d. two-tailed
   e. rejecting the hypothesis that the mean amount is 650 mg when it really is 650 mg
   f. failing to reject the hypothesis that the mean amount is 650 mg when it really is not 650 mg
   g. there is sufficient evidence to support the claim that the mean amount is unequal to 650 mg
   h. there is not sufficient evidence to support the claim that the mean amount is unequal to 650 mg

3. a. $p > .5$
   b. $H_o:p \leq .5$
   c. $H_1:p > .5$
   d. right-tailed
   e. rejecting the hypothesis that the proportion is at most .5 when it really is at most .5
   f. failing to reject the hypothesis that the proportion is at most .5 when it really is greater than .5
   g. there is sufficient evidence to support the claim that the proportion is greater than .5
   h. there is not sufficient evidence to support the claim that the proportion is more than .5

5. a. $p = .15$
   b. $H_o:p = .15$
   c. $H_1:p \neq .15$
   d. two-tailed
   e. rejecting the claim that the proportion is .15 when it really is .15
   f. failing to reject the claim that the proportion is .15 when it really is not .15
   g. there is sufficient evidence to reject the claim that the proportion is .15
   h. there is not sufficient evidence to reject the claim that the proportion is .15

7. a. $\mu > 30$
   b. $H_o:\mu \leq 30$
   c. $H_1:\mu > 30$
   d. right-tailed
   e. rejecting the hypothesis that the mean mpg is no more than 30 when it really is no more than 30
   f. failing to reject the hypothesis that the mean mpg is no more than 30 when it really is greater than 30
   g. there is sufficient evidence to support the claim that the mean mpg is greater than 30
   h. there is not sufficient evidence to support the claim that the mean mpg is greater than 30

NOTE: The manual continues to use the last row of Table A-3 as the most convenient and most accurate choice for obtaining frequently used z scores. Reading backwards from Table A-2 is done only when necessary.

9. the critical z value is $z_{.05} = 1.645$
    the critical region is z > 1.645

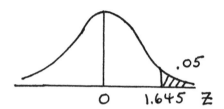

11. the critical z value is $-z_{.01} = -2.327$
    the critical region is z < -2.327

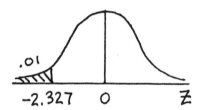

13. the critical z values are $\pm z_{.05} = \pm 1.645$
    the critical region is z < -1.645
                           z > 1.645

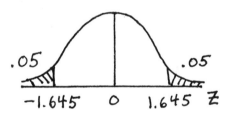

15. the critical z value is $-z_{.025} = -1.960$
    the critical region is z < -1.960

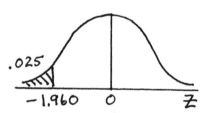

17. Mathematically, in order for α to equal 0 the magnitude of the critical value would have to be infinite. Practically, the only way never to make a type I error is to always fail to reject H_o. From either perspective, the only way to achieve $\alpha = 0$ is to never reject H_o no matter how extreme the sample data might be.

7-3 Testing a Claim about a Mean: Large Samples

1. original claim: $\mu = 75$ [n > 30, use z (with s for σ)]
   H_o: $\mu = 75$
   H_1: $\mu \neq 75$
   $\alpha = .05$
   C.R. $z < -z_{.025} = -1.960$
   $\quad\quad z > z_{.025} = 1.960$
   calculations:

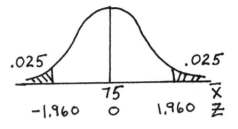

   $\quad z_{\bar{x}} = (\bar{x} - \mu)/\sigma_{\bar{x}}$
   $\quad\quad = (78 - 75)/(15/\sqrt{100})$
   $\quad\quad = 3/1.50$
   $\quad\quad = 2.000$
   conclusion:
      Reject H_o; there is sufficient evidence to reject the claim that $\mu = 75$ and conclude that $\mu \neq 75$ (in fact, $\mu > 75$).
   P-value $= 2 \cdot P(z > 2.00) = 2 \cdot (.5000 - .4772) = 2 \cdot (.0228) = .0456$

3. original claim: $\mu < 2.50$ [n > 30, use z (with s for σ)]
   H_o: $\mu \geq 2.50$
   H_1: $\mu < 2.50$
   $\alpha = .02$
   C.R. $z < -z_{.02} = -2.05$
   calculations:

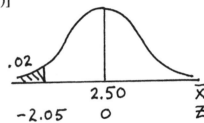

   $\quad z_{\bar{x}} = (\bar{x} - \mu)/\sigma_{\bar{x}}$
   $\quad\quad = (2.45 - 2.50)/(.80/\sqrt{64})$
   $\quad\quad = -.05/.1$
   $\quad\quad = -.500$
   conclusion:
      Do not reject H_o; there is not sufficient evidence to conclude that $\mu < 2.50$.
   P-value $= P(z < -.50) = .5000 - .1915 = .3085$

5. original claim: $\mu \neq 92.84$ [n > 30, use z (with s for σ)]
   H_o: $\mu = 92.84$
   H_1: $\mu \neq 92.84$
   $\alpha = .05$
   C.R. $z < -z_{.025} = -1.960$
   $\quad\quad z > z_{.025} = 1.960$
   calculations:

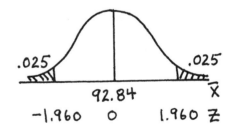

   $\quad z_{\bar{x}} = (\bar{x} - \mu)/\sigma_{\bar{x}}$
   $\quad\quad = (92.67 - 92.84)/(1.79/\sqrt{40})$
   $\quad\quad = -.17/.283$
   $\quad\quad = -.601$
   conclusion:
      Do not reject H_o; there is not sufficient evidence to conclude that $\mu \neq 92.84$.
   P-value $= 2 \cdot P(z < -.60) = 2 \cdot (.5000 - .2257) = 2 \cdot (.2743) = .5486$
   No; the new balls do not appear to be "juiced."

7. original claim: $\mu > .21$ [n > 30, use z (with s for σ)]
   $H_o: \mu \leq .21$
   $H_1: \mu > .21$
   $\alpha = .01$
   C.R. $z > z_{.01} = 2.327$
   calculations:

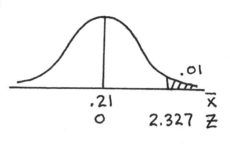

$$z_{\bar{x}} = (\bar{x} - \mu)/\sigma_{\bar{x}}$$
$$= (.83 - .21)/(.24/\sqrt{32})$$
$$= .62/.0424$$
$$= 14.614$$

   conclusion:
   Reject H_o; there is sufficient evidence to conclude that $\mu > .21$.
   P-value = $P(z > 14.61) = .5000 - .4999 = .0001$

The questionnaire may or may not be effective at "identifying compulsive buyers" -- there are at least two concerns that should be addressed. First, the test may miss certain compulsive shoppers -- the data analyzed comes from a sample of self-identified compulsive shoppers and may not be representative of compulsive shoppers who are unwilling to acknowledge themselves as such. Second, the test may also give high scores to certain others who are not compulsive shoppers -- i.e., while compulsive shoppers score high on the test, so might other subpopulations (e.g., compulsive gamblers).

9. original claim: $\mu < 1.39$ [n > 30, use z (with s for σ)]
   $H_o: \mu \geq 1.39$
   $H_1: \mu < 1.39$
   $\alpha = .01$
   C.R. $z < -z_{.01} = -2.327$
   calculations:

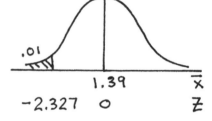

$$z_{\bar{x}} = (\bar{x} - \mu)/\sigma_{\bar{x}}$$
$$= (.83 - 1.39)/(.16/\sqrt{123})$$
$$= -.56/.0144$$
$$= -38.817$$

   conclusion:
   Reject H_o; there is sufficient evidence to conclude that $\mu < 1.39$.
   P-value = $P(z < -38.82) = .5000 - .4999 = .0001$
   Yes; wearing seat belts seems to help.

11. original claim: $\mu > 0$ [n > 30, use z (with s for σ)]
   $H_o: \mu \leq 0$
   $H_1: \mu > 0$
   $\alpha = .05$
   C.R. $z > z_{.05} = 1.645$
   calculations:

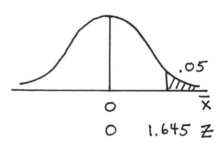

$$z_{\bar{x}} = (\bar{x} - \mu)/\sigma_{\bar{x}}$$
$$= (.6 - 0)/(3.8/\sqrt{75})$$
$$= .6/.439$$
$$= 1.367$$

   conclusion:
   Do not reject H_o; there is not sufficient evidence to conclude that $\mu > 0$.
   P-value = $P(z > 1.37) = .5000 - .4147 = .0853$

No; people should not take the course. Even if the mean increase of .6 had been statistically significant, it would not be of practical significance.

13. original claim: $\mu < 7.5$ [n > 30, use z (with s for σ)]
   H_o: $\mu \geq 7.5$
   H_1: $\mu < 7.5$
   $\alpha = .05$
   C.R. $z < -z_{.05} = -1.645$
   calculations:

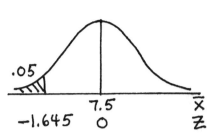

$$z_{\bar{x}} = (\bar{x} - \mu)/\sigma_{\bar{x}}$$
$$= (7.01 - 7.5)/(3.74/\sqrt{100})$$
$$= -.49/.374$$
$$= -1.310$$

   conclusion:
   Do not reject H_o; there is not sufficient evidence to conclude that $\mu < 7.5$.
   P-value = $P(z < -1.31)$ = .5000 - .4049 = .0951

15. original claim: $\mu = 600$ [n > 30, use z (with s for σ)]
   H_o: $\mu = 600$
   H_1: $\mu \neq 600$
   $\alpha = .01$
   C.R. $z < -z_{.005} = -2.575$
       $z > z_{.005} = 2.575$
   calculations:

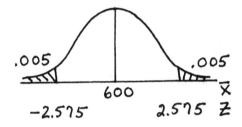

$$z_{\bar{x}} = (\bar{x} - \mu)/\sigma_{\bar{x}}$$
$$= (589 - 600)/(21/\sqrt{65})$$
$$= -11/2.605$$
$$= -4.223$$

   conclusion:
   Reject H_o; there is sufficient evidence to reject the claim that $\mu = 600$ and to conclude that
   $\mu \neq 600$ (in fact, $\mu < 600$).
   P-value = $2 \cdot P(z < -4.22)$ = $2 \cdot (.5000 - .4999)$ = $2 \cdot (.0001)$ = .0002
   The fact that the product has less than the advertised amount of a certain ingredient does not
   necessarily make it ineffective, but it does raise questions about the company's quality control
   and/or honesty and thwart efforts to monitor the intake of users. It would probably be wise not
   to buy the product.

17. original claim: $\mu > 40,000$ [n > 30, use z (with s for σ)]
   H_o: $\mu \leq 40,000$
   H_1: $\mu > 40,000$
   $\alpha = .005$
   C.R. $z > z_{.005} = 2.575$
   calculations:

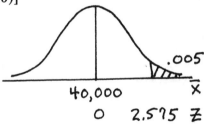

$$z_{\bar{x}} = (\bar{x} - \mu)/\sigma_{\bar{x}}$$
$$= (41,182 - 40,000)/(19,990/\sqrt{1700})$$
$$= 1182/484.829$$
$$= 2.438$$

   conclusion:
   Do not reject H_o; there is not sufficient evidence to conclude that $\mu > 40,000$.
   P-value = $P(z > 2.44)$ = .5000 - .4927 = .0073

19. original claim: $\mu > 420$ [n > 30, use z (with s for σ)]
    $H_o: \mu \leq 420$
    $H_1: \mu > 420$
    $\alpha = .05$
    C.R. $z > z_{.05} = 1.645$
    calculations:

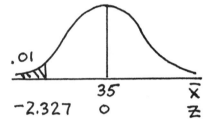

$$z_{\bar{x}} = (\bar{x} - \mu)/\sigma_{\bar{x}}$$
$$= (385 - 420)/(24/\sqrt{35}\ )$$
$$= -35/4.057$$
$$= -8.628$$

    conclusion:
      Do not reject H_o; there is not sufficient evidence to conclude that $\mu > 420$.
    P-value = $P(z > -8.63) = .4999 + .5000 = .9999$

21. original claim: $\mu < 35$ [n > 30, use z (with s for σ)]
    $H_o: \mu \geq 35$
    $H_1: \mu < 35$
    $\alpha = .01$
    C.R. $z < -z_{.01} = -2.327$
    calculations:

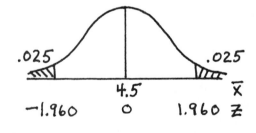

$$z_{\bar{x}} = (\bar{x} - \mu)/\sigma_{\bar{x}}$$
$$= (27.44 - 35)/(12.46/\sqrt{62}\ )$$
$$= -7.56/1.582$$
$$= -4.777$$

    conclusion:
      Reject H_o; there is sufficient evidence to conclude that $\mu < 35$.
    P-value = $P(z < -4.78) = .5000 - .4999 = .0001$
    No; based on this result, there is no cause for concern that there might be too much garbage.

23. NOTE: Minitab calls the calculated statistic T instead of Z because it uses the notation and techniques of the next section, 7-4.
    original claim: $\mu = 4.5$ [n > 30, use z (with s for σ)]
    $H_o: \mu = 4.5$
    $H_1: \mu \neq 4.5$
    $\alpha = .05$
    C.R. $z < -z_{.025} = -1.960$
         $z > z_{.025} = 1.960$
    calculations:

$$z_{\bar{x}} = (\bar{x} - \mu)/\sigma_{\bar{x}}$$
$$= (2.319 - 4.500)/(2.899/\sqrt{141})$$
$$= -2.181/.244$$
$$= -8.93\ \text{[from Minitab]}$$

    conclusion:
      Reject H_o; there is sufficient evidence to reject the claim that $\mu = 4.5$ and to conclude that $\mu \neq 4.5$ (in fact, $\mu < 4.5$).
    P-value = $2 \cdot P(z < -8.93) = 2 \cdot (.5000 - .4999) = 2 \cdot (.0001) = .0002$ [NOTE: Minitab gives P-value = .0000, rounded to 4 decimal places, which is more accurate.]
    As $\bar{x} \approx 2.3$ is so low, there are probably a disproportionate number of lengths ending in 0. A reasonable guess would be that most of the lengths have been rounded off to end in 0 or 5, with slightly more 0's than 5's.

25. original claim: $\mu = 100$ [n > 30, use z (with s for σ)]
    H_o: $\mu = 100$
    H_1: $\mu \neq 100$
    $\alpha = .01$
    C.R. z < $-z_{.005}$ = -2.575
         z > $z_{.005}$ = 2.575
    calculations:
       $z_{\bar{x}} = (\bar{x} - \mu)/\sigma_{\bar{x}}$
       2.575 < (103.6 - 100)/(s/$\sqrt{62}$)
          s < (103.6 - 100)/(2.575/$\sqrt{62}$)
          s < 3.6/.327
          s < 11.008

NOTE: There is no conclusion because this exercise is not asking for the completion of a test of hypotheses.

27. original claim: $\mu = 98.6$ [n > 30, use z (with s for σ)]
    H_o: $\mu = 98.6$
    H_1: $\mu \neq 98.6$
    $\alpha = .05$
    C.R. z < $-z_{.025}$ = -1.960
         z > $z_{.025}$ = 1.960
    calculations:

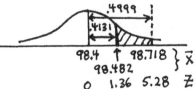

   $z_{\bar{x}} = (\bar{x} - \mu)/\sigma_{\bar{x}}$ $z_{\bar{x}} = (\bar{x} - \mu)/\sigma_{\bar{x}}$
   -1.960 = ($\bar{x}$ - 98.6)/(.62/$\sqrt{106}$)   1.960 = ($\bar{x}$ - 98.6)/(.62/$\sqrt{106}$)
         $\bar{x}$ = 98.482 $\bar{x}$ = 98.718

NOTE: There is no conclusion because this exercise is not asking for the completion of a test of hypotheses.

a. β = P(98.482 < $\bar{x}$ < 98.718 | μ = 98.7)
     = P(-3.62 < z < .30)
     = .4999 + .1179
     = .6178

b. β = P(98.482 < $\bar{x}$ < 98.718 | μ = 98.4)
     = P(1.36 < z < 5.28)
     = .4999 - .4131
     = .0868

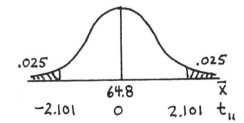

7-4 Testing a Claim about a Mean: Small Samples

1. a. $\pm t_{6,.025} = \pm 2.447$
   b. $-t_{11,.05} = -1.796$
   c. $t_{8,.01} = 2.896$

3. original claim: $\mu \neq 64.8$ [n ≤ 30 and σ unknown, use t]
   H_o: $\mu = 64.8$
   H_1: $\mu \neq 64.8$
   $\alpha = .05$
   C.R. t < $-t_{11,.025}$ = -2.201
        t > $t_{11,.025}$ = 2.201
   calculations:
      $t_{\bar{x}} = (\bar{x} - \mu)/s_{\bar{x}}$
          = (59.8 - 64.8)/(8.7/$\sqrt{12}$)
          = -5.0/2.511
          = -1.991

conclusion:
   Do not reject H_o; there is not sufficient evidence to conclude that $\mu \neq 64.8$.

5. original claim: $\mu > 12.00$ [$n \leq 30$ and σ unknown, use t]
   H_o: $\mu \leq 12.00$
   H_1: $\mu > 12.00$
   $\alpha = .01$
   C.R. t > $t_{10,.01} = 2.764$
   calculations:

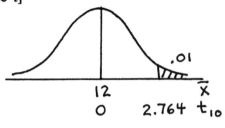

$$t_{\bar{x}} = (\bar{x} - \mu)/s_{\bar{x}}$$
$$= (13.286 - 12.00)/(1.190/\sqrt{11})$$
$$= 1.286/.576$$
$$= 2.23 \quad \text{[from Minitab]}$$
conclusion:
   Do not reject H_o; there is not sufficient evidence to conclude that $\mu > 12.00$.
   NOTE: The decision could also be made using P-values. Since P-value = .025 [from Minitab] is greater than $\alpha = .01$, we fail to reject H_o.

7. original claim: $\mu > 40$ [$n \leq 30$ and σ unknown, use t]
   H_o: $\mu \leq 40$
   H_1: $\mu > 40$
   $\alpha = .01$
   C.R. t > $t_{9,.01} = 2.821$
   calculations:

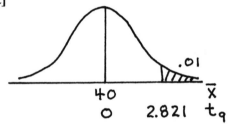

$$t_{\bar{x}} = (\bar{x} - \mu)/s_{\bar{x}}$$
$$= (43.3 - 40)/(3.8/\sqrt{10})$$
$$= 3.3/1.202$$
$$= 2.746$$
conclusion:
   Do not reject H_o; there is not sufficient evidence to conclude that $\mu > 40$.

9. original claim: $\mu = 98.6$ [$n \leq 30$ and σ unknown, use t]
   H_o: $\mu = 98.6$
   H_1: $\mu \neq 98.6$
   $\alpha = .05$
   C.R. t < $-t_{24,.025} = -2.064$
       t > $t_{24,.025} = 2.064$
   calculations:

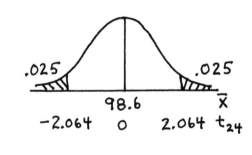

$$t_{\bar{x}} = (\bar{x} - \mu)/s_{\bar{x}}$$
$$= (98.24 - 98.6)/(.56/\sqrt{25})$$
$$= -.36/.112$$
$$= -3.214$$
conclusion:
   Reject H_o; there is sufficient evidence to reject the claim that $\mu = 98.6$ and to conclude that $\mu \neq 98.6$ (in fact, $\mu < 98.6$).

11. original claim: $\mu > 1800$ [$n \leq 30$ and σ unknown, use t]
   H_o: $\mu \leq 1800$
   H_1: $\mu > 1800$
   $\alpha = .01$
   C.R. $t > t_{11,.01} = 2.718$
   calculations:

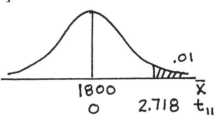

   $t_{\bar{x}} = (\bar{x} - \mu)/s_{\bar{x}}$
   $= (2133 - 1800)/(345/\sqrt{12})$
   $= 333/99.593$
   $= 3.344$
   conclusion:
   Reject H_o; there is sufficient evidence to conclude that $\mu > 1800$.

13. original claim: $\mu > 20000$ [$n > 30$, use z (with s for σ)]
   H_o: $\mu \leq 2000$
   H_1: $\mu > 2000$
   $\alpha = .025$
   C.R. $z > z_{.025} = 1.960$
   calculations:
   $z_{\bar{x}} = (\bar{x} - \mu)/\sigma_{\bar{x}}$
   $= (2177 - 2000)/(1257/\sqrt{50})$
   $= 177/177.8$
   $= .996$
   conclusion:
   Do not reject H_o; there is not sufficient evidence to conclude that $\mu > 2000$.
   No; based on this sample, the monitoring system will not be implemented.

15. original claim: $\mu < 10.00$ [$n \leq 30$ and σ unknown, use t]
   H_o: $\mu \geq 10.00$
   H_1: $\mu < 10.00$
   $\alpha = .01$
   C.R. $t < -t_{15,.01} = -2.602$
   calculations:
   $t_{\bar{x}} = (\bar{x} - \mu)/s_{\bar{x}}$
   $= (8.33 - 10.00)/(1.96/\sqrt{16})$
   $= -1.67/.49$
   $= -3.408$
   conclusion:
   Reject H_o; there is sufficient evidence to conclude that $\mu < 10.00$.

17. original claim: $\mu > 5$ [$n > 30$, use z (with s for σ)]
   H_o: $\mu \leq 5$
   H_1: $\mu > 5$
   $\alpha = .10$
   C.R. $z > z_{.10} = 1.282$
   calculations:
   $t_{\bar{x}} = (\bar{x} - \mu)/s_{\bar{x}}$
   $= (5.15 - 5)/(1.68/\sqrt{80})$
   $= .15/.188$
   $= .799$
   conclusion:
   Do not reject H_o; there is not sufficient evidence to conclude that $\mu > 5$.

19. original claim: $\mu \geq .9085$ [$n \leq 30$ and σ unknown, use t]
   H_o: $\mu \geq .9085$
   H_1: $\mu < .9085$
   $\alpha = .05$
   C.R. $t < -t_{4,.05} = -2.132$
   calculations:

   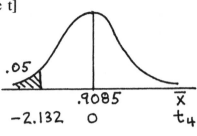

   $t_{\bar{x}} = (\bar{x} - \mu)/s_{\bar{x}}$
   $\qquad = (.9014 - .9085)/(.0573/\sqrt{5})$
   $\qquad = -.0071/.0256$
   $\qquad = -.277$
   conclusion:
   Do not reject H_o; there is not sufficient evidence to reject the claim that $\mu \geq .9085$ and to conclude that $\mu < .9085$.

   No; based on this sample we cannot conclude that the package contents do not agree with the claimed weight printed on the label.

   NOTE: Even rejecting H_o would not mean the package weight is less than claimed, since the test results apply only to blue M&M's and not to all M&M's. Also, the notion that M&M's bags are filled with 1498 candies is probably incorrect: it is more likely that the bags are filled by weight and not by count, so the mean weight per candy would not be relevant for determining whether a bag of unknown count had full weight.

21. original claim: $\mu = 11,000$ [$n \leq 30$ and σ unknown, use t]
   summary statistics: $n = 7$, $\Sigma x = 74,209$, $\Sigma x^2 = 793,035,705$, $\bar{x} = 10,601.3$, $s = 1026.7$
   H_o: $\mu = 11,000$
   H_1: $\mu \neq 11,000$
   $\alpha = .05$
   C.R. $t < -t_{6,.025} = -2.447$
   $\qquad t > t_{6,.025} = 2.447$
   calculations:

   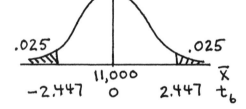

   $t_{\bar{x}} = (\bar{x} - \mu)/s_{\bar{x}}$
   $\qquad = (10601.3 - 11000)/(1026.7/\sqrt{7})$
   $\qquad = -398.7/388.1$
   $\qquad = -1.027$
   conclusion:
   Do not reject H_o; there is not sufficient evidence to reject the claim that $\mu = 11000$.

23. original claim: $\mu > 63.2$ [$n \leq 30$ and σ unknown, use t]
   summary statistics: $n = 15$, $\Sigma x = 1043$, $\Sigma x^2 = 77739$, $\bar{x} = 69.53$, $s = 19.30$
   H_o: $\mu \leq 63.2$
   H_1: $\mu > 63.2$
   $\alpha = .05$ [assumed]
   C.R. $t > t_{14,.05} = 1.761$
   calculations:

   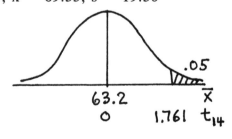

   $t_{\bar{x}} = (\bar{x} - \mu)/s_{\bar{x}}$
   $\qquad = (69.53 - 63.2)/(19.30/\sqrt{15})$
   $\qquad = 6.333/4.984$
   $\qquad = 1.271$
   conclusion:
   Do not reject H_o; there is not sufficient evidence to conclude that $\mu > 63.2$.

   No; while the new mean is higher, Rita cannot be 95% confident that the new show is not just reflecting chance variations from the former distribution.

25. a. Since the calculated t = 2.746 is between $t_{9,.01}$ = 2.821 and $t_{9,.025}$ = 2.262,
     the P-value for this one-tailed test is between .01 and .025
     -- i.e., .01 < P-value < .025.

  b. Since the calculated t = -1.320 is between $-t_{11,.10}$ = -1.363 and $-t_{11,.25}$ = -.697,
     the P-value for this one-tailed test is between .10 and .25
     -- i.e., .10 < P-value < .25.

  c. Since the calculated t = -3.214 is to the left of $-t_{24,.005}$ = 2.797,
     the P-value for this two-tailed test is less than 2·(.005)
     -- i.e., P-value < .01.

  d. Since the calculated z = -3.004 is between z = -3.00 and = -3.01,
     and since both of these scores are $z_{.0013}$ [A = .4987],
     the P-value for this two-tailed test is 2·(.0013)
     -- i.e., P-value = .0026.

27. $A = z_{.05}·[(8·df + 3)/(8·df + 1)]$
    $= 1.645·[(8·9 + 3)/(8·9 + 1)]$
    $= 1.645·[75/73] = 1.690$
    $t = \sqrt{df·(e^{A·A/df} - 1)}$
    $= \sqrt{9·(e^{(1.690)(1.690)/9} - 1)}$
    $= \sqrt{9·(e^{.317} -1)}$
    $= \sqrt{9·(.3735)} = 1.833$
    This agrees exactly with $t_{9,.05}$ = 1.833 given in Table A-3.

7-5 Testing a Claim about a Proportion

NOTE: To be consistent with the notation of the previous sections, and thereby reinforcing the patterns and concepts presented in those sections, the manual uses the "usual" z formula written to apply to $\hat{p}$'s
$$z_{\hat{p}} = (\hat{p} - \mu_{\hat{p}})/\sigma_{\hat{p}}$$
When the normal approximation to the binomial applies, the $\hat{p}$'s are normally distributed
    with $\mu_{\hat{p}} = p$
    and $\sigma_{\hat{p}} = \sqrt{pq/n}$
And so the formula for the z statistic may also be written as
$$z_{\hat{p}} = (\hat{p} - p)/\sqrt{pq/n}$$

1. original claim: p = .01 [normal approximation to the binomial, use z]
   H_o: p = .01 $\hat{p} = x/n = 20/1234 = .0162$
   H_1: p ≠ .01
   α = .05
   C.R. z < $-z_{.025}$ = -1.960
        z > $z_{.025}$ = 1.960
   calculations:
   $z_{\hat{p}} = (\hat{p} - \mu_{\hat{p}})/\sigma_{\hat{p}}$
       $= (.0162 - .01)/\sqrt{(.01)(.99)/1234}$
       $= .0062/.0028$
       $= 2.192$

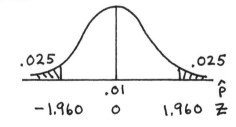

   conclusion:
   Reject H_o; there is sufficient evidence reject the claim that p = .01 and to conclude that p ≠ .01 (in fact, p > .01).
   P-value = 2·P(z > 2.19) = 2·(.5000 - .4857) = 2·(0143) = .0286
   No; scanners do not appear to help consumers avoid overcharges.

3. original claim: $p \leq .04$ [normal approximation to the binomial, use z]
   H_o: $p \leq .04$ $\hat{p} = x/n = 9/150 = .06$
   H_1: $p > .04$
   $\alpha = .05$
   C.R. $z > z_{.05} = 1.645$
   calculations:
   $\quad z_{\hat{p}} = (\hat{p} - \mu_{\hat{p}})/\sigma_{\hat{p}}$
   $\qquad = (.06 - .04)/\sqrt{(.04)(.96)/150}$
   $\qquad = .02/.016$
   $\qquad = 1.25$

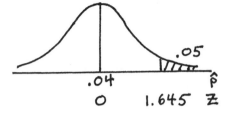

   conclusion:
   $\quad$ Do not reject H_o; there is not sufficient evidence to reject the claim that $p \leq .04$.
   P-value $= P(z > 1.25) = .5000 - .3944 = .1056$
   No; we cannot be 95% certain that corrective action is needed.

5. original claim: $p > .50$ [normal approximation to the binomial, use z]
   NOTE: A "success" is <u>not</u> being knowledgeable about the Holocaust.
   H_o: $p \leq .50$ $\hat{p} = x/n = 268/506 = .530$
   H_1: $p > .50$
   $\alpha = .05$
   C.R. $z > z_{.05} = 1.645$
   calculations
   $\quad z_{\hat{p}} = (\hat{p} - \mu_{\hat{p}})/\sigma_{\hat{p}}$
   $\qquad = (.530 - .50)/\sqrt{(.50)(.50)/506}$
   $\qquad = .0296/.0222$
   $\qquad = 1.334$

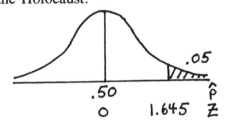

   conclusion:
   $\quad$ Do not reject H_o; there is not sufficient evidence to conclude that $p > .50$.
   P-value $= P(z > 1.33) = .5000 - .4082 = .0918$
   No; we cannot be 95% certain that $p > .50$ and that the curriculum needs to be revised.

7. original claim: $p > .043$ [normal approximation to the binomial, use z]
   H_o: $p \leq .043$ $\hat{p} = x/n = 56/400 = .140$
   H_1: $p > .043$
   $\alpha = .005$
   C.R. $z > z_{.005} = 2.575$
   calculations:
   $\quad z_{\hat{p}} = (\hat{p} - \mu_{\hat{p}})/\sigma_{\hat{p}}$
   $\qquad = (.140 - .043)/\sqrt{(.043)(.957)/400}$
   $\qquad = .0970/.0101$
   $\qquad = 9.563$

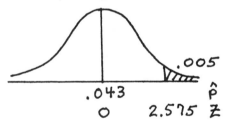

   conclusion:
   $\quad$ Reject H_o; there is sufficient evidence to conclude that $p > .043$.
   P-value $= P(z > 9.56) = .5000 - .4999 = .0001$
   No; it appears that having the return prepared by a different company would decrease the chance of being audited.

9. original claim: p < .50   [normal approximation to the binomial, use z]

H_o: p ≥ .50 $\hat{p}$ = x/n = x/1998 = .48

H_1: p < .50

α = .05

C.R. z < $-z_{.05}$ = -1.645

calculations:

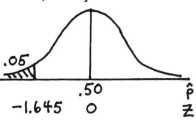

$$z_{\hat{p}} = (\hat{p} - \mu_{\hat{p}})/\sigma_{\hat{p}}$$
$$= (.48 - .50)/\sqrt{(.50)(.50)/1998}$$
$$= -.02/.0112$$
$$= -1.788$$

conclusion:

  Reject H_o; there is sufficient evidence to conclude that p < .50.

P-value = P(z < -1.79) = .5000 - .4633 = .0367

NOTE: In reality, the information given is not precise enough to perform the test of hypothesis. Any value of x between 950 and 969 inclusive rounds to x/1998 = 48%, the accuracy to which $\hat{p}$ is given. For x = 950, $z_{\hat{p}}$ = -2.192 and the conclusion is to reject H_o; for x = 969, $z_{\hat{p}}$ = -1.342 and the conclusion is not to reject H_o. As the problem is stated, only 1 digit accuracy is possible for the difference in the numerator of $z_{\hat{p}}$.

11. original claim: p = .10   [normal approximation to the binomial, use z]

  H_o: p = .10 $\hat{p}$ = x/n = 90/500 = .180

  H_1: p ≠ .10

  α = .02

  C.R. z < $-z_{.01}$ = -2.327

      z > $z_{.01}$ = 2.327

  calculations:

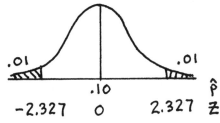

$$z_{\hat{p}} = (\hat{p} - \mu_{\hat{p}})/\sigma_{\hat{p}}$$
$$= (.180 - .10)/\sqrt{(.10)(.90)/500}$$
$$= .08/.0134$$
$$= 5.962$$

conclusion:

  Reject H_o; there is sufficient evidence to reject the claim that p = .10 and to conclude that p ≠ .10 (in fact, p > .10).

P-value = 2·P(z > 5.96) = 2·(.5000 - .4999) = 2·(.0001) = .0002

13. original claim: p = .08   [normal approximation to the binomial, use z]

  H_o: p = .08 $\hat{p}$ = x/n = 70/781 = .0896

  H_1: p ≠ .08

  α = .10

  C.R. z < $-z_{.05}$ = -1.645

      z > $z_{.05}$ = 1.645

  calculations

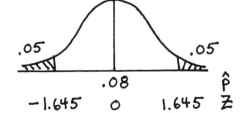

$$z_{\hat{p}} = (\hat{p} - \mu_{\hat{p}})/\sigma_{\hat{p}}$$
$$= (.0896 - .08)/\sqrt{(.08)(.92)/781}$$
$$= .00962/.00971$$
$$= .992$$

conclusion:

  Do not reject H_o; there is not sufficient evidence to reject the claim that p = .08.

P-value = 2·P(z > .99) = 2·(.5000 - .3389) = 2·(.1611) = .3222

15. original claim: $p < .10$ [normal approximation to the binomial, use z]
   $H_o: p \geq .10$ $\hat{p} = x/n = 64/1068 = .0599$
   $H_1: p < .10$
   $\alpha = .01$
   C.R. $z < -z_{.01} = -2.327$
   calculations
   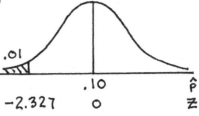

$$z_{\hat{p}} = (\hat{p} - \mu_{\hat{p}})/\sigma_{\hat{p}}$$
$$= (.0599 - .10)/\sqrt{(.10)(.90)/1068}$$
$$= -.0401/.00918$$
$$= -4.366$$

   conclusion:
   Reject H_o; there is sufficient evidence to conclude that $p < .10$.
   P-value $= P(z < -4.37) = .5000 - .4999 = .0001$

17. original claim: $p > .50$ [normal approximation to the binomial, use z]
   NOTE: A "success" is <u>not</u> having quit smoking one year later.
   $H_o: p \leq .50$ $\hat{p} = x/n = 39/71 = .5493$
   $H_1: p > .50$
   $\alpha = .10$
   C.R. $z > z_{.10} = 1.282$
   calculations
   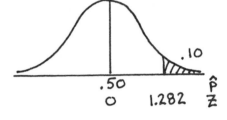

$$z_{\hat{p}} = (\hat{p} - \mu_{\hat{p}})/\sigma_{\hat{p}}$$
$$= (.5493 - .50)/\sqrt{(.50)(.50)/71}$$
$$= .0493/.0593$$
$$= .831$$

   conclusion:
   Do not reject H_o; there is not sufficient evidence to conclude that $p > .50$.
   P-value $= P(z > .83) = .5000 - .2967 = .2033$
   No; while the sample success rate is less than ½, that value must be seen in context -- would you call a major league baseball player with a .475 batting average a failure because his success rate is less than ½? The success rate must be compared to success rates of other programs and to the fact that helping even a few is worth it for those few.

19. original claim: $p = .50$ [normal approximation to the binomial, use z]
   $H_o: p = .50$ $\hat{p} = x/n = x/1012 = .28$
   $H_1: p \neq .50$
   $\alpha = .01$
   C.R. $z < -z_{.005} = -2.575$
        $z > z_{.005} = 2.575$
   calculations:
   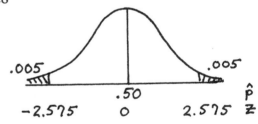

$$z_{\hat{p}} = (\hat{p} - \mu_{\hat{p}})/\sigma_{\hat{p}}$$
$$= (.28 - .50)/\sqrt{(.50)(.50)/1012}$$
$$= -.22/.0157$$
$$= -13.997$$

   conclusion:
   Reject H_o; there is sufficient evidence to reject the claim that $p = .50$ and to conclude that $p \neq .50$ (in fact, $p < .50$).
   P-value $= 2 \cdot P(z < -14.00) = 2 \cdot (.5000 - .4999) = 2 \cdot (.0001) = .0002$
   The sample results are extremely unlikely to have occurred by chance if $p = .50$. It appears that the fruitcake producers have a problem.

21. original claim: $p = .10$ [$np = (15)(.10) = 1.5 < 5$; use binomial distribution]

    $x = 0$

    H_o: $p = .10$

    H_1: $p \neq .10$

    $\alpha = .05$

    C.R. $x < 0$ NOTE: obtained by placing .025 (or

        $x > 4$ much as possible) in each tail

    calculations

      $x = 0$

    conclusion:

      Do not reject H_o; there is not sufficient evidence

      to conclude that $p \neq .10$.

    NOTE: P-value = P(a value as extreme as our value)

               = P($x=0$ or $x \geq 3$)

               = .206 + (.129 + .043 + .010 + .002 + ...)

               = .206 + .184

               = .390

NOTE: Refer to this probability distribution for n=15 and p=.10.

x	P(x)
0	.206
1	.343
2	.267
3	.129
4	.043
5	.010
6	.002
7	0^+
.	.
.	.
.	.
	1.000

23. a. original claim: $p = .20$ [normal approximation to the binomial, use z]

    H_o: $p = .20$ $\hat{p} = x/n = 21/100 = .21$

    H_1: $p \neq .20$

    $\alpha = .05$

    C.R. $z < -z_{.025} = -1.960$

         $z > z_{.025} = 1.960$

    calculations

      $z_{\hat{p}} = (\hat{p} - \mu_{\hat{p}})/\sigma_{\hat{p}}$

          $= (.21 - .20)/\sqrt{(.20)(.80)/100}$

          $= .01/.04$

          $= .25$

    conclusion:

      Do not reject H_o; there is not sufficient evidence to reject the claim that $p = .20$.

  b. original claim: $p = .20$ [normal approximation to the binomial, use z]

    H_o: $p = .20$ $\hat{p} = x/n = 21/100 = .21$

    H_1: $p \neq .20$

    $\alpha = .05$

    C.R. P-value $< .05$

    calculations

      $z_{\hat{p}} = (\hat{p} - \mu_{\hat{p}})/\sigma_{\hat{p}}$

          $= (.21 - .20)/\sqrt{(.20)(.80)/100}$

          $= .01/.04$

          $= .25$

    P-value $= 2 \cdot P(z > .25) = 2 \cdot (.5000 - .0987) = 2 \cdot (.4013) = .8026$

    conclusion:

      Do not reject H_o; there is not sufficient evidence to reject the claim that $p = .20$.

c. There are 21 of the 100 M&M's that were red.

H_o: p = .20 $\hat{p}$ = x/n = 21/100 = .210

H_1: p ≠ .20

α = .05

calculations:

$\hat{p} \pm z_{.025}\sqrt{\hat{p}\hat{q}/n}$

.210 ± 1.960$\sqrt{(.210)(.790)/100}$

.210 ± .080

.130 < p < .290

conclusion:

Since .20 is within the confidence interval, do not reject H_o; there is not sufficient evidence to reject the claim that p = .20.

7-6 Testing a Claim about a Standard Deviation or Variance

NOTE: Following the pattern used with the z and t distributions, this manual uses the closest entry from Table A-4 for χ^2 as if it were the precise value necessary and does not use interpolation. This procedure sacrifices very little accuracy -- and even interpolation does not yield precise values. When extreme accuracy is needed in practice, statisticians refer either to more accurate tables or to computer-produced values.

1. a. $\chi^2_{9,.995}$ = 1.735

$\chi^2_{9,.005}$ = 23.589

b. $\chi^2_{26,.01}$ = 45.642

c. $\chi^2_{20,.95}$ = 10.851

3. original claim: $\sigma \neq$ 43.7

H_o: σ = 43.7

H_1: $\sigma \neq$ 43.7

α = .05

C.R. $\chi^2 < \chi^2_{80,.975}$ = 57.153

$\chi^2 > \chi^2_{80,.025}$ = 106.629

calculations:

$\chi^2 = (n-1)s^2/\sigma^2$

$= (80)(52.3)^2/(43.7)^2$

$= 114.586$

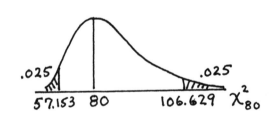

conclusion:

Reject H_o; there is sufficient evidence to conclude that $\sigma \neq$ 43.7 (in fact, $\sigma >$ 43.7).

Since a larger standard deviation indicates more variability it appears that the new production method is worse than the old one.

5. original claim: $\sigma < 6.2$
   H_o: $\sigma \geq 6.2$
   H_1: $\sigma < 6.2$
   $\alpha = .05$
   C.R. $\chi^2 < \chi^2_{24,.95} = 13.848$
   calculations:
   $\quad \chi^2 = (n-1)s^2/\sigma^2$
   $\quad\quad = (24)(3.8)^2/(6.2)^2$
   $\quad\quad = 9.016$

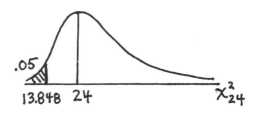

   conclusion:
       Reject H_o; there is sufficient evidence to conclude that $\sigma < 6.2$.
   No; the smaller variability in times does <u>not</u> indicate a smaller mean time. There should be no
   more very short or very long waits, but the mean wait time is not affected.

7. original claim: $\sigma < .75$
   H_o: $\sigma \geq .75$
   H_1: $\sigma < .75$
   $\alpha = .05$
   C.R. $\chi^2 < \chi^2_{60,.95} = 43.188$
   calculations:
   $\quad \chi^2 = (n-1)s^2/\sigma^2$
   $\quad\quad = (60)(.48)^2/(.75)^2$
   $\quad\quad = 24.576$

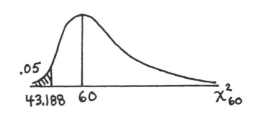

   conclusion:
       Reject H_o; there is sufficient evidence to conclude that $\sigma < .75$.
   If the weights are "more consistent," they tend to cluster more closely about their typical value
   and have a smaller standard deviation.

9. original claim: $\sigma < .15$
   H_o: $\sigma \geq .15$
   H_1: $\sigma < .15$
   $\alpha = .05$
   C.R. $\chi^2 < \chi^2_{70,.95} = 51.739$
   calculations:
   $\quad \chi^2 = (n-1)s^2/\sigma^2$
   $\quad\quad = (70)(.12)^2/(.15)^2$
   $\quad\quad = 44.800$

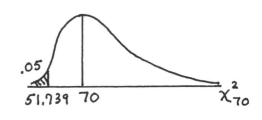

   conclusion:
       Reject H_o; there is sufficient evidence to conclude that $\sigma < 15$.
   Yes; based on this sample, purchase of the machine should be considered.

11. original claim: $\sigma > 19.7$
    H_o: $\sigma \leq 19.7$
    H_1: $\sigma > 19.7$
    $\alpha = .05$
    C.R. $\chi^2 > \chi^2_{49,.05} = 67.505$
    calculations:
    $\quad \chi^2 = (n-1)s^2/\sigma^2$
    $\quad\quad = (49)(23.4)^2/(19.7)^2$
    $\quad\quad = 69.135$

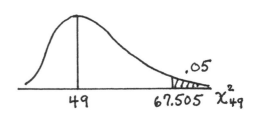

    conclusion:
        Reject H_o; there is sufficient evidence to conclude that $\sigma > 19.7$.

13. summary statistics: $n = 25$, $\Sigma x = 1721.07$, $\Sigma x^2 = 118614.7057$, $\bar{x} = 68.843$, $s^2 = 5.476$
    original claim: $\sigma \neq 2.9$
    H_o: $\sigma = 2.9$
    H_1: $\sigma \neq 2.9$
    $\alpha = .05$ [assumed]
    C.R. $\chi^2 < \chi^2_{24,.975} = 12.401$
    $\quad\;\; \chi^2 > \chi^2_{24,.025} = 39.364$
    calculations:
    $\quad \chi^2 = (n-1)s^2/\sigma^2$
    $\qquad = (24)(5.476)/(2.9)^2$
    $\qquad = 15.628$
    conclusion:
    Do not reject H_o; there is not sufficient evidence to conclude that $\sigma \neq 2.9$.

15. a. To find the P-value for $\chi^2_{80} = 114.586$ in a two-tailed test,
       note that $\chi^2_{80,.01} = 112.329$ and $\chi^2_{80,.005} = 116.321$.
       Therefore, $.005 < \tfrac{1}{2}$(P-value) $< .01$ and $.01 <$ P-value $< .02$.

    b. To find the P-value for $\chi^2_{24} = 9.016$ in a one-tailed less than test,
       note that $\chi^2_{24,.995} = 9.886$ is the smallest χ^2 in the row.
       Therefore, P-value $< .005$.

    c. To find the P-value for $\chi^2_{70} = 44.800$ in a one-tailed less than test,
       note that $\chi^2_{70,.995} = 43.275$ and $\chi^2_{70,.99} = 45.442$.
       Therefore, $.005 <$ P-value $< .01$

17. a. upper $\chi^2 = df \cdot [1 - 2/(9 \cdot df) + z_{.025}\sqrt{2/(9 \cdot df)}\,]^3$
    $\qquad\qquad = 100 \cdot [1 - 2/(9 \cdot 100) + 1.960\sqrt{2/(9 \cdot 100)}\,]^3$
    $\qquad\qquad = 100 \cdot (1.0902)^3$
    $\qquad\qquad = 129.565 \quad$ [compare to $\chi^2_{100,.025} = 129.561$ from Table A-4]
    $\quad$ lower $\chi^2 = df \cdot [1 - 2/(9 \cdot df) - z_{.025}\sqrt{2/(9 \cdot df)}\,]^3$
    $\qquad\qquad = 100 \cdot [1 - 2/(9 \cdot 100) - 1.960\sqrt{2/(9 \cdot 100)}\,]^3$
    $\qquad\qquad = 100 \cdot (.9054)^3$
    $\qquad\qquad = 74.216 \quad$ [compare to $\chi^2_{100,.975} = 74.222$ from Table A-4]
    b. upper $\chi^2 = df \cdot [1 - 2/(9 \cdot df) + z_{.025}\sqrt{2/(9 \cdot df)}\,]^3$
    $\qquad\qquad = 149 \cdot [1 - 2/(9 \cdot 149) + 1.960\sqrt{2/(9 \cdot 149)}\,]^3$
    $\qquad\qquad = 149 \cdot (1.0742)^3$
    $\qquad\qquad = 184.690$
    $\quad$ lower $\chi^2 = df \cdot [1 - 2/(9 \cdot df) - z_{.025}\sqrt{2/(9 \cdot df)}\,]^3$
    $\qquad\qquad = 149 \cdot [1 - 2/(9 \cdot 149) - 1.960\sqrt{2/(9 \cdot 149)}\,]^3$
    $\qquad\qquad = 149 \cdot (.9228)^3$
    $\qquad\qquad = 117.093$

Review Exercises

1. a. concerns μ:
   n ≤ 30 and σ unknown, use t
   $-t_{9,.05} = -1.833$
   b. concerns p:
   normal approx. to the binomial, use z
   $\pm z_{.005} = \pm 2.575$
   c. concerns σ:
   use χ^2
   $\chi^2_{19,.975} = 8.907$ and $\chi^2_{19,.025} = 32.852$

2. a. concerns μ:
   n > 30, use z (with s for σ)
   $\pm z_{.05} = \pm 1.645$
   b. concerns σ:
   use χ^2
   $\chi^2_{9,.025} = 19.023$
   c. concerns p:
   normal approx. to the binomial, use z
   $-z_{.01} = -2.327$

3. original claim: $\mu = 90,000$
   a. $H_o: \mu = 90,000$
   b. two-tailed
   c. rejecting the claim that the mean income is $90,000 when it really is $90,000
   d. failing to reject the claim that the mean income is $90,000 when is really is not $90,000
   e. $\alpha = .05$

4. original claim: $\sigma < 15$
   a. $H_o: \sigma \geq 15$
   b. left-tailed
   c. concluding that the standard deviation is less than 15 seconds when it really is 15 seconds or more
   d. failing to conclude that the standard deviation is less than 15 seconds when it really is less than 15 seconds
   e. $\alpha = .01$

5. original claim: p > .50 [concerns p: normal approximation to the binomial, use z]
   $H_o: p \leq .50$ $\hat{p} = x/n = x/504 = .57$
   $H_1: p > .50$
   $\alpha = .01$
   C.R. z > $z_{.01} = 2.327$
   calculations:

   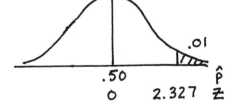

   $z_{\hat{p}} = (\hat{p} - \mu_{\hat{p}})/\sigma_{\hat{p}}$
   $= (.57 - .50)/\sqrt{(.50)(.50)/504}$
   $= .07/.0223$
   $= 3.143$
   conclusion:
   Reject H_o; there is sufficient evidence to conclude that p > .50.

6. original claim: $\mu < 5.00$ [concerns μ: n > 30, use z (with s for σ)]
H_o: $\mu \geq 5.00$
H_1: $\mu < 5.00$
$\alpha = .01$
C.R. $z < -z_{.01} = -2.327$
calculations:

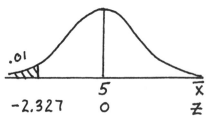

$$z_{\bar{x}} = (\bar{x} - \mu)/\sigma_{\bar{x}}$$
$$= (4.13 - 5.00)/(1.91/\sqrt{36})$$
$$= -.87/.3183$$
$$= -2.733$$
conclusion:
Reject H_o; there is sufficient evidence to conclude that $\mu < 5.00$.

7. original claim: $\sigma < 2.50$ [concerns σ: use χ^2]
H_o: $\sigma \geq 2.50$
H_1: $\sigma < 2.50$
$\alpha = .01$
C.R. $\chi^2 < \chi^2_{35,.99}$
$\quad\quad < (14.954 + 22.164)/2 = 18.559$
calculations:

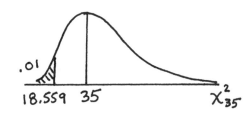

$$\chi^2 = (n-1)s^2/\sigma^2$$
$$= (35)(1.91)^2/(2.50)^2$$
$$= 20.429$$
conclusion:
Do not reject H_o; there is not sufficient evidence to conclude that $\sigma < 2.50$.
Even if the mean level of radiation is acceptable, a large standard deviation could mean that there is an unacceptably high probability of receiving too much radiation on any one trial.
NOTE: Since the desired 35 df is exactly ½ way between the tabled entries of 30 and 40, and because the method of compensating for this can affect the conclusion of the test, we deviate from our usual practice and use interpolation to determine the C.R. Other consistent approaches [using 30 for both the df in the C.R. and the n-1 in the calculations, or using 40 for both the df in the C.R. and the n-1 in the calculations] also lead to the same conclusion.

8. original claim: $p < .10$ [concerns p: normal approximation to the binomial, use z]
H_o: $p \geq .10$ $\quad\quad\quad\quad\quad \hat{p} = x/n = x/1248 = .08$
H_1: $p < .10$
$\alpha = .01$
C.R. $z < -z_{.01} = -2.327$
calculations:

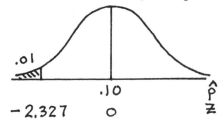

$$z_{\hat{p}} = (\hat{p} - \mu_{\hat{p}})/\sigma_{\hat{p}}$$
$$= (.08 - .10)/\sqrt{(.10)(.90)/1248}$$
$$= -.02/.00849$$
$$= -2.356$$
conclusion:
Reject H_o; there is sufficient evidence to conclude that $p < .10$.
Yes; based on this result, the phrase "almost 1 out of 10" is justified. While "almost" is not well-defined, 10% is the closest round number to 8%. The test indicates we are confident that $p < .10$, but does not specify how much less.

9. original claim: $\mu < 12$ [concerns μ: $n \leq 30$ and σ unknown, use t]
   H_o: $\mu \geq 12$
   H_1: $\mu < 12$
   $\alpha = .01$
   C.R. $t < -t_{23,.01} = -2.500$
   calculations:
   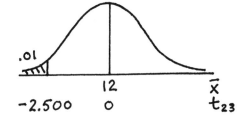

$$t_{\bar{x}} = (\bar{x} - \mu)/s_{\bar{x}}$$
$$= (11.82 - 12.00)/(.38/\sqrt{24})$$
$$= -.18/.0776$$
$$= -2.321$$

   conclusion:
       Do not reject H_o; there is not sufficient evidence to conclude that $\mu < 12$.

10. summary statistics: $n = 20$, $\Sigma x = 465.7$, $\Sigma x^2 = 11234.37$, $\bar{x} = 23.285$, $s = 4.534$
    original claim: $\mu = 20.0$ [concerns μ: $n \leq 30$ and σ unknown, use t]
    H_o: $\mu = 20.0$
    H_1: $\mu \neq 20.0$
    $\alpha = .01$
    C.R. $t < -t_{19,.005} = -2.861$
         $t > t_{19,.005} = 2.861$
    calculations:
    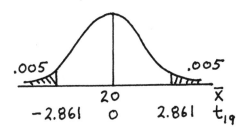

$$t_{\bar{x}} = (\bar{x} - \mu)/\sigma_{\bar{x}}$$
$$= (23.285 - 20.0)/(4.534/\sqrt{20})$$
$$= 3.285/1.014$$
$$= 3.240$$

    conclusion:
        Reject H_o; there is sufficient evidence to reject the claim that $\mu = 20.0$ and to conclude that $\mu \neq 20.0$ (in fact, $\mu > 20.0$).
    No; these pills are not acceptable at the .01 level because their mean amount appears to be higher than the stated 20.0 mg. Even if the hypothesis that $\mu = 20.0$ had not been rejected, the pills may be unacceptable because of their variability. While no specific criteria for variability were stated, note that the pills range from 165.0% (33 mg.) to 79.9% (15.9 mg.) of their stated amount.

Cumulative Review Exercises

1. a. normal distribution: $\mu = 114.8$ and $\sigma = 13.1$
      $P(x > 124.23)$
        $= P(z > .72)$
        $= .5000 - .2642$
        $= .2358$

   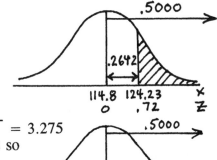

   b. $n = 16$, $\mu_{\bar{x}} = \mu = 114.8$, $\sigma_{\bar{x}} = \sigma/\sqrt{n} = 13.1/\sqrt{16} = 3.275$
      normal distribution, since the original distribution is so
      $P(\bar{x} > 124.23)$
        $= P(z > 2.88)$
        $= .5000 - .4980$
        $= .0020$

   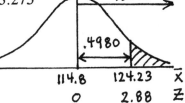

2. summary statistics: n = 16, Σx = 1987.7, Σx² = 254543.43

   a. $\bar{x}$ = 124.23, s = 22.52

   b. original claim: μ = 114.8 [concerns μ: n ≤ 30 and σ unknown, use t]

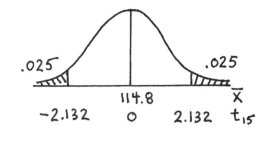

      H_o: μ = 114.8
      H_1: μ ≠ 114.8
      α = .05
      C.R. t < -$t_{15,.025}$ = -2.132
           t > $t_{15,.025}$ = 2.132
      calculations:

$$t_{\bar{x}} = (\bar{x} - \mu)/\sigma_{\bar{x}}$$
$$= (124.23 - 114.8)/(22.52/\sqrt{16})$$
$$= 9.431/5.631$$
$$= 1.675$$

      conclusion:

      Do not reject H_o; there is not sufficient evidence to reject the claim that μ = 114.8. NOTE: The given value of σ = 13.1 was for healthy women and may not apply. It's that large value of s = 22.52 that prevents rejection of H_o and creates the seeming discrepancy with part (b) of exercise #1 above -- which used σ = 13.1.

   c. $\bar{x}$ = ±$t_{15,.025}s_{\bar{x}}$
      124.23 ± 2.132·(22.52/$\sqrt{16}$)
      124.23 ± 12.00
      112.23 < μ < 136.24
      Yes; the interval contains 114.8

   d. original claim: σ = 13.1

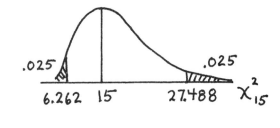

      H_o: σ = 13.1
      H_1: σ ≠ 13.1
      α = .05
      C.R. χ^2 < $\chi^2_{15,.975}$ = 6.262
           χ^2 > $\chi^2_{15,.025}$ = 27.488
      calculations:

$$\chi^2 = (n-1)s^2/\sigma^2$$
$$= (15)(22.52)^2/(13.1)^2$$
$$= 44.339$$

      conclusion:

      Reject H_o; there is sufficient evidence to reject the claim that σ = 13.1 and to conclude that σ ≠ 13.1 (in fact, σ > 13.1).

   e. Yes; the new strain of viral infection seems to be associated with an increase in the variability among systolic blood pressures.

3. binomial distribution: n = 25 and p = .25

   a. μ = np = (25)(.25) = 6.25

   b. σ = $\sqrt{np(1-p)}$ = $\sqrt{25(.25)(.75)}$ = 2.165

   c. Use the normal approximation (with μ = 6.25 and σ = 2.165) to the binomial, since np = 25(.25) = 6.25 ≥ 5 and n(1-p) = 25(.75) = 18.75 ≥ 5.

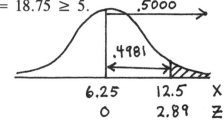

      P(x > 12) = P_c(x > 12.5)
           = P(z > 2.89)
           = .5000 - .4981
           = .0019

d. original claim: p = .25   [concerns p: normal approximation to the binomial, use z]
   H_o: p = .25 $\hat{p}$ = (more than 12)/n = 13/25 = .52
   H_1: p ≠ .25
   α = .05
   C.R. z < -$z_{.025}$ = -1.960
       z > $z_{.025}$ = 1.960
   calculations:

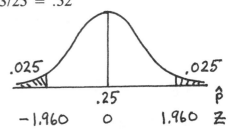

   $z_{\hat{p}}$ = ($\hat{p}$ - μ_p)/$\sigma_{\hat{p}}$
       = (.52 - .25)/$\sqrt{(.25)(.75)/25}$
       = .27/.0866
       = 3.118
   conclusion:
       Reject H_o; there is sufficient evidence to reject the claim that p = .25 and to conclude that
       p ≠ .25 (in fact, p > .25).
   NOTE: The above test of hypothesis follows the form and notation used in chapter 7. One
   could also consider the .0019 of part (c) above, calculate a P-value of 2·(.0019) = .0038,
   and reject H_o because .0038 < .05. Part (c) uses the correction for continuity while the
   section on testing hypotheses concerning p ignores the correction.

e. $\hat{p}$ unknown, use $\hat{p}$ = .5
   n = [($z_{.05}$)²$\hat{p}$(1-$\hat{p}$)]/E^2 = [(1.645)²(.5)(.5)]/(.04)² = 422.8 rounded up to 423

4. a. The stem-and-plot is given below.

9.	0005
10.	00000
11.	00055555
12.	05555
13.	000000005555555
14.	00555
15.	055555
16.	00055
17.	0

   b. Yes; the distribution is approximately normal -- at least the mode is near the middle and
      the smallest frequencies are at the extremes.
   c. Graphically, a histogram could be used. Mathematically, one could calculate $\bar{x}$ and s to see
      if there are about 68% of the scores within 1·s of $\bar{x}$ and about 95% of the scores within 2·s
      of $\bar{x}$.
   d. All the measurements end in 0 or 5. This is probably due to the fact that measurements
      were taken to the nearest half inch instead of the nearest tenth inch. Such decisions are
      arbitrary, usually related to the nature of the measurement, and do not reflect negatively on
      the data.

Chapter 8

Inferences from Two Samples

8-2 Inferences about Two Means: Dependent Samples

NOTE: To be consistent with the notation of the previous sections, and thereby reinforcing the patterns and concepts presented in those sections, the manual uses the "usual" t formula written to apply to $\bar{d}$'s
$$t_{\bar{d}} = (\bar{d} - \mu_{\bar{d}})/s_{\bar{d}}$$
with $\mu_{\bar{d}} = \mu_d$
and $s_{\bar{d}} = s_d/\sqrt{n}$
And so the formula for the t statistic may also be written as
$$t_{\bar{d}} = (\bar{d} - \mu_d)/(s_d/\sqrt{n})$$

1. d = x - y: 5 6 0 5 -2
    summary statistics: n = 5, $\Sigma d = 14$, $\Sigma d^2 = 90$
    a. $\bar{d} = (\Sigma d)/n = 14/5 = 2.8$
    b. $s_d^2 = [n \cdot \Sigma d^2 - (\Sigma d)^2]/[n(n-1)]$
       $= [5 \cdot 90 - (14)^2]/[5(4)] = 254/20 = 12.7$
       $s_d = 3.6$
    c. $t_{\bar{d}} = (\bar{d} - \mu_{\bar{d}})/s_{\bar{d}}$
       $= (2.8 - 0)/(3.564/\sqrt{5})$
       $= 2.8/1.594$
       $= 1.757$
    d. $\pm t_{4,.025} = \pm 2.776$

3. $\bar{d} \pm t_{4,.025} \cdot s_d/\sqrt{n}$
    $2.8 \pm 2.776 \cdot 3.564/\sqrt{5}$
    2.8 ± 4.4
    $-1.6 < \mu_d < 7.2$

5. original claim: $\mu_d < 0$ [n ≤ 30 and σ_d unknown, use t]
    d = x - y: -20 0 10 -40 -30 -10 30 -20 -20 -10
        n = 10
        $\Sigma d = -110$ $\bar{d} = -11.0$
        $\Sigma d^2 = 4900$ $s_d = 20.248$
    $H_o: \mu_d \geq 0$
    $H_1: \mu_d < 0$
    $\alpha = .05$
    C.R. $t < -t_{9,.05} = -1.833$
    calculations:

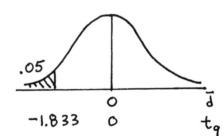

    $t_{\bar{d}} = (\bar{d} - \mu_{\bar{d}})/s_{\bar{d}}$
       $= (-11.0 - 0)/(20.248/\sqrt{10})$
       $= -11.0/6.40$
       $= -1.718$
    conclusion:
        Do not reject H_o; there is not sufficient evidence to conclude that $\mu_d < 0$.
        NOTE: The original claim is that the course "has an effect" (i.e., $\mu_d \neq 0$) on SAT scores. In

this case, it seems reasonable to assume that the reference is to the desired effect of improving scores (i.e., $\mu_d < 0$) and to use a one-tailed test. Technically, the lack of specificity calls for a two-tailed test with $H_o : \mu_d = 0$ and critical values $t = \pm 2.262$. The conclusion is then "Do not reject H_o; there is not sufficient evidence to conclude that $\mu_d \neq 0$."

7. asks for confidence interval for μ_d [$n \leq 30$ and σ_d unknown, use t]
   $d = x_B - x_A$: 9 4 21 3 20 31 17 26 26 10 23 33
   $\quad$ n = 12
   $\quad \Sigma d = 223 \qquad\qquad \bar{d} = 18.58$
   $\quad \Sigma d^2 = 5267 \qquad\quad s_d = 10.104$
   $\bar{d} \pm t_{11,.005} \cdot s_d / \sqrt{n}$
   $18.58 \pm 3.106 \cdot 10.104 / \sqrt{12}$
   18.58 ± 9.06
   $9.5 < \mu_d < 27.6$

9. original claim: $\mu_d > 0$ [$n \leq 30$ and σ_d unknown, use t]
   $d = x_B - x_A$: -.2 4.1 1.6 1.8 3.2 2.0 2.9 9.6
   $\quad$ n = 8
   $\quad \Sigma d = 25.0 \qquad\qquad \bar{d} = 3.125$
   $\quad \Sigma d^2 = 137.46 \qquad\; s_d = 2.9114$
   $H_o : \mu_d \leq 0$
   $H_1 : \mu_d > 0$
   $\alpha = .05$
   C.R. $t > t_{7,.05} = 1.895$
   calculations:

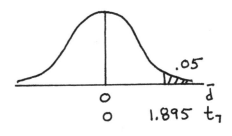

   $\quad t_{\bar{d}} = (\bar{d} - \mu_{\bar{d}})/s_{\bar{d}}$
   $\qquad = (3.125 - 0)/(2.9114/\sqrt{8})$
   $\qquad = 3.125/1.029$
   $\qquad = 3.036$
   conclusion:
   $\quad$ Reject H_o; there is sufficient evidence to conclude that $\mu_d > 0$.
   Yes; hypnotism appears to be effective in reducing pain.

11. original claim: $\mu_d = 0$ [$n \leq 30$ and σ_d unknown, use t]
    $d = x_1 - x_2$: 0 13 5 15 15 126 28 -2 -5 31 3 51 3 14 37
    $\quad$ n = 15
    $\quad \Sigma d = 334 \qquad\qquad \bar{d} = 22.27$
    $\quad \Sigma d^2 = 22478 \qquad\; s_d = 32.777$
    $H_o : \mu_d = 0$
    $H_1 : \mu_d \neq 0$
    $\alpha = .01$
    C.R. $t < -t_{14,.005} = -2.977$
    $\qquad\; t > t_{14,.005} = 2.977$
    calculations:

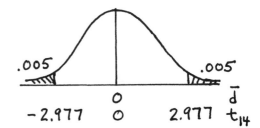

    $\quad t_{\bar{d}} = (\bar{d} - \mu_{\bar{d}})/s_{\bar{d}}$
    $\qquad = (22.27 - 0)/(32.777/\sqrt{15})$
    $\qquad = 22.27/8.463$
    $\qquad = 2.631$
    conclusion:
    $\quad$ Do not reject H_o; there is not sufficient evidence to reject the claim that $\mu_d = 0$.

13. original claim: $\mu_d = 0$ [n $\leq$ 30 and σ_d unknown, use t]
  d = x_1 - x_2: 5 0 0 0 8 1 1 4 0 1
    n = 10
    $\Sigma d = 20$ $\bar{d} = 2.00$
    $\Sigma d^2 = 108$ $s_d = 2.749$
  H_o: $\mu_d = 0$
  H_1: $\mu_d \neq 0$
  $\alpha = .05$
  C.R. t < $-t_{9,.025}$ = -2.262
      t > $t_{9,.025}$ = 2.262
  calculations:
    $t_{\bar{d}} = (\bar{d} - \mu_{\bar{d}})/s_{\bar{d}}$
        $= (2.00 - 0)/(2.749/\sqrt{10}\,)$
        $= 2.00/.869$
        $= 2.301$

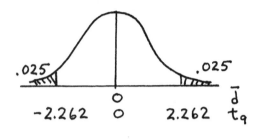

  conclusion:
    Reject H_o; there is sufficient evidence to reject the claim that $\mu_d = 0$ and to conclude that $\mu_d \neq 0$ (in fact, $\mu_d > 0$).

15. asks for confidence interval for μ_d [n $\leq$ 30 and σ_d unknown, use t]
  d = x_8 - x_{12}: given at the right
    n = 11
    $\Sigma d = -8.6$ $\bar{d} = -.782$
    $\Sigma d^2 = 15.06$ $s_d = .913$
  $\bar{d} \pm t_{10,.025} \cdot s_d/\sqrt{n}$
  $-.782 \pm 2.228 \cdot .913/\sqrt{11}$
  $-.782 \pm .613$
  $-1.40 < \mu_d < -.17$

8am	12am	diff
97.0	97.7	-0.7
98.0	98.8	-0.8
96.4	98.0	-1.6
98.2	98.7	-0.5
98.8	98.0	0.8
98.6	98.5	0.1
97.8	98.3	-0.5
98.7	98.7	0.0
97.8	99.1	-1.3
96.4	98.2	-1.8
96.9	99.2	-2.3

17. original claim: $\mu_d \neq 0$ [n $\leq$ 30 and σ_d unknown, use t]
  d = x_B - x_A:
  H_o: $\mu_d = 0$
  H_1: $\mu_d \neq 0$
  $\alpha = .05$
  C.R. t < $-t_{9,.025}$ = -2.262
      t > $t_{9,.025}$ = 2.262
  calculations:
    $t_{\bar{d}} = (\bar{d} - \mu_{\bar{d}})/s_{\bar{d}}$
        $= (-7.5 - 0)/(57.7/\sqrt{10}\,)$
        $= -7.5/18.2$
        $= -.41$ [from Minitab]

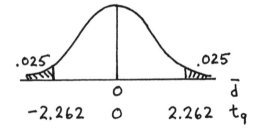

  conclusion:
    Do not reject H_o; there is not sufficient evidence to conclude that $\mu_d \neq 0$.
  No; based on this result, do not spend the money for the drug.
  NOTE: The conclusion could also have been reached by the P-value method. Since the Minitab's P-value =.69 is greater than $\alpha = .05$, there is not sufficient evidence to reject H_o.

19. Refer to the notation and data summary for exercise #17

$\bar{d} \pm t_{9,.025} \cdot s_d/\sqrt{n}$

$-7.5 \pm 2.262 \cdot 57.7/\sqrt{10}$

-7.5 ± 41.3

$-48.8 < \mu_d < 33.8$

Yes; the confidence interval includes the value 0, indicating that the drug does not have a significant effect on vulnerability to motion sickness as measured.

21. In general, a test of hypothesis is not affected by a change in units. When hypothesizing about a mean, the calculated test statistic is a ratio of the difference between the observed and expected values to the spread within the problem. Both the numerator and denominator are in the units of the problem, and the final ratio will be the same unit-free value for any choice of units.

Since the confidence interval endpoints have the units of the problem, a change in units will change the numerical value of the endpoints. But any specific amount that is in one confidence interval for one unit of measure will remain in the confidence interval for any other unit of measure. The new end points may always be found by converting the old end points to the new units.

8-3 Inferences about Two Means: Independent and Large Samples

NOTE: To be consistent with the notation of the previous sections, and thereby reinforcing the patterns and concepts presented in those sections, the manual uses the "usual" z formula written to apply to $\bar{x}_1 - \bar{x}_2$'s

$$z_{\bar{x}_1-\bar{x}_2} = (\bar{x}_1-\bar{x}_2 - \mu_{\bar{x}_1-\bar{x}_2})/\sigma_{\bar{x}_1-\bar{x}_2}$$

with $\mu_{\bar{x}_1-\bar{x}_2} = \mu_1 - \mu_2$

and $\sigma_{\bar{x}_1-\bar{x}_2} = \sqrt{\sigma_1^2/n_1 + \sigma_2^2/n_2}$

And so the formula for the z statistic may also be written as

$$z_{\bar{x}_1-\bar{x}_2} = ((\bar{x}_1-\bar{x}_2) - (\mu_1-\mu_2))/\sqrt{\sigma_1^2/n_1 + \sigma_2^2/n_2}$$

1. original claim: $\mu_1-\mu_2 = 0$ [$n_1 > 30$ and $n_2 > 30$, use z (with s's for σ's)]

$\bar{x}_1-\bar{x}_2 = 75 - 73 = 2$

$H_o: \mu_1-\mu_2 = 0$

$H_1: \mu_1-\mu_2 \neq 0$

$\alpha = .05$

C.R. $z < -z_{.025} = -1.960$

$z > z_{.025} = 1.960$

calculations:

$z_{\bar{x}_1-\bar{x}_2} = (\bar{x}_1-\bar{x}_2 - \mu_{\bar{x}_1-\bar{x}_2})/\sigma_{\bar{x}_1-\bar{x}_2}$

$= (2 - 0)/\sqrt{(15)^2/50 + (14)^2/100}$

$= 2/2.542$

$= .787$

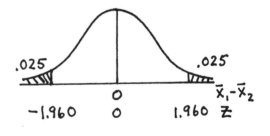

conclusion:

Do not reject H_o; there is not sufficient evidence to reject the claim that $\mu_1-\mu_2 = 0$.

3. $(\bar{x}_1-\bar{x}_2) \pm z_{.025}\sqrt{\sigma_1^2/n_1 + \sigma_2^2/n_2}$

$2 \pm 1.960 \cdot \sqrt{(15)^2/50 + (14)^2/100}$

2 ± 5

$-3 < \mu_1-\mu_2 < 7$

Since the confidence interval limits include 0, we cannot conclude there is a real difference between the control group and the experimental group.

NOTE: Since the sample means were given as whole numbers, decimal accuracy in the limits is not possible.

5. original claim: $\mu_1-\mu_2 > 0$ [$n_1 > 30$ and $n_2 > 30$, use z (with s's for σ's)]
   $\overline{x}_1-\overline{x}_2 = 53.3 - 45.3 = 8.0$
   $H_o: \mu_1-\mu_2 \le 0$
   $H_1: \mu_1-\mu_2 > 0$
   $\alpha = .01$
   C.R. $z > z_{.01} = 2.327$
   calculations:

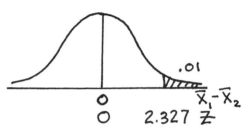

   $z_{\overline{x}_1-\overline{x}_2} = (\overline{x}_1-\overline{x}_2 - \mu_{\overline{x}_1-\overline{x}_2})/\sigma_{\overline{x}_1-\overline{x}_2}$
   $= (8.0 - 0)/\sqrt{(11.6)^2/40 + (13.2)^2/40}$
   $= 8.0/2.778$
   $= 2.879$
   conclusion:
   Reject H_o; there is sufficient evidence to conclude that $\mu_1-\mu_2 > 0$.

7. original claim: $\mu_1-\mu_2 = 0$ [$n_1 > 30$ and $n_2 > 30$, use z (with s's for σ's)]
   $\overline{x}_1-\overline{x}_2 = 203.4 - 189.4 = 14.0$
   $H_o: \mu_1-\mu_2 = 0$
   $H_1: \mu_1-\mu_2 \ne 0$
   $\alpha = .01$
   C.R. $z < -z_{.005} = -2.575$
   $z > z_{.005} = 2.575$
   calculations:

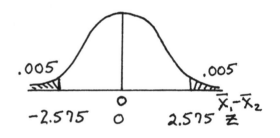

   $z_{\overline{x}_1-\overline{x}_2} = (\overline{x}_1-\overline{x}_2 - \mu_{\overline{x}_1-\overline{x}_2})/\sigma_{\overline{x}_1-\overline{x}_2}$
   $= (14.0 - 0)/\sqrt{(39.4)^2/50 + (39.0)^2/100}$
   $= 14.0/6.801$
   $= 2.058$
   conclusion:
   Do not reject H_o; there is not sufficient evidence to reject the claim that $\mu_1-\mu_2 = 0$.
   While this study does not provide conclusive evidence that the product does affect blood pressure, it certainly does not provide evidence that the product does not have an effect.

9. Let the students be group 1.
   original claim: $\mu_1-\mu_2 > 0$ [$n_1 > 30$ and $n_2 > 30$, use z (with s's for σ's)]
   $\overline{x}_1-\overline{x}_2 = 7.89 - 5.99 = 1.90$
   $H_o: \mu_1-\mu_2 \le 0$
   $H_1: \mu_1-\mu_2 > 0$
   $\alpha = .05$
   C.R. $z > z_{.05} = 1.645$
   calculations:

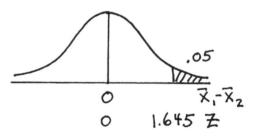

   $z_{\overline{x}_1-\overline{x}_2} = (\overline{x}_1-\overline{x}_2 - \mu_{\overline{x}_1-\overline{x}_2})/\sigma_{\overline{x}_1-\overline{x}_2}$
   $= (1.90 - 0)/\sqrt{(3.67)^2/217 + (3.65)^2/152}$
   $= 1.90/.387$
   $= 4.910$
   conclusion:
   Reject H_o; there is sufficient evidence to conclude that $\mu_1-\mu_2 > 0$.

11. Let the younger men be group 1.
    original claim: $\mu_1-\mu_2 > 0$ [$n_1 > 30$ and $n_2 > 30$, use z (with s's for σ's)]
    $\overline{x}_1-\overline{x}_2 = 176 - 164 = 12$
    H_o: $\mu_1-\mu_2 \le 0$
    H_1: $\mu_1-\mu_2 > 0$
    $\alpha = .01$
    C.R. $z > z_{.01} = 2.327$
    calculations:

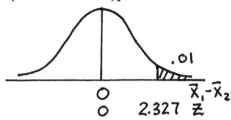

$$z_{\overline{x}_1-\overline{x}_2} = (\overline{x}_1-\overline{x}_2 - \mu_{\overline{x}_1-\overline{x}_2})/\sigma_{\overline{x}_1-\overline{x}_2}$$
$$= (12 - 0)/\sqrt{(35.0)^2/804 + (27.0)^2/1657}$$
$$= 12/1.401$$
$$= 8.563$$

conclusion:
    Reject H_o; there is sufficient evidence to conclude that $\mu_1-\mu_2 > 0$.

13. Let American be group 1.
    original claim: $\mu_1-\mu_2 = 0$ [$n_1 > 30$ and $n_2 > 30$, use z (with s's for σ's)]
    $\overline{x}_1-\overline{x}_2 = 23,870 - 22,025 = 1845$
    H_o: $\mu_1-\mu_2 = 0$
    H_1: $\mu_1-\mu_2 \ne 0$
    $\alpha = .10$
    C.R. $z < -z_{.05} = -1.645$
    $z > z_{.05} = 1.645$
    calculations:

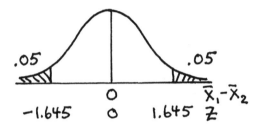

$$z_{\overline{x}_1-\overline{x}_2} = (\overline{x}_1-\overline{x}_2 - \mu_{\overline{x}_1-\overline{x}_2})/\sigma_{\overline{x}_1-\overline{x}_2}$$
$$= (1845 - 0)/\sqrt{(2960)^2/40 + (3065)^2/35}$$
$$= 1845/698.17$$
$$= 2.643$$

conclusion:
    Reject H_o; there is sufficient evidence to conclude that $\mu_1-\mu_2 \ne 0$ (in fact, $\mu_1-\mu_2 > 0$).
    The conclusion that the mean pay is more at American than TWA is not necessarily relevant for an individual considering employment. Both companies might have exactly the same pay scale, for example, but the American attendants might have more accumulated experience and hence be paid more.

15. Let last edition's values be group 1. This exercise can be completed either by testing $H_o: \mu_1-\mu_2 = 0$ or by checking whether a confidence interval constructed for $\mu_1-\mu_2$ contains the value 0. We choose the former technique.
    original claim: $\mu_1-\mu_2 \ne 0$ [$n_1 > 30$ and $n_2 > 30$, use z (with s's for σ's)]
    $\overline{x}_1-\overline{x}_2 = .9160 - .9147 = .0013$
    H_o: $\mu_1-\mu_2 = 0$
    H_1: $\mu_1-\mu_2 \ne 0$
    $\alpha = .05$ [assumed]
    C.R. $z < -z_{.025} = -1.960$
    $z > z_{.025} = 1.960$
    calculations:

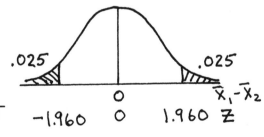

$$z_{\overline{x}_1-\overline{x}_2} = (\overline{x}_1-\overline{x}_2 - \mu_{\overline{x}_1-\overline{x}_2})/\sigma_{\overline{x}_1-\overline{x}_2}$$
$$= (.0013 - 0)/\sqrt{(.0433)^2/100 + (.0369)^2/100}$$
$$= .0013/.00569$$
$$= .229$$

conclusion:
    Do not reject H_o; there is not sufficient evidence to conclude that $\mu_1-\mu_2 \ne 0$.

17. Let the 18-24 year olds be group 1.
original claim: $\mu_1 - \mu_2 = 0$ [$n_1 > 30$ and $n_2 > 30$, use z (with s's for σ's)]
    group 1: 18-24 inclusive (n = 37)   group 2: 25 and older (n = 56)
       $\Sigma x = 3634.3$ $\Sigma x = 5491.2$
       $\Sigma x^2 = 356{,}992.32$ $\Sigma x^2 = 538{,}473.56$
       $\bar{x} = 98.2243$ $\bar{x} = 98.0571$
       $s^2 = .4349$ $s^2 = .4032$
   $\bar{x}_1 - \bar{x}_2 = 98.2243 - 98.0571 = .1672$

a. H_o: $\mu_1 - \mu_2 = 0$
   H_1: $\mu_1 - \mu_2 \neq 0$
   $\alpha = .05$
   C.R. $z < -z_{.025} = -1.960$
         $z > z_{.025} = 1.960$
   calculations:

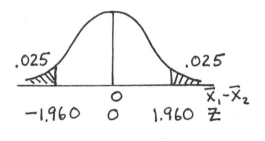

      $z_{\bar{x}_1 - \bar{x}_2} = (\bar{x}_1 - \bar{x}_2 - \mu_{\bar{x}_1 - \bar{x}_2})/\sigma_{\bar{x}_1 - \bar{x}_2}$
                $= (.1672 - 0)/\sqrt{.4349/37 + .4032/56}$
                $= .1672/.1377$
                $= 1.214$
   conclusion:
      Do not reject H_o; there is not sufficient evidence to reject the claim that $\mu_1 - \mu_2 = 0$.

b. H_o: $\mu_1 - \mu_2 = 0$
   H_1: $\mu_1 - \mu_2 \neq 0$
   $\alpha = .05$
   C.R. P-value $< .05$
   calculations:

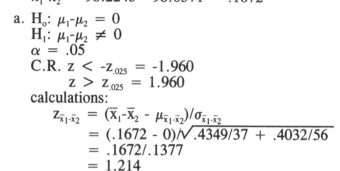

      $z_{\bar{x}_1 - \bar{x}_2} = (\bar{x}_1 - \bar{x}_2 - \mu_{\bar{x}_1 - \bar{x}_2})/\sigma_{\bar{x}_1 - \bar{x}_2}$
                $= (.1672 - 0)/\sqrt{.4349/37 + .4032/56}$
                $= .1672/.1377$
                $= 1.214$
   P-value $= 2 \cdot P(z > 1.21) = 2 \cdot (.5000 - .3869) = 2 \cdot (.1131) = .2262$
   conclusion:
      Do not reject H_o; there is not sufficient evidence to reject the claim that $\mu_1 - \mu_2 = 0$.

c. $(\bar{x}_1 - \bar{x}_2) \pm z_{.025}\sqrt{\sigma_1^2/n_1 + \sigma_2^2/n_2}$
   $(98.2243 - 98.0571) \pm 1.960 \cdot \sqrt{.4349/37 + .4032/56}$
   $.17 \pm .27$
   $-.10 < \mu_1 - \mu_2 < .44$
   Since the confidence interval includes the value zero, there is not sufficient evidence to reject the claim that $\mu_1 - \mu_2 = 0$.

d. Parts (a), (b) and (c) are equivalent techniques, and the results agree.

19. First find the summary statistics for the 175 axial loads of the .0111 cans including the outlier of 504. If we had the orignal Σx and Σx^2 values, the new ones could be obtained by subtracting 504 and 504^2.
   Usually we use n, Σx and Σx^2 to solve for $\bar{x}$ and s^2 using
   $\bar{x} = (\Sigma x)/n$
   $s^2 = [n(\Sigma x^2) - (\Sigma x)^2]/[n(n-1)]$
   Solving the above equations in reverse,
   $\Sigma x = n \cdot \bar{x}$
   $\Sigma x^2 = [n(n-1)s^2 + (\Sigma x)^2]/n$
   The original values were, therefore,
   $\Sigma x = 175 \cdot (281.8) = 49315$
   $\Sigma x^2 = [175(174)(27.8)^2 + (49315)^2]/175 = 14{,}031{,}441$

And so the new summary statistics for the .0111 cans are

$n = 174$ $\bar{x} = 280.5$

$\Sigma x = 49315 - 504 = 48811$ $s^2 = 490.27$

$\Sigma x^2 = 14,031,441 - 504^2 = 13,777,425$ $s = 22.1$

The new mean is very slightly smaller than the old one, and the new standard deviation is more noticeably smaller -- just the types of changes one should expect when removing a large outlier.

a. <u>old 5-number boxplot summary</u>   <u>new 5-number boxplot summary</u>

   $L = x_1 = 205$ $L = x_1 = 205$

   $.25(175) = 43.74; P_{25} = x_{44} = 275$ $.25(174) = 43.5; P_{25} = x_{44} = 275$

   $.50(175) = 87.50; P_{50} = x_{88} = 285$ $.50(174) = 87; P_{50} = (x_{87}+x_{88})/2 = (285+285)/2 = 285$

   $.75(175) = 131.25; P_{75} = x_{132} = 295$ $.75(174) = 130.5; P_{75} = x_{132} = 294$

   $H = x_{175} = 504$ $H = x_{174} = 317$

The new boxplot is given below. The only signifcant difference is the smaller length of the extension from the third quartile to the maximum value.

        205   275 285 294   317

b. original claim: $\mu_1-\mu_2 < 0$ [$n_1 > 30$ and $n_2 > 30$, use z (with s's for σ's)]

   $H_o: \mu_1-\mu_2 \geq 0$

   $H_1: \mu_1-\mu_2 < 0$

   $\alpha = .01$

   C.R. $z < -z_{.01} = -2.327$

   calculations:

      $z_{\bar{x}_1-\bar{x}_2} = (\bar{X}_1-\bar{X}_2 - \mu_{\bar{x}_1-\bar{x}_2})/\sigma_{\bar{x}_1-\bar{x}_2}$

          $= [(267.1 - 280.5)- 0]/\sqrt{(22.1)^2/175 + (22.1)^2/174}$

          $= -13.4/2.366$

          $= -5.664$

   conclusion:

      Reject H_o; there is sufficient evidence to conclude that $\mu_1-\mu_2 < 0$.

The calculated statistic changed slightly from -5.48 to -5.66, but the conclusion was the same.

c. $(\bar{X}_1-\bar{X}_2) \pm z_{.005}\sqrt{\sigma_1^2/n_1 + \sigma_2^2/n_2}$

   $(267.1 - 280.5) \pm 2.575 \cdot\sqrt{(22.1)^2/175 + (22.1)^2/174}$

   -13.4 ± 6.1

   $-19.5 < \mu_1-\mu_2 < -7.3$

The decrease in variability realized by dropping the outlier created a slightly narrower confidence interval. In gerenal, the sample is so large that the single outlier does not have a major effect on the test of hypothesis or the confidence interval.

21. a. $x = 5,10,15$

   $\mu = \Sigma x/n = 30/3 = 10$

   $\sigma^2 = \Sigma(x-\mu)^2/n = [(-5)^2 + (0)^2 + (5)^2]/3 = 50/3$

   b. $y = 1,2,3$

   $\mu = \Sigma y/n = 6/3 = 2$

   $\sigma^2 = \Sigma(y-\mu)^2/n = [(-1)^2 + (0)^2 + (1)^2]/3 = 2/3$

c. $z = x-y = 4,3,2,9,8,7,14,13,12$
   $\mu = \Sigma z/n = 72/9 = 8$
   $\sigma^2 = \Sigma(z-\mu)^2/n$
   $= [(-4)^2 + (-5)^2 + (-6)^2 + (1)^2 + (0)^2 + (-1)^2 + (6)^2 + (5)^2 + (4)^2]/9 = 156/9 = 52/3$

d. $\sigma^2_{x-y} = \sigma^2_x + \sigma^2_y$
   $52/3 = 50/3 + 2/3$
   $52/3 = 52/3$

e. Let R stand for range.
   $R_{x-y} = \text{highest}_{x-y} - \text{lowest}_{x-y}$
   $= (\text{highest x} - \text{lowest y}) - (\text{lowest x} - \text{highest y})$
   $= \text{highest x} - \text{lowest y} - \text{lowest x} + \text{highest y}$
   $= (\text{highest x} - \text{lowest x}) + (\text{highest y} - \text{lowest y})$
   $= R_x + R_y$
   That is, the range of all possible x-y values is the sum of the individual ranges of x and y.
   NOTE: The problem refers to all possible x-y differences (where n_x and n_y might even be different) and not to x-y differences for paired data.

8-4   Comparing Two Variances

NOTE: The following conventions are used in this manual regarding the F test.
* The set of scores with the larger sample variance is designated group 1.
* Even though always designating the scores with the larger sample variance as group 1 makes lower critical values are unnecessary in two-tailed tests, the lower critical value will be calculated (using the method given in exercise #15) and included (in brackets) for completeness and for consistency with the other tests.
* The degrees of freedom for group 1 (numerator) and group 2 (denominator) will be given with the F as a superscript and subscript respectively.
* If the desired degrees of freedom does not appear in Table A-5, the closest entry will be used. If the desired degrees of freedom is exactly halfway between two tabled values, the conservative approach of using the smaller degrees of freedom is employed. Since any finite number is closer to 120 than ∞, 120 is used for all degrees of freedom larger than 120.
* Since all hypotheses in the text question the equality of σ^2_1 and σ^2_2, the calculation of F [which is statistically defined to be $F = (s^2_1/\sigma^2_1)/(s^2_2/\sigma^2_2)$] is shortened to $F = s^2_1/s^2_2$.

1. Let sample B be group 1. original claim: $\sigma^2_1 \neq \sigma^2_2$
   H_o: $\sigma^2_1 = \sigma^2_2$
   H_1: $\sigma^2_1 \neq \sigma^2_2$
   $\alpha = .05$
   C.R. $F < F^{30}_{24,.975} = [.4682]$
   $F > F^{30}_{24,.025} = 2.2090$
   calculations:
   $F = s^2_1/s^2_2$
   $= 2000/900$
   $= 2.2222$
   conclusion:
   Reject H_o; there is sufficient evidence to conclude that $\sigma^2_1 \neq \sigma^2_2$ (in fact, $\sigma^2_1 > \sigma^2_2$).

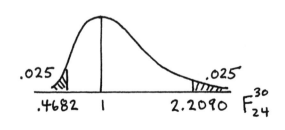

3. Let the stress sample be group 1. original claim: $\sigma_1 \neq \sigma_2$ [i.e., $\sigma_1^2 \neq \sigma_2^2$]
   H_o: $\sigma_1^2 = \sigma_2^2$
   H_1: $\sigma_1^2 \neq \sigma_2^2$
   $\alpha = .10$
   C.R. $F < F_{39,.95}^{39} = [.5907]$
   $\qquad F > F_{39,.05}^{39} = 1.6928$
   calculations:
   $\qquad F = s_1^2/s_2^2$
   $\qquad\quad = (13.2)^2/(11.6)^2$
   $\qquad\quad = 1.2949$
   conclusion:
   $\qquad$ Do not reject H_o; there is not sufficient evidence to conclude that $\sigma_1^2 \neq \sigma_2^2$.

5. Let the 25 to 34 year olds be group 1. original claim: $\sigma_1^2 > \sigma_2^2$
   H_o: $\sigma_1^2 \leq \sigma_2^2$
   H_1: $\sigma_1^2 > \sigma_2^2$
   $\alpha = .01$
   C.R. $F > F_{1656,.01}^{803} = 1.5330$
   calculations:
   $\qquad F = s_1^2/s_2^2$
   $\qquad\quad = (35.0)^2/(27.0)^2$
   $\qquad\quad = 1.6804$
   conclusion:
   $\qquad$ Reject H_o; there is sufficient evidence to conclude that $\sigma_1^2 > \sigma_2^2$.

7. Let TWA be group 1. original claim: $\sigma_1^2 = \sigma_2^2$
   H_o: $\sigma_1^2 = \sigma_2^2$
   H_1: $\sigma_1^2 \neq \sigma_2^2$
   $\alpha = .10$
   C.R. $F < F_{39,.95}^{34} = [.5581]$
   $\qquad F > F_{39,.05}^{34} = 1.7444$
   calculations:
   $\qquad F = s_1^2/s_2^2$
   $\qquad\quad = (3065)^2/(2960)^2$
   $\qquad\quad = 1.0722$
   conclusion:
   $\qquad$ Do not reject H_o; there is not sufficient evidence to reject the claim that $\sigma_1^2 = \sigma_2^2$.

9. <u>experimental</u>   <u>control</u>
   $\quad n = 20$ $\quad n = 16$
   $\quad \Sigma x = 1327.61$ $\quad \Sigma x = 1660.28$
   $\quad \Sigma x^2 = 96477.3313$ $\quad \Sigma x^2 = 181167.5442$
   $\quad \bar{x} = 66.3805$ $\quad \bar{x} = 103.7675$
   $\quad s^2 = 439.469$ $\quad s^2 = 592.296$

Let the controls be group 1. original claim: $\sigma_1^2 = \sigma_2^2$

H_o: $\sigma_1^2 = \sigma_2^2$

H_1: $\sigma_1^2 \neq \sigma_2^2$

$\alpha = .05$

C.R. $F < F_{19,.975}^{15} = [.3629]$

   $F > F_{19,.025}^{15} = 2.6171$

calculations:

   $F = s_1^2/s_2^2$

   $= 592.296/439.469$

   $= 1.3478$

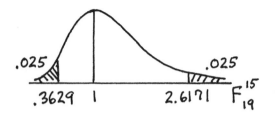

conclusion:

   Do not reject H_o; there is not sufficient evidence to reject the claim that $\sigma_1^2 = \sigma_2^2$.

11. placebo   calcium

   n = 13   n = 15

   $\Sigma x = 1490.5$ $\Sigma x = 1740.4$

   $\Sigma x^2 = 171965.47$ $\Sigma x^2 = 202936.92$

   $\bar{x} = 114.65$ $\bar{x} = 116.03$

   $s^2 = 89.493$ $s^2 = 71.722$

Let the placebos be group 1. original claim: $\sigma_1^2 = \sigma_2^2$

H_o: $\sigma_1^2 = \sigma_2^2$

H_1: $\sigma_1^2 \neq \sigma_2^2$

$\alpha = .10$

C.R. $F < F_{14,.95}^{12} = [.3821]$

   $F > F_{14,.05}^{12} = 2.5342$

calculations:

   $F = s_1^2/s_2^2$

   $= 89.493/71.722$

   $= 1.2478$

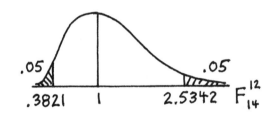

conclusion:

   Do not reject H_o; there is not sufficient evidence to reject the claim that $\sigma_1^2 = \sigma_2^2$.

13. a. The upper critical value is $F_{34,.025}^{49} = 2.0089$. Since 2.0933 > 2.0089, the conclusion is:
      Reject H_o, there is sufficient evidence to reject the claim that $\sigma_1^2 = \sigma_2^2$ and to conclude that $\sigma_1^2 \neq \sigma_2^2$ (in fact, $\sigma_1^2 > \sigma_2^2$).

   b. The upper critical value is $F_{34,.025}^{49} = 2.0089$. Since 1.8025 < 2.0089, the conclusion is:
      Do not reject H_o, there is not sufficient evidence to reject the claim that $\sigma_1^2 = \sigma_2^2$.

   c. The upper critical value is $F_{19,.025}^{39} = 2.3329$. Since 2.3935 > 2.3329, the conclusion is:
      Reject H_o, there is sufficient evidence to reject the claim that $\sigma_1^2 = \sigma_2^2$ and to conclude that $\sigma_1^2 \neq \sigma_2^2$ (in fact, $\sigma_1^2 > \sigma_2^2$).

15. a. $F_L = F_{9,.975}^9 = 1/F_{9,.025}^9 = 1/4.0260 = .2484$

      $F_R = F_{9,.025}^9 = 4.0260$

   b. $F_L = F_{6,.975}^9 = 1/F_{9,.025}^6 = 1/4.3197 = .2315$

      $F_R = F_{6,.025}^9 = 5.5234$

   c. $F_L = F_{9,.975}^6 = 1/F_{6,.025}^9 = 1/5.5234 = .1810$

      $F_R = F_{9,.025}^6 = 4.3197$

   d. $F_L = F_{9,.99}^{24} = 1/F_{24,.025}^9 = 1/3.2560 = .3071$

      $F_R = F_{9,.01}^{24} = 4.7290$

   e. $F_L = F_{24,.99}^9 = 1/F_{9,.025}^{24} = 1/4.7290 = .2115$

      $F_R = F_{24,.01}^9 = 3.2560$

17. a. Adding a constant to each score does not affect the spread of the scores. All the standard deviations and variances remain the same, and so the F statistic (i.e., the ratio of the variances) is unchanged.

b. Multiplying each score by a constant multiplies the standard deviation of those scores by that constant and the variance of those scores by the square of that constant. If this is done to both groups, so that each variance (i.e., the numerator and the denominator of the F statistic) is multiplied by the square of that constant, then the F statistic is unchanged.

c. The change from Fahrenheit to Celsius is done by multiplication and addition. As noted in part (a), the additive constant does not affect the variances and, therefore, does not affect the F statistic. As noted in part (b), the multiplicative constant affects both the numerator and denominator of the F statistic in the same manner and, therefore, does not affect the value of the F statistic.

8-5 Inferences about Two Means: Independent and Small samples

1. Let Sample A be group 1.
   original claim: $\mu_1 - \mu_2 = 0$ [small samples and σ unknown, first test $H_o: \sigma_1^2 = \sigma_2^2$]
   $H_o: \sigma_1^2 = \sigma_2^2$
   $H_1: \sigma_1^2 \neq \sigma_2^2$
   $\alpha = .05$
   C.R. $F < F^9_{19,.975} = [.2727]$
   $\quad\quad F > F^9_{19,.025} = 2.8801$

   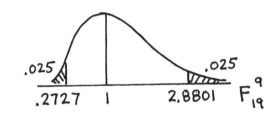

   calculations:
   $\quad F = s_1^2/s_2^2$
   $\quad\quad = (15)^2/(12)^2$
   $\quad\quad = 1.5625$
   conclusion:
   $\quad$ Do not reject H_o; there is not sufficient evidence to reject the claim that $\sigma_1^2 = \sigma_2^2$.
   Now proceed using s_p^2 for both s_1^2 and s_2^2
   $\quad \bar{x}_1 - \bar{x}_2 = 75 - 80 = -5$
   $\quad s_p^2 = (df_1 \cdot s_1^2 + df_2 \cdot s_2^2)/(df_1 + df_2)$
   $\quad\quad = (9 \cdot 15^2 + 19 \cdot 12^2)/(9 + 19) = 4761/28 = 170.03$
   $H_o: \mu_1 - \mu_2 = 0$
   $H_1: \mu_1 - \mu_2 \neq 0$
   $\alpha = .05$
   C.R. $t < -t_{28,.025} = -2.048$
   $\quad\quad t > t_{28,.025} = 2.048$

   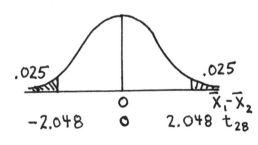

   calculations:
   $\quad t_{\bar{x}_1 - \bar{x}_2} = (\bar{x}_1 - \bar{x}_2 - \mu_{\bar{x}_1 - \bar{x}_2})/s_{\bar{x}_1 - \bar{x}_2}$
   $\quad\quad = (-5 - 0)/\sqrt{170.03/10 + 170.03/20}$
   $\quad\quad = -5/5.05$
   $\quad\quad = -.990$
   conclusion:
   $\quad$ Do not reject H_o; there is not sufficient evidence to reject the claim that $\mu_1 - \mu_2 = 0$.

3. Refer to exercise #1.
   $(\bar{x}_1 - \bar{x}_2) \pm t_{28,.025}\sqrt{s_p^2/n_1 + s_p^2/n_2}$
   $-5 \pm 2.048 \cdot \sqrt{170.03/10 + 170.03/20}$
   -5 ± 10
   $-15 < \mu_1 - \mu_2 < 5$

5. Since the F test failed to reject H_o: $\sigma_1^2 = \sigma_2^2$, proceed using s_p^2 for both s_1^2 and s_2^2
   $$\bar{x}_1 - \bar{x}_2 = 2985 - 4788 = -1803$$
   $$s_p^2 = (df_1 \cdot s_1^2 + df_2 \cdot s_2^2)/(df_1 + df_2)$$
   $$= (5 \cdot 1352^2 + 5 \cdot 2948^2)/(5 + 5) = 52593040/10 = 5259304$$
   $$(\bar{x}_1 - \bar{x}_2) \pm t_{10,.025}\sqrt{s_p^2/n_1 + s_p^2/n_2}$$
   $$-1803 \pm 2.228 \cdot \sqrt{5259304/6 + 5259304/6}$$
   $$-1803 \pm 2950$$
   $$-4753 < \mu_1 - \mu_2 < 1147$$

   Yes, the confidence interval limits include 0. The mean ozone concentrations for Maine and Rhode Island are not significantly different.

   NOTE: Reading directly from the Minitab output, $-4754 < \mu_1 - \mu_2 < 1148$. These limits are computed from the raw data (i.e., not from summary statistics) and are more accurate.

7. Let the O-C patients be group 1.
   original claim: $\mu_1 - \mu_2 = 0$ [small samples and σ unknown, first test H_o: $\sigma_1^2 = \sigma_2^2$]
   H_o: $\sigma_1^2 = \sigma_2^2$
   H_1: $\sigma_1^2 \neq \sigma_2^2$
   $\alpha = .05$ [assumed]
   C.R. $F < F^9_{9,.975} = [.2484]$
   $\quad\quad F > F^9_{9,.025} = 4.0260$
   calculations:
   $\quad\quad F = s_1^2/s_2^2$
   $\quad\quad\quad = (.08)^2/(.08)^2$
   $\quad\quad\quad = 1.0000$

   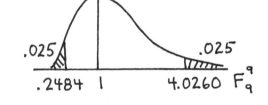

   conclusion:
   $\quad$ Do not reject H_o; there is not sufficient evidence to conclude that $\sigma_1^2 \neq \sigma_2^2$.
   Now proceed using s_p^2 for both s_1^2 and s_2^2
   $$\bar{x}_1 - \bar{x}_2 = .34 - .45 = -.11$$
   $$s_p^2 = (df_1 \cdot s_1^2 + df_2 \cdot s_2^2)/(df_1 + df_2)$$
   $$= (9 \cdot (.08)^2 + 9 \cdot (.08)^2)/(9 + 9) = .1152/18 = .0064$$
   H_o: $\mu_1 - \mu_2 = 0$
   H_1: $\mu_1 - \mu_2 \neq 0$
   $\alpha = .01$
   C.R. $t < -t_{18,.005} = -2.878$
   $\quad\quad t > t_{18,.005} = 2.878$
   calculations:

   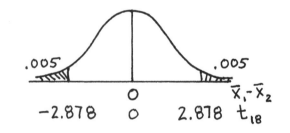

   $\quad t_{\bar{x}_1-\bar{x}_2} = (\bar{x}_1 - \bar{x}_2 - \mu_{\bar{x}_1-\bar{x}_2})/s_{\bar{x}_1-\bar{x}_2}$
   $\quad\quad\quad = (-.11 - 0)/\sqrt{.0064/10 + .0064/10}$
   $\quad\quad\quad = -.11/.0358$
   $\quad\quad\quad = -3.075$
   conclusion:
   $\quad$ Reject H_o; there is sufficient evidence to reject the claim that $\mu_1 - \mu_2 = 0$ and to conclude that $\mu_1 - \mu_2 \neq 0$ (in fact, $\mu_1 - \mu_2 < 0$).
   Based on this result, it appears that obsessive-compulsive disorders have a biological indicator.

9. Let the O-C patients be group 1.
   confidence interval: $\mu_1-\mu_2$ [small samples and σ unknown, first test H_o: $\sigma_1^2 = \sigma_2^2$]
   H_o: $\sigma_1^2 = \sigma_2^2$
   H_1: $\sigma_1^2 \neq \sigma_2^2$
   $\alpha = .05$
   C.R. $F < F_{9,.975}^9 = [.2484]$
       $F > F_{9,.025}^9 = 4.0260$
   calculations:

   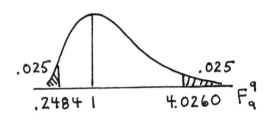

   $F = s_1^2/s_2^2$
     $= (156.84)^2/(137.97)^2$
     $= 1.2922$
   conclusion:
       Do not reject H_o; there is not sufficient evidence to conclude that $\sigma_1^2 \neq \sigma_2^2$.
   Now proceed using s_p^2 for both s_1^2 and s_2^2
   $\bar{x}_1-\bar{x}_2 = 1390.03 - 1268.41 = 121.62$
   $s_p^2 = (df_1 \cdot s_1^2 + df_2 \cdot s_2^2)/(df_1 + df_2)$
     $= (9 \cdot (156.84)^2 + 9 \cdot (137.97)^2)/(9 + 9) = 43634.5065/18 = 21817.3$
   $(\bar{x}_1-\bar{x}_2) \pm t_{18,.025}\sqrt{s_p^2/n_1 + s_p^2/n_2}$
   $121.62 \pm 2.101 \cdot \sqrt{21817.3/10 + 21817.3/10}$
   121.62 ± 138.78
   $-17.16 < \mu_1-\mu_2 < 260.40$

11. Let the orange ones be group 1.
    original claim: $\mu_1-\mu_2 = 0$ [small samples and σ unknown, first test H_o: $\sigma_1^2 = \sigma_2^2$]
    H_o: $\sigma_1^2 = \sigma_2^2$
    H_1: $\sigma_1^2 \neq \sigma_2^2$
    $\alpha = .05$ [assumed]
    C.R. $F < F_{20,.975}^7 = [.2239]$
        $F > F_{20,.025}^7 = 3.0074$
    calculations:

    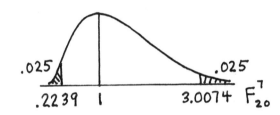

    $F = s_1^2/s_2^2$
      $= (.0472)^2/(.0275)^2$
      $= 2.9459$
    conclusion:
        Do not reject H_o; there is not sufficient evidence to reject the claim that $\sigma_1^2 = \sigma_2^2$.
    Now proceed using s_p^2 for both s_1^2 and s_2^2
    $\bar{x}_1-\bar{x}_2 = .9251 - .9097 = .0154$
    $s_p^2 = (df_1 \cdot s_1^2 + df_2 \cdot s_2^2)/(df_1 + df_2)$
      $= (7 \cdot (.0472)^2 + 20 \cdot (.0275)^2)/(7 + 20) = .0307/27 = .00114$
    H_o: $\mu_1-\mu_2 = 0$
    H_1: $\mu_1-\mu_2 \neq 0$
    $\alpha = .05$
    C.R. $t < -t_{27,.025} = -2.052$
        $t > t_{27,.025} = 2.052$
    calculations:

    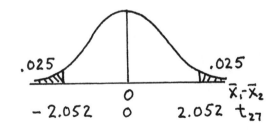

    $t_{\bar{x}_1-\bar{x}_2} = (\bar{x}_1-\bar{x}_2 - \mu_{\bar{x}_1-\bar{x}_2})/s_{\bar{x}_1-\bar{x}_2}$
      $= (.0154 - 0)/\sqrt{.00114/8 + .00114/21}$
      $= .0154/.0140$
      $= 1.099$
    conclusion:
        Do not reject H_o; there is not sufficient evidence to reject the claim that $\mu_1-\mu_2 = 0$.
    No; no corrective action is necessary.

13. <u>R rated</u>　　　　　　　　<u>G/PG/PG-13 Rated</u>
　　　n = 35　　　　　　　　　n = 25
　　　$\Sigma x = 3890$　　　　　　　$\Sigma x = 2740$
　　　$\Sigma x^2 = 446620$　　　　　$\Sigma x^2 = 315574$
　　　$\bar{x} = 111.14$　　　　　　$\bar{x} = 109.60$
　　　$s^2 = 419.83$　　　　　　　$s^2 = 636.25$
Let the G/PG/PG-13 films be group 1.
original claim: $\mu_1 - \mu_2 = 0$ [small n_1 and σ unknown, first test H_o: $\sigma_1^2 = \sigma_2^2$]
　　H_o: $\sigma_1^2 = \sigma_2^2$
　　H_1: $\sigma_1^2 \neq \sigma_2^2$
　　$\alpha = .05$ [assumed]
　　C.R. $F < F_{34,.975}^{24} = [.4527]$
　　　　$F > F_{34,.025}^{24} = 2.1359$
　　calculations:
　　　　$F = s_1^2 / s_2^2$
　　　　　$= 636.25/419.83$
　　　　　$= 1.5155$

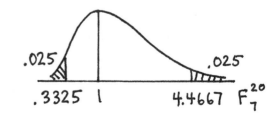

　　conclusion:
　　　　Do not reject H_o; there is not sufficient evidence to reject the claim that $\sigma_1^2 = \sigma_2^2$.
Now proceed using s_p^2 for both s_1^2 and s_2^2
　　$\bar{x}_1 - \bar{x}_2 = 109.60 - 111.14 = -1.543$
　　$s_p^2 = (df_1 \cdot s_1^2 + df_2 \cdot s_2^2)/(df_1 + df_2)$
　　　$= (24 \cdot (636.25) + 34 \cdot (419.83))/(24 + 34) = 29544.29/58 = 509.38$
H_o: $\mu_1 - \mu_2 = 0$
H_1: $\mu_1 - \mu_2 \neq 0$
$\alpha = .05$ [assumed]
C.R. $t < -t_{58,.025} = -1.960$
　　$t > t_{58,.025} = 1.960$
calculations:
　　$t_{\bar{x}_1 - \bar{x}_2} = (\bar{x}_1 - \bar{x}_2 - \mu_{\bar{x}_1 - \bar{x}_2})/s_{\bar{x}_1 - \bar{x}_2}$
　　　　$= (-1.543 - 0)/\sqrt{509.38/25 + 509.38/35}$
　　　　$= -1.543/5.910$
　　　　$= .261$

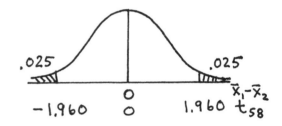

conclusion:
　　Do not reject H_o; there is not sufficient evidence to reject the claim that $\mu_1 - \mu_2 = 0$.

15. <u>filtered</u>　　　　　　　　<u>non-filtered</u>
　　　n = 21　　　　　　　　　n = 8
　　　$\Sigma x = 270$　　　　　　　$\Sigma x = 125$
　　　$\Sigma x^2 = 3660$　　　　　　$\Sigma x^2 = 1963$
　　　$\bar{x} = 12.857$　　　　　　$\bar{x} = 15.625$
　　　$s^2 = 9.429$　　　　　　　$s^2 = 1.411$
Let the filtered ones be group 1.
original claim: $\mu_1 - \mu_2 < 0$ [small samples and σ unknown, first test H_o: $\sigma_1^2 = \sigma_2^2$]
　　H_o: $\sigma_1^2 = \sigma_2^2$
　　H_1: $\sigma_1^2 \neq \sigma_2^2$
　　$\alpha = .05$ [assumed]
　　C.R. $F < F_{7,.975}^{20} = [.3325]$
　　　　$F > F_{7,.025}^{20} = 4.4667$
　　calculations:
　　　　$F = s_1^2 / s_2^2$
　　　　　$= 9.429/1.411$
　　　　　$= 6.6835$

conclusion:
    Reject H_o; there is sufficient evidence to reject the claim that $\sigma_1^2 = \sigma_2^2$ and to conclude
    that $\sigma_1^2 \neq \sigma_2^2$ (in fact, $\sigma_1^2 > \sigma_2^2$.).
Now proceed using s_1^2 and s_2^2 and the smaller degrees of freedom (viz., $df_2 = 7$).
    $\bar{x}_1-\bar{x}_2 = 12.857 - 15.625 = -2.768$
H_o: $\mu_1-\mu_2 \geq 0$
H_1: $\mu_1-\mu_2 < 0$
$\alpha = .05$ [assumed]
C.R. $t < -t_{7,.05} = -1.895$
calculations:
    $t_{\bar{x}_1-\bar{x}_2} = (\bar{x}_1-\bar{x}_2 - \mu_{\bar{x}_1-\bar{x}_2})/s_{\bar{x}_1-\bar{x}_2}$

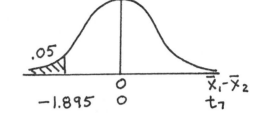

$\qquad = (-2.768 - 0)/\sqrt{9.429/21 + 1.411/8}$
$\qquad = -2.768/.791$
$\qquad = -3.500$
conclusion:
    Reject H_o; there is sufficient evidence to conclude that $\mu_1-\mu_2 < 0$.
Based on this result and the examples of this section, it appears that cigarette filters are truly
effective. The ultimate filter, which reduces harmful first-hand intake to zero, is to refrain
from smoking. Even then, there is legitimate concern about second-hand smoke -- and cigarette
filters probably do not reduce the harmful levels of second-hand smoke.

17. Let the treatment sample be group 1.
    original claim: $\mu_1-\mu_2 = 0$ [small samples and σ unknown, first test H_o: $\sigma_1^2 = \sigma_2^2$]
        H_o: $\sigma_1^2 = \sigma_2^2$
        H_1: $\sigma_1^2 \neq \sigma_2^2$
        $\alpha = .05$
        C.R. $F < F_{21,.975}^{21} = [.4124]$
            $F > F_{21,.025}^{21} = 2.4247$
        calculations:
            $F = s_1^2/s_2^2$

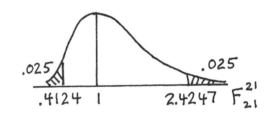

            $\qquad = (.015)^2/(.000)^2$
            $\qquad = \infty$
        conclusion:
            Reject H_o; there is sufficient evidence to reject the claim that $\sigma_1^2 = \sigma_2^2$ and to conclude that
            $\sigma_1^2 \neq \sigma_2^2$ (in fact, $\sigma_1^2 > \sigma_2^2$).
Now proceed using s_1^2 and s_2^2 and the smaller degrees of freedom (viz., $df_1 = df_2 = 21$).
    $\bar{x}_1-\bar{x}_2 = .049 - .000 = .049$
H_o: $\mu_1-\mu_2 = 0$
H_1: $\mu_1-\mu_2 \neq 0$
$\alpha = .05$
C.R. $t < -t_{21,.025} = -2.080$
    $t > t_{21,.025} = 2.080$
calculations:
    $t_{\bar{x}_1-\bar{x}_2} = (\bar{x}_1-\bar{x}_2 - \mu_{\bar{x}_1-\bar{x}_2})/s_{\bar{x}_1-\bar{x}_2}$

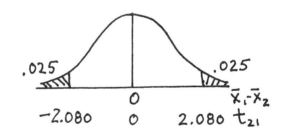

    $\qquad = (.049 - 0)/\sqrt{(.015)^2/22 + (0)^2/22}$
    $\qquad = .049/.00319$
    $\qquad = 15.322$
conclusion:
    Reject H_o; there is sufficient evidence to reject the claim that $\mu_1-\mu_2 = 0$ and to conclude that
    $\mu_1-\mu_2 \neq 0$ (in fact, $\mu_1-\mu_2 > 0$).

19. $A = s_1^2/n_1 = (3.7)^2/21 = .65190$
    $B = s_2^2/n_2 = (1.7)^2/8 = .36125$
    $df = (A + B)^2/(A^2/df_1 + B^2/df_2)$
    $= (1.01315)^2/(.42498/20 + .13050/7)$
    $= 1.02648/.039892$
    $= 25.7$

For the test of hypothesis:
    The C.R. changes from $t < -t_{7,.05} = -1.895$ to $t < -t_{26,.05} = -1.706$. The calculated test statistic of $t = -10.630$ is not affected. In this case, the conclusion remains the same.

For the confidence interval:
    $(\overline{x}_1-\overline{x}_2) \pm t_{25.7,.025}\sqrt{s_1^2/n_1 + s_2^2/n_2}$
    $-10.7 \pm 2.060\cdot\sqrt{(3.7)^2/21 + (1.7)^2/8}$
    -10.7 ± 2.1
    $-12.8 < \mu_1-\mu_2 < -8.6$
    This is slightly narrower than the given confidence interval based on df=7.

8-6 Inferences about Two Proportions

NOTE: To be consistent with the notation of the previous sections, thereby reinforcing the patterns and concepts presented there, the manual uses the "usual" z formula written to apply to $\hat{p}_1-\hat{p}_2$'s
    $z_{\hat{p}1-\hat{p}2} = (\hat{p}_1-\hat{p}_2 - \mu_{\hat{p}1-\hat{p}2})/\sigma_{\hat{p}1-\hat{p}2}$
with $\mu_{\hat{p}1-\hat{p}2} = p_1 - p_2$
    and $\sigma_{\hat{p}1-\hat{p}2} = \sqrt{\overline{pq}/n_1 + \overline{pq}/n_2}$ [when H_o includes $p_1=p_2$]
    where $\overline{p} = (x_1 + x_2)/(n_1 + n_2)$
And so the formula for the z statistic may also be written as
    $z_{\hat{p}1-\hat{p}2} = ((\hat{p}_1-\hat{p}_2) - (p_1-p_2))/\sqrt{\overline{pq}/n_1 + \overline{pq}/n_2}$

1. $\hat{p}_1 = x_1/n_1 = 25/50 = .500$
   $\hat{p}_2 = x_2/n_2 = 55/100 = .550$
   $\hat{p}_1-\hat{p}_2 = .500 - .550 = -.050$
   a. $\overline{p} = (x_1 + x_2)/(n_1 + n_2) = (25 + 55)/(50 + 100) = 80/150 = .533$
   b. $z_{\hat{p}1-\hat{p}2} = (\hat{p}_1-\hat{p}_2 - \mu_{\hat{p}1-\hat{p}2})/\sigma_{\hat{p}1-\hat{p}2}$
      $= (-.050 - 0)/\sqrt{(.533)(.467)/50 + (.533)(.467)/100}$
      $= -.050/.0864 = -.579$
   c. $\pm z_{.025} = \pm 1.960$
   d. P-value $= 2\cdot P(z < -.58) = 2\cdot(.5000 - .2190) = 2\cdot(.2810) = .5620$

NOTE: Since $\overline{p}$ is the weighted average of $\hat{p}_1$ and $\hat{p}_2$, it must always fall between those two values. If it does not, then an error has been made that must be corrected before proceeding. Calculation of $\sigma_{\hat{p}1-\hat{p}2} = \sqrt{\overline{pq}/n_1 + \overline{pq}/n_2}$ can be accomplished with no round-off loss on most calculators by calculating $\overline{p}$ and proceeding as follows: STORE 1-RECALL = * RECALL = STORE RECALL $\div$ n_1 + RECALL $\div$ n_2 = $\sqrt{}$. The quantity $\sigma_{\hat{p}1-\hat{p}2}$ may then be STORED for future use. Each calculator is different -- learn how your calculator works, and do the homework on the same calculator you will use for the exam. If you have any questions about performing/storing calculations on your calculator, check with your instructoror class assistant.

3. Let those receiving warmth be group 1.
  original claim: $p_1-p_2 < 0$
    $\hat{p}_1 = x_1/n_1 = 6/104 = .0577$
    $\hat{p}_2 = x_2/n_2 = 18/96 = .1875$
    $\hat{p}_1-\hat{p}_2 = .0577 - .1875 = -.1298$
    $\bar{p} = (x_1 + x_2)/(n_1 + n_2)$
       $= (6 + 18)/(104 + 96) = 24/200 = .12$
  H_o: $p_1-p_2 \geq 0$
  H_1: $p_1-p_2 < 0$
  $\alpha = .05$
  C.R. $z < -z_{.05} = -1.645$
  calculations:

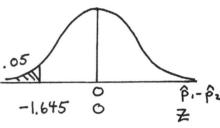

    $z_{\hat{p}_1-\hat{p}_2} = (\hat{p}_1-\hat{p}_2 - \mu_{\hat{p}_1-\hat{p}_2})/\sigma_{\hat{p}_1-\hat{p}_2}$
       $= (-.1298 - 0)/\sqrt{(.12)(.88)/104 + (.12)(.88)/96}$
       $= -.1298/.0460$
       $= -2.822$
  conclusion:
    Reject H_o; there is sufficient evidence to conclude that $p_1-p_2 < 0$.
  Yes; if these results are verified, surgical patients should be routinely warmed.

5. Let the 18-24 year olds be group 1.
    $\hat{p}_1 = x_1/n_1 = .0425$
    $\hat{p}_2 = x_2/n_2 = .0455$
    $\hat{p}_1-\hat{p}_2 = .0425 - .0455 = -.0029$
  $(\hat{p}_1-\hat{p}_2) \pm z_{.025}\sqrt{\hat{p}_1\hat{q}_1/n_1 + \hat{p}_2\hat{q}_2/n_2}$
  $-.0029 \pm 1.960 \cdot \sqrt{(.0425)(.9575)/2750 + (.0455)(.9545)/2200}$
  $-.0029 \pm .0115$
  $-.0144 < p_1-p_2 < .0086$
  Yes; the interval contains zero, indicating no significant difference between the two rates of crime.

7. Let the vinyl gloves be group 1.
  original claim: $p_1-p_2 > 0$
    $\hat{p}_1 = x_1/n_1 = .63$
    $\hat{p}_2 = x_2/n_2 = .07$
    $\hat{p}_1-\hat{p}_2 = .63 - .07 = .56$
    $\bar{p} = (x_1 + x_2)/(n_1 + n_2)$
       $= (.63 \cdot 240 + .07 \cdot 240)/(240 + 240) = 168/480 = .350$
  H_o: $p_1-p_2 \leq 0$
  H_1: $p_1-p_2 > 0$
  $\alpha = .005$
  C.R. $z > z_{.005} = 2.575$
  calculations:
    $z_{\hat{p}_1-\hat{p}_2} = (\hat{p}_1-\hat{p}_2 - \mu_{\hat{p}_1-\hat{p}_2})/\sigma_{\hat{p}_1-\hat{p}_2}$
       $= (.56 - 0)/\sqrt{(.350)(.650)/240 + (.350)(.650)/240}$
       $= .56/.0435$
       $= 12.861$
  conclusion:
    Reject H_o; there is sufficient evidence to conclude that $p_1-p_2 > 0$.

9. Let the Democrats be group 1.
   original claim: $p_1 - p_2 = 0$
   $\hat{p}_1 = x_1/n_1 = .35$ [see NOTE below]
   $\hat{p}_2 = x_2/n_2 = .41$ [see NOTE below]
   $\hat{p}_1 - \hat{p}_2 = .35 - .41 = -.06$
   $\bar{p} = (x_1 + x_2)/(n_1 + n_2)$
   $= (.35 \cdot 552 + .41 \cdot 417)/(552 + 417) = 364.17/969 = .376$
   $H_o: p_1 - p_2 = 0$
   $H_1: p_1 - p_2 \neq 0$
   $\alpha = .05$
   C.R. $z < -z_{.025} = -1.960$
   $z > z_{.025} = 1.960$
   calculations:
   $z_{\hat{p}_1 - \hat{p}_2} = (\hat{p}_1 - \hat{p}_2 - \mu_{\hat{p}_1 - \hat{p}_2})/\sigma_{\hat{p}_1 - \hat{p}_2}$
   $= (-.06 - 0)/\sqrt{(.376)(.624)/552 + (.376)(.624)/417}$
   $= -.06/.0314$
   $= -1.909$

   conclusion:
   Do not reject H_o; there is not sufficient evidence to conclude that $p_1 - p_2 \neq 0$.

   No; based on this result it does not appear appropriate to invest in different approaches for Republicans and Democrats.

NOTE: In the preceding problem x_1 and x_2 were not given and must be deduced from the values of $\hat{p}_1 = 35\%$ and $\hat{p}_2 = 41\%$. Unfortunately, any x_1 between 191 and 195 inclusive and any x_2 between 169 and 173 inclusive produces the given percents. Whenever this occurs (i.e., x_1 and x_2 cannot be determined exactly), the estimate $x = \hat{p} \cdot n$ will be employed without further manipulation -- even if that estimate is not a whole number.

11. Let the convicted arsonists be group 1.
    original claim: $p_1 - p_2 > 0$
    $\hat{p}_1 = x_1/n_1 = 50/(50 + 43) = 50/93 = .538$
    $\hat{p}_2 = x_2/n_2 = 63/(63 + 144) = 63/207 = .304$
    $\hat{p}_1 - \hat{p}_2 = .538 - .304 = .233$
    $\bar{p} = (x_1 + x_2)/(n_1 + n_2)$
    $= (50 + 63)/(93 + 207) = 113/300 = .377$
    $H_o: p_1 - p_2 \leq 0$
    $H_1: p_1 - p_2 > 0$
    $\alpha = .01$
    C.R. $z > z_{.01} = 2.327$
    calculations:
    $z_{\hat{p}_1 - \hat{p}_2} = (\hat{p}_1 - \hat{p}_2 - \mu_{\hat{p}_1 - \hat{p}_2})/\sigma_{\hat{p}_1 - \hat{p}_2}$
    $= (.233 - 0)/\sqrt{(.377)(.623)/93 + (.377)(.623)/207}$
    $= .233/.0605$
    $= 3.857$

    conclusion:
    Reject H_o; there is sufficient evidence to conclude that $p_1 - p_2 > 0$.

    Yes; it does seem reasonable that drinking might be related to the type of crime. The kinds of problems and personalities associated with drinking may well be more likely to be associated with some crimes more than others.

13. Let the 18-24 year olds be group 1.
    original claim: $p_1 - p_2 = 0$
        $\hat{p}_1 = x_1/n_1 = .360$ [x_1 must be 72]
        $\hat{p}_2 = x_2/n_2 = .540$ [x_2 must be 135]
        $\hat{p}_1 - \hat{p}_2 = .360 - .540 = -.180$
        $\bar{p} = (x_1 + x_2)/(n_1 + n_2)$
            $= (72 + 135)/(200 + 250) = 207/450 = .460$
    H_o: $p_1 - p_2 = 0$
    H_1: $p_1 - p_2 \neq 0$
    $\alpha = .05$
    C.R. $z < -z_{.025} = -1.960$
        $z > z_{.025} = 1.960$
    calculations:

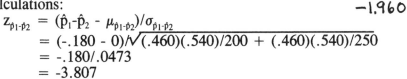

        $z_{\hat{p}_1 - \hat{p}_2} = (\hat{p}_1 - \hat{p}_2 - \mu_{\hat{p}_1 - \hat{p}_2})/\sigma_{\hat{p}_1 - \hat{p}_2}$
            $= (-.180 - 0)/\sqrt{(.460)(.540)/200 + (.460)(.540)/250}$
            $= -.180/.0473$
            $= -3.807$
    conclusion:
        Reject H_o; there is sufficient evidence to reject that claim that $p_1 - p_2 = 0$ and to conclude that $p_1 - p_2 \neq 0$ (in fact, $p_1 - p_2 < 0$).

15. Let those not wearing seat belts be group 1.
    original claim: $p_1 - p_2 > 0$
        $\hat{p}_1 = x_1/n_1 = 50/290 = .1724$
        $\hat{p}_2 = x_2/n_2 = 16/123 = .1301$
        $\hat{p}_1 - \hat{p}_2 = .1724 - .1301 = .0423$
        $\bar{p} = (x_1 + x_2)/(n_1 + n_2)$
            $= (50 + 16)/(290 + 123) = 66/413 = .160$
    H_o: $p_1 - p_2 \leq 0$
    H_1: $p_1 - p_2 > 0$
    $\alpha = .05$
    C.R. $z > z_{.05} = 1.645$
    calculations:
        $z_{\hat{p}_1 - \hat{p}_2} = (\hat{p}_1 - \hat{p}_2 - \mu_{\hat{p}_1 - \hat{p}_2})/\sigma_{\hat{p}_1 - \hat{p}_2}$
            $= (.0423 - 0)/\sqrt{(.160)(.840)/290 + (.160)(.840)/123}$
            $= .0423/.0394$
            $= 1.074$
    conclusion:
        Do not reject H_o; there is not sufficient evidence to conclude that $p_1 - p_2 > 0$.
    Based on these results, no specific action should be taken.
    NOTE: This test involves only children who were hospitalized. It is probably true that a much greater percentage of children wearing seat belts avoided having to be hospitalized in the first place.

17. Let the women exposed to glycol be group 1.
original claim: $p_1-p_2 = .10$
$$\hat{p}_1 = x_1/n_1 = 10/30 = .333$$
$$\hat{p}_2 = x_2/n_2 = 120/750 = .160$$
$$\hat{p}_1-\hat{p}_2 = .333 - .160 = .173$$
H_o: $p_1-p_2 = .10$
H_1: $p_1-p_2 \neq .10$
$\alpha = .05$
C.R. $z < -z_{.025} = -1.960$
$z > z_{.025} = 1.960$
calculations:

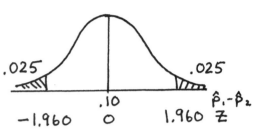

$$z_{\hat{p}1-\hat{p}2} = (\hat{p}_1-\hat{p}_2 - \mu_{\hat{p}1-\hat{p}2})/\sigma_{\hat{p}1-\hat{p}2}$$
$$= (.173 - .100)/\sqrt{(.333)(.667)/30 + (.160)(.840)/750}$$
$$= .0733/.0871 = .842$$
conclusion:
   Do not reject H_o; there is not sufficient evidence to reject the claim that $p_1-p_2 = .10$.

19. $E^2 = (z_{\alpha/2})^2(p_1q_1/n_1 + p_2q_2/n_2)$ squaring the original equation
$E^2 = (z_{\alpha/2})^2(.25/n_1 + .25/n_2)$ setting the unknown proportions to .5
$E^2 = (z_{\alpha/2})^2(.25/n + .25/n)$ requiring $n_1 = n_2 = n$
$E^2 = (z_{\alpha/2})^2(.50/n)$ addition
$n = (z_{\alpha/2})^2(.50)/E^2$ solving for n
$\quad = (1.960)^2(.50)/(.03)^2$ for $\alpha = .05$ and $E = .03$
$\quad = 2134.2$ rounded up to 2135

Review Exercises

1. Let those who saw someone else getting help be group 1.
original claim: $p_1-p_2 > 0$
$$\hat{p}_1 = x_1/n_1 = 58/2000 = .0290 \quad [x_1 = .0290 \cdot 2000 = 58]$$
$$\hat{p}_2 = x_2/n_2 = 35/2000 = .0175 \quad [x_2 = .0175 \cdot 2000 = 35]$$
$$\hat{p}_1-\hat{p}_2 = .0290 - .0175 = .0115$$
$$\bar{p} = (x_1 + x_2)/(n_1 + n_2)$$
$$= (58 + 35)/(2000 + 2000) = 93/4000 = .02325$$
H_o: $p_1-p_2 \leq 0$
H_1: $p_1-p_2 > 0$
$\alpha = .05$
C.R. $z > z_{.05} = 1.645$
calculations:

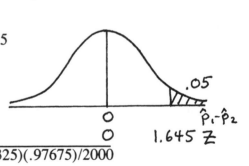

$$z_{\hat{p}1-\hat{p}2} = (\hat{p}_1-\hat{p}_2 - \mu_{\hat{p}1-\hat{p}2})/\sigma_{\hat{p}1-\hat{p}2}$$
$$= (.0115 - 0)/\sqrt{(.02325)(.97675)/2000 + (.02325)(.97675)/2000}$$
$$= .0115/.00477$$
$$= 2.413$$
conclusion:
   Reject H_o; there is sufficient evidence to conclude that $p_1-p_2 > 0$.

2. a. original claim $\mu_d > 0$ [$n \leq 30$ and σ_d unknown, use t]
       d = $x_{after} - x_{before}$: 90 25 17 16 65 10 5 9 29 6 30 46
         n = 12
         $\Sigma d = 348$ $\bar{d} = 29.0$
         $\Sigma d^2 = 17594$ $s_d = 26.1$
       H_o: $\mu_d \leq 0$
       H_1: $\mu_d > 0$
       $\alpha = .05$
       C.R. $t > t_{11,.05} = 1.796$

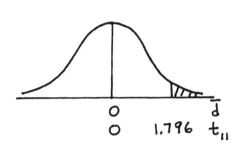

       calculations:
         $t_{\bar{d}} = (\bar{d} - \mu_{\bar{d}})/s_{\bar{d}}$
            $= (29.0 - 0)/(26.1/\sqrt{12})$
            $= 29.0/7.539$
            $= 3.847$
       conclusion:
         Reject H_o; there is sufficient evidence to conclude that $\mu_d > 0$.
     b. $\bar{d} \pm t_{11,.025} \cdot s_{\bar{d}}$
         $29.0 \pm 2.201 \cdot 26.1/\sqrt{12}$
         29.0 ± 16.6
         $12.4 < \mu_d < 45.6$

3.   <u>Dozenol</u>   <u>Niteze</u>
       n = 12   n = 12
       $\Sigma x = 5936$ $\Sigma x = 6284$
       $\Sigma x^2 = 2,938,906$ $\Sigma x^2 = 3,294,058$
       $\bar{x} = 494.667$ $\bar{x} = 523.667$
       $s^2 = 233.152$ $s^2 = 303.333$
     Let Niteze be group 1.
     a. original claim: $\mu_1 - \mu_2 = 0$ [small samples and σ unknown, first test H_o: $\sigma_1^2 = \sigma_2^2$]
        H_o: $\sigma_1^2 = \sigma_2^2$
        H_1: $\sigma_1^2 \neq \sigma_2^2$
        $\alpha = .05$
        C.R. $F < F_{11,.975}^{11} = [.2836]$ [OR .2916]
             $F > F_{11,.025}^{11} = 3.5257$ [OR 3.4296]

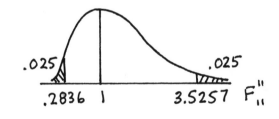

        calculations:
          $F = s_1^2/s_2^2$
             $= 303.333/233.152$
             $= 1.3010$
        conclusion:
          Do not reject H_o; there is not sufficient evidence to conclude that $\sigma_1^2 \neq \sigma_2^2$.
     Now proceed using s_p^2 for both s_1^2 and s_2^2
       $\bar{x}_1 - \bar{x}_2 = 523.667 - 494.667 = 29.00$
       $s_p^2 = (df_1 \cdot s_1^2 + df_2 \cdot s_2^2)/(df_1 + df_2)$
          $= (11 \cdot (303.333) + 11 \cdot (233.152))/(11 + 11) = 5901.33/22 = 268.24$

H_o: $\mu_1-\mu_2 = 0$
H_1: $\mu_1-\mu_2 \neq 0$
$\alpha = .05$
C.R. $t < -t_{22,.025} = -2.074$
    $t > t_{22,.025} = 2.074$
calculations:

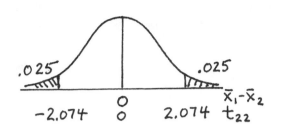

$t_{\bar{x}_1-\bar{x}_2} = (\bar{X}_1-\bar{X}_2 - \mu_{\bar{x}_1-\bar{x}_2})/s_{\bar{x}_1-\bar{x}_2}$
        $= (29.00 - 0)/\sqrt{268.24/12 + 268.24/12}$
        $= 29.00/6.686$
        $= 4.337$
conclusion:
    Reject H_o; there is sufficient evidence to reject the claim that $\mu_1-\mu_2 = 0$ and to conclude
    that $\mu_1-\mu_2 \neq 0$ (in fact, $\mu_1-\mu_2 > 0$).
b. $(\bar{X}_1-\bar{X}_2) \pm t_{22,.025}\sqrt{s_p^2/n_1 + s_p^2/n_2}$
    $29.00 \pm 2.074\sqrt{268.24/12 + 268.24/12}$
    29.00 ± 13.87
    $15.1 < \mu_1-\mu_2 < 42.9$

4. Let the women be group 1. original claim: $\sigma_1 = \sigma_2$ [i.e., $\sigma_1^2 = \sigma_2^2$]
   H_o: $\sigma_1^2 = \sigma_2^2$
   H_1: $\sigma_1^2 \neq \sigma_2^2$
   $\alpha = .02$
   C.R. $F < F_{85,.99}^{67} = [.5446]$
       $F > F_{85,.01}^{67} = 1.8363$
   calculations:

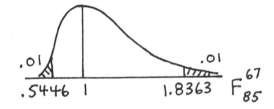

   $F = s_1^2/s_2^2$
     $= (114.16)^2/(97.23)^2$
     $= 1.3786$
   conclusion:
       Do not reject H_o; there is not sufficient evidence to reject the claim that $\sigma_1^2 = \sigma_2^2$.

5. Let the women be group 1.
   original claim: $\mu_1-\mu_2 = 0$ [$n_1 > 30$ and $n_2 > 30$, use z (with s's for σ's)]
     $\bar{X}_1-\bar{X}_2 = 538.82 - 525.23 = 13.59$
   H_o: $\mu_1-\mu_2 = 0$
   H_1: $\mu_1-\mu_2 \neq 0$
   $\alpha = .02$
   C.R. $z < -z_{.01} = -2.327$
       $z > z_{.01} = 2.327$
   calculations:

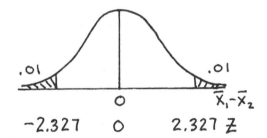

   $z_{\bar{x}_1-\bar{x}_2} = (\bar{X}_1-\bar{X}_2 - \mu_{\bar{x}_1-\bar{x}_2})/\sigma_{\bar{x}_1-\bar{x}_2}$
        $= (13.59 - 0)/\sqrt{(114.16)^2/68 + (97.23)^2/86}$
        $= 13.59/17.366$
        $= .783$
   conclusion:
       Do not reject H_o; there is not sufficient evidence to reject the claim that $\mu_1-\mu_2 = 0$.

6. Let the prepared students be group 1.
    original claim: $p_1-p_2 > 0$ [concerns p's, use z]
        $\hat{p}_1 = x_1/n_1 = 62/80 = .775$
        $\hat{p}_2 = x_2/n_2 = 23/50 = .460$
        $\hat{p}_1-\hat{p}_2 = .775 - .460 = .315$
        $\bar{p} = (x_1 + x_2)/(n_1 + n_2)$
            $= (62 + 23)/(80 + 50) = 85/130 = .654$
    H_o: $p_1-p_2 \leq 0$
    H_1: $p_1-p_2 > 0$
    $\alpha = .05$
    C.R. $z > z_{.05} = 1.645$
    calculations:

$z_{\hat{p}_1-\hat{p}_2} = (\hat{p}_1-\hat{p}_2 - \mu_{\hat{p}_1-\hat{p}_2})/\sigma_{\hat{p}_1-\hat{p}_2}$
$= (.315 - 0)/\sqrt{(.654)(.346)/80 + (.654)(.346)/50}$
$= .315/.0858$
$= 3.673$

conclusion:
    Reject H_o; there is sufficient evidence to conclude that $p_1-p_2 > 0$.

7. original claim $\mu_d = 0$ [$n \leq 30$ and σ_d unknown, use t]
    $d = x_{sitting} - x_{supine}$: .03 -.64 -.35 .11 .17 .47 .49 .40 .29 .44
        $n = 10$
        $\Sigma d = 1.41$ $\bar{d} = .141$
        $\Sigma d^2 = 1.4727$ $s_d = .376$
    H_o: $\mu_d = 0$
    H_1: $\mu_d \neq 0$
    $\alpha = .05$
    C.R. $t < -t_{9,.025} = -2.262$
        $t > t_{9,.025} = 2.262$
    calculations:

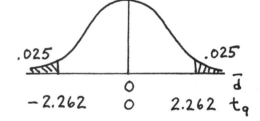

$t_{\bar{d}} = (\bar{d} - \mu_{\bar{d}})/s_{\bar{d}}$
$= (.141 - 0)/(.376/\sqrt{10})$
$= .141/.119$
$= 1.185$

conclusion:
    Do not reject H_o; there is not sufficient evidence to reject the claim that $\mu_d = 0$.

8. Let the six-week program be group 1.
    original claim: $\mu_1-\mu_2 > 0$ [$n_1 > 30$ and $n_2 > 30$, use z (with s's for σ's)]
        $\bar{x}_1-\bar{x}_2 = 83.5 - 79.8 = 3.7$
    H_o: $\mu_1-\mu_2 \leq 0$
    H_1: $\mu_1-\mu_2 > 0$
    $\alpha = .01$
    C.R. $z > z_{.01} = 2.327$
    calculations:

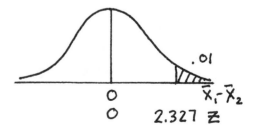

$z_{\bar{x}_1-\bar{x}_2} = (\bar{x}_1-\bar{x}_2 - \mu_{\bar{x}_1-\bar{x}_2})/\sigma_{\bar{x}_1-\bar{x}_2}$
$= (3.7 - 0)/\sqrt{(16.3)^2/60 + (19.2)^2/35}$
$= 3.7/3.868$
$= .957$

conclusion:
    Do not reject H_o; there is not sufficient evidence to conclude that $\mu_1-\mu_2 > 0$.

9. Let Barrington be group 1.
   original claim: $\mu_1-\mu_2 = 0$ [small samples and σ unknown, first test H_o: $\sigma_1^2 = \sigma_2^2$]
   H_o: $\sigma_1^2 = \sigma_2^2$
   H_1: $\sigma_1^2 \neq \sigma_2^2$
   $\alpha = .05$
   C.R. $F < F_{17,.975}^{23} = [.4054]$
   $\quad\quad F > F_{17,.025}^{23} = 2.5598$
   calculations:
   $\quad F = s_1^2/s_2^2$
   $\quad\quad = (9.7)^2/(2.8)^2$
   $\quad\quad = 12.0013$

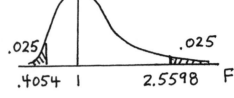

   conclusion:
   $\quad$ Reject H_o; there is sufficient evidence to reject the claim that $\sigma_1^2 = \sigma_2^2$ and to conclude that $\sigma_1^2 \neq \sigma_2^2$ (in fact, $\sigma_1^2 > \sigma_2^2$).
   Now proceed using s_1^2 and s_2^2 and the smaller degrees of freedom (viz., $df_2 = 17$).
   $\bar{x}_1-\bar{x}_2 = 80.6 - 85.7 = -5.1$
   H_o: $\mu_1-\mu_2 = 0$
   H_1: $\mu_1-\mu_2 \neq 0$
   $\alpha = .05$
   C.R. $t < -t_{17,.025} = -2.110$
   $\quad\quad t > t_{17,.025} = 2.110$
   calculations:
   $t_{\bar{x}_1-\bar{x}_2} = (\bar{x}_1-\bar{x}_2 - \mu_{\bar{x}_1-\bar{x}_2})/s_{\bar{x}_1-\bar{x}_2}$
   $\quad\quad = (-5.1 - 0)/\sqrt{(9.7)^2/24 + (2.8)^2/18}$
   $\quad\quad = -5.1/2.087$
   $\quad\quad = -2.444$

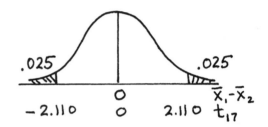

   conclusion:
   $\quad$ Reject H_o; there is sufficient evidence to reject the claim that $\mu_1-\mu_2 = 0$ and to conclude that $\mu_1-\mu_2 \neq 0$ (in fact, $\mu_1-\mu_2 < 0$).

10. a. Let the first time period be group 1.
   original claim: $\mu_1-\mu_2 = 0$ [small samples and σ unknown, first test H_o: $\sigma_1^2 = \sigma_2^2$]
   H_o: $\sigma_1^2 = \sigma_2^2$
   H_1: $\sigma_1^2 \neq \sigma_2^2$
   $\alpha = .05$
   C.R. $F < F_{11,.975}^{11} = [.2836]$ [OR .2916]
   $\quad\quad F > F_{11,.025}^{11} = 3.5257$ [OR 3.4296]
   calculations:
   $\quad F = s_1^2/s_2^2$
   $\quad\quad = (11.07)^2/(10.39)^2$
   $\quad\quad = 1.1352$

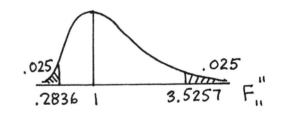

   conclusion:
   $\quad$ Do not reject H_o; there is not sufficient evidence to reject the claim that $\sigma_1^2 = \sigma_2^2$.
   Now proceed using s_p^2 for both s_1^2 and s_2^2
   $\bar{x}_1-\bar{x}_2 = 46.42 - 51.00 = -4.58$
   $s_p^2 = (df_1 \cdot s_1^2 + df_2 \cdot s_2^2)/(df_1 + df_2)$
   $\quad\quad = (11 \cdot (11.07)^2 + 11 \cdot (10.39)^2)/(11 + 11) = 2535.467/22 = 115.25$

H_o: $\mu_1-\mu_2 = 0$
H_1: $\mu_1-\mu_2 \neq 0$
$\alpha = .05$
C.R. t $< -t_{22,.025} = -2.074$
      t $> t_{22,.025} = 2.074$
calculations:

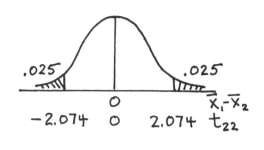

$$t_{\bar{x}_1-\bar{x}_2} = (\bar{x}_1-\bar{x}_2 - \mu_{\bar{x}_1-\bar{x}_2})/s_{\bar{x}_1-\bar{x}_2}$$
$$= (-4.58 - 0)/\sqrt{115.25/12 + 115.25/12}$$
$$= -4.58/4.383$$
$$= -1.045$$

conclusion:
   Do not reject H_o; there is not sufficient evidence to reject the claim that $\mu_1-\mu_2 = 0$.

b. $(\bar{x}_1-\bar{x}_2) \pm t_{22,.025}\sqrt{s_p^2/n_1 + s_p^2/n_2}$
   $-4.58 \pm 2.074\sqrt{115.25/12 + 115.25/12}$
   -4.58 ± 9.09
   $-13.67 < \mu_1-\mu_2 < 4.51$

Cumulative Review Exercises

1. Refer to the summary table at the right.

   a. $P(Y) = 53/750$
         $= .0707$

   b. $P(M$ or $Y) = P(M) + P(Y) - P(M$ and $Y)$
         $= 250/750 + 53/750 - 26/750$
         $= 277/750 = .3693$

   c. $P(Y \mid M) = 26/250$
          $= .104$

   d. $P(Y \mid F) = 27/500$
          $= .054$

		TICKET?		
		Y	N	
SEX	M	26	224	250
	F	27	473	500
		53	697	750

   e. Let the males be group 1.
      original claim: $p_1-p_2 > 0$
         $\hat{p}_1 = x_1/n_1 = 26/250 = .104$
         $\hat{p}_2 = x_2/n_2 = 27/500 = .054$
         $\hat{p}_1-\hat{p}_2 = .104 - .054 = .050$
         $\bar{p} = (x_1 + x_2)/(n_1 + n_2)$
             $= (26 + 27)/(250 + 500) = 53/750 = .0707$
      H_o: $p_1-p_2 \leq 0$
      H_1: $p_1-p_2 > 0$
      $\alpha = .05$
      C.R. z $> z_{.05} = 1.645$
      calculations:

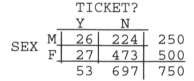

         $z_{\hat{p}_1-\hat{p}_2} = (\hat{p}_1-\hat{p}_2 - \mu_{\hat{p}_1-\hat{p}_2})/\sigma_{\hat{p}_1-\hat{p}_2}$
             $= (.050 - 0)/\sqrt{(.0707)(.9293)/250 + (.0707)(.9293)/500}$
             $= .050/.0199$
             $= 2.519$

conclusion:
   Reject H_o; there is sufficient evidence to conclude that $p_1-p_2 > 0$.

No; we cannot conclude that men speed more, only that they are ticketed more. It is possible, for example, that men drive more.

2. a. original claim $\mu = 0$ [n > 30, use z (with s for σ)]
    H_o: $\mu = 0$
    H_1: $\mu \neq 0$
    $\alpha = .05$
    C.R. $z < -z_{.025} = -1.960$
    $\quad\quad z > z_{.025} = 1.960$
    calculations:
    $\quad z_{\bar{x}} = (\bar{x} - \mu_{\bar{x}})/\sigma_{\bar{x}}$
    $\quad\quad = (1.2 - 0)/(3.9/\sqrt{40})$
    $\quad\quad = 1.2/.617$
    $\quad\quad = 1.946$

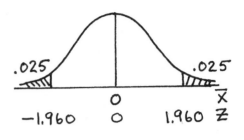

conclusion:
    Do not reject H_o; there is not sufficient evidence to reject the claim that $\mu = 0$.
    NOTE: See the NOTE for part (d).

  b. original claim $\mu = 0$ [n > 30, use z (with s for σ)]
    H_o: $\mu = 0$
    H_1: $\mu \neq 0$
    $\alpha = .05$
    C.R. $z < -z_{.025} = -1.960$
    $\quad\quad z > z_{.025} = 1.960$
    calculations:
    $\quad z_{\bar{x}} = (\bar{x} - \mu_{\bar{x}})/\sigma_{\bar{x}}$
    $\quad\quad = (-1.4 - 0)/(4.3/\sqrt{33})$
    $\quad\quad = -1.4/.749$
    $\quad\quad = -1.870$

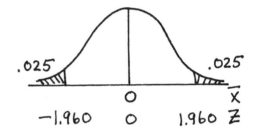

conclusion:
    Do not reject H_o; there is not sufficient evidence to reject the claim that $\mu = 0$.

  c. Let the day shift be group 1.
    original claim: $\mu_1-\mu_2 = 0$ [$n_1 > 30$ and $n_2 > 30$, use z (with s's for σ's)]
    $\bar{x}_1-\bar{x}_2 = 1.2 - (-1.4) = 2.6$
    H_o: $\mu_1-\mu_2 = 0$
    H_1: $\mu_1-\mu_2 \neq 0$
    $\alpha = .05$
    C.R. $z < -z_{.025} = -1.960$
    $\quad\quad z > z_{.025} = 1.960$
    calculations:
    $\quad z_{\bar{x}_1-\bar{x}_2} = (\bar{x}_1-\bar{x}_2 - \mu_{\bar{x}_1-\bar{x}_2})/\sigma_{\bar{x}_1-\bar{x}_2}$
    $\quad\quad = (2.6 - 0)/\sqrt{(3.9)^2/40 + (4.3)^2/33}$
    $\quad\quad = 2.6/.970$
    $\quad\quad = 2.681$

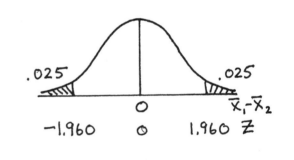

conclusion:
    Reject H_o; there is sufficient evidence to reject the claim that $\mu_1-\mu_2 = 0$ and to conclude that $\mu_1-\mu_2 \neq 0$ (in fact, $\mu_1-\mu_2 > 0$).
    NOTE: There was not enough evidence in part (a) to conclude that μ_1 was significantly different from 0, and there was not enough evidence in part (b) to conclude that μ_2 was significantly different from 0. Yet in part (c) we conclude that μ_1 and μ_2 are significantly different from each other. This is not a contradiction, but another example of the principle illustrated in exercise #18 of section 8-6.
    Consider this analogy: Young children Arney, Barney and Charlie are each allowed by their parents to travel no more than 4 blocks from home. Barney lives in the middle, Arney lives 3 blocks due west of Barney, and Charlie lives 3 blocks due east of Barney. Arney and Barney can visit each other's homes, Barney and Charlie can visit each other's homes, but

Arney and Charlie cannot visit each other's homes.

d. Refer to part (a).
$$\bar{x} \pm z_{.025} \cdot \sigma_{\bar{x}}$$
$$1.2 \pm 1.960 \cdot 3.9/\sqrt{40}$$
$$1.2 \pm 1.2$$
$$0.0 < \mu < 2.4 \text{ (grams)}$$
NOTE: This seems to contradict the conclusion in part (a). Given the 2 significant digit accuracy of $\bar{x}$ and s, however, this is the proper confidence interval. Using (inappropriately) more accuracy gives 1.2 ± 1.209 -- i.e., $-.009 < \mu < 2.409$. This latter interval includes the value 0 and seems to agree with the conclusion in part (a) that failed to reject $H_o : \mu = 0$. In truth, the 2 significant digit accuracy of the given values do not justify more than 2 significant digits in the calculated z statistic of part (a). In short, more accuracy is needed to make a definitive statement about the relationship between μ and 0.

e. Refer to part (b).
$$\bar{x} \pm z_{.025} \cdot \sigma_{\bar{x}}$$
$$-1.4 \pm 1.960 \cdot 4.3/\sqrt{33}$$
$$-1.4 \pm 1.5$$
$$-2.9 < \mu < 0.1 \text{ (grams)}$$

f. Refer to part (c)
$$\bar{x}_1 - \bar{x}_2 \pm z_{.025} \cdot \sigma_{\bar{x}_1 - \bar{x}_2}$$
$$2.6 \pm 1.960 \cdot \sqrt{(3.9)^2/40 + (4.3)^2/33}$$
$$2.6 \pm 1.9$$
$$0.7 < \mu_1 - \mu_2 < 4.5 \text{ (grams)}$$

g. Refer to part (a).
$$n = [z_{.025} \cdot \sigma/E]^2$$
$$= [1.960 \cdot 3.9/.5]$$
$$= 233.7, \text{ rounded up to } 234$$

Chapter 9

Correlation and Regression

9-2 Correlation

1. The critical values below are taken from Table A-6.
   a. CV $= \pm.361$; r $= .992$ indicates a significant (positive) linear correlation
   b. CV $= \pm.279$; r $= -.333$ indicates a significant (negative) linear correlation
   c. CV $= \pm.482$; r $= .456$ indicates no significant linear correlation

NOTE: In addition to the value of n, calculation of r requires five sums: Σx, Σy, Σx^2, Σy^2 and Σxy. The next problem shows the chart prepared to find these sums. As the sums can usually be found conveniently using a calculator and without constructing the chart, subsequent problems typically give only the values of the sums and do not show a chart.

Also, calculation of r also involves three subcalculations.

(1) $n(\Sigma xy) - (\Sigma x)(\Sigma y)$ determines the sign of r. If large values of x are associated with large values of y, it will be positive. If large values of x are associated with small values of y, it will be negative. If not, a mistake has been made.

(2) $n(\Sigma x^2) - (\Sigma x)^2$ cannot be negative. If it is, a mistake has been made.

(3) $n(\Sigma y^2) - (\Sigma y)^2$ cannot be negative. If it is, a mistake has been made.

Finally, r must be between -1 and 1 inclusive. If not, a mistake has been made. If this or any of the previous mistakes occurs, stop immediately and find the error; continuing will be a fruitless waste of effort.

3.

x	y	xy	x^2	y^2
2	6	12	4	36
3	9	27	9	81
5	14	70	25	196
5	16	80	25	256
10	30	300	100	900
25	75	489	163	1469

$n(\Sigma xy) - (\Sigma x)(\Sigma y) = 5(489) - (25)(75) = 570$
$n(\Sigma x^2) - (\Sigma x)^2 = 5(163) - (25)^2 = 190$
$n(\Sigma y^2) - (\Sigma y)^2 = 5(1469) - (75)^2 = 1720$

a. According to the scatter diagram, there appears to be a significant (positive) linear correlation between x and y.

b. n = 5 $\Sigma x = 25$ $\Sigma x^2 = 163$ $(\Sigma x)^2 = (25)^2 = 625$ $\Sigma xy = 489$

$r = [n(\Sigma xy) - (\Sigma x)(\Sigma y)]/[\sqrt{n(\Sigma x^2) - (\Sigma x)^2} \cdot \sqrt{n(\Sigma y^2) - (\Sigma y)^2}]$
$= 570/[\sqrt{190} \cdot \sqrt{1720}]$
$= .997$

NOTE: In each of exercises 5-18, the first variable listed is given the designation x, and the second variable listed is given the designation y. In correlation problems, the designation of x and y is arbitrary -- so long as a person remains consistent after making the designation. For part (d) of each problem, the following summary statistics should be saved: n, Σx, Σy, Σx^2, Σy^2, Σxy.

5. a. weight (lbs)

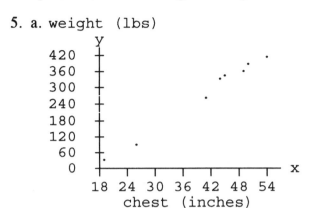

       n = 8
       Σx = 327
       Σy = 2186
       Σx^2 = 14397
       Σy^2 = 730220
       Σxy = 100972

b. $n(\Sigma xy) - (\Sigma x)(\Sigma y) = 8(100972) - (327)(2186) = 92954$
   $n(\Sigma x^2) - (\Sigma x)^2 = 8(14397) - (327)^2 = 8247$
   $n(\Sigma y^2) - (\Sigma y)^2 = 8(730220) - (2186)^2 = 1063164$
   $r = [n(\Sigma xy) - (\Sigma x)(\Sigma y)]/[\sqrt{n(\Sigma x^2) - (\Sigma x)^2} \cdot \sqrt{n(\Sigma y^2) - (\Sigma y)^2}]$
   $= 92954/[\sqrt{8247} \cdot \sqrt{1063164}]$
   $= .993$

c. $H_o: \rho = 0$
   $H_1: \rho \neq 0$
   $\alpha = .05$
   C.R. r < -.707   OR C.R. $t < -t_{6,.025} = -2.447$
        r > .707 $t > t_{6,.025} = 2.447$
   calculations: calculations:
        r = .993 $t_r = (r - \mu_r)/s_r$
                                   $= (.993 - 0)/\sqrt{(1-(.993)^2)/6}$
                                   $= .993/.0492$
                                   $= 20.167$

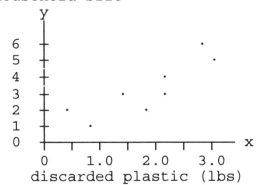

   conclusion:
       Reject H_o; there is sufficient evidence to reject the claim that $\rho = 0$ and to conclude that $\rho \neq 0$ (in fact, $\rho > 0$).
   Yes; based on these results, a bear's weight does seem to be related to its chest size. No; the results do not change if the chest measurements are changed to feet.

7. a. household size

       n = 8
       Σx = 14.60
       Σy = 26
       Σx^2 = 32.9632
       Σy^2 = 104
       Σxy = 56.80

b. $n(\Sigma xy) - (\Sigma x)(\Sigma y) = 8(56.80) - (14.60)(26) = 74.80$
   $n(\Sigma x^2) - (\Sigma x)^2 = 8(32.9632) - (14.60)^2 = 50.5456$
   $n(\Sigma y^2) - (\Sigma y)^2 = 8(104) - (26)^2 = 156$
   $r = [n(\Sigma xy) - (\Sigma x)(\Sigma y)]/[\sqrt{n(\Sigma x^2) - (\Sigma x)^2} \cdot \sqrt{n(\Sigma y^2) - (\Sigma y)^2}]$
   $= 74.80/[\sqrt{50.5456} \cdot \sqrt{156}]$
   $= .842$

c. $H_o: \rho = 0$
   $H_1: \rho \neq 0$
   $\alpha = .05$
   C.R. $r < -.707$ OR C.R. $t < -t_{6,.025} = -2.447$
       $r > .707$ $t > t_{6,.025} = 2.447$
   calculations: calculations:
      $r = .842$ $t_r = (r - \mu_r)/s_r$
                              $= (.842 - 0)/\sqrt{(1-(.842)^2)/6}$
                              $= .842/.220$
                              $= 3.829$

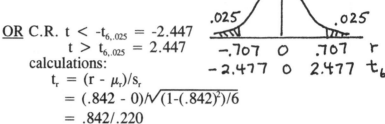

conclusion:
   Reject H_o; there is sufficient evidence to reject the claim that $\rho = 0$ and to conclude that $\rho \neq 0$ (in fact, $\rho > 0$).

Yes; there does appear to be a linear relationship between the amount of discarded plastic and household size.

9. a. household size

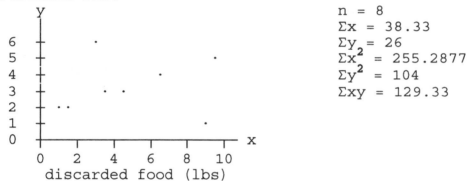

   discarded food (lbs)

   $n = 8$
   $\Sigma x = 38.33$
   $\Sigma y = 26$
   $\Sigma x^2 = 255.2877$
   $\Sigma y^2 = 104$
   $\Sigma xy = 129.33$

b. $n(\Sigma xy) - (\Sigma x)(\Sigma y) = 8(129.33) - (38.33)(26) = 38.06$
   $n(\Sigma x^2) - (\Sigma x)^2 = 8(255.2877) - (38.33)^2 = 573.1127$
   $n(\Sigma y^2) - (\Sigma y)^2 = 8(104) - (26)^2 = 156$
   $r = [n(\Sigma xy) - (\Sigma x)(\Sigma y)]/[\sqrt{n(\Sigma x^2) - (\Sigma x)^2} \cdot \sqrt{n(\Sigma y^2) - (\Sigma y)^2}]$
   $= 38.06/[\sqrt{573.1127} \cdot \sqrt{156}]$
   $= .127$

c. $H_o: \rho = 0$
   $H_1: \rho \neq 0$
   $\alpha = .05$
   C.R. $r < -.707$ OR C.R. $t < -t_{6,.025} = -2.447$
       $r > .707$ $t > t_{6,.025} = 2.447$
   calculations: calculations:
      $r = .127$ $t_r = (r - \mu_r)/s_r$
                               $= (.127 - 0)/\sqrt{(1-(.127)^2)/6}$
                              $= .127/.405$
                              $= .314$
   conclusion:
   Do not reject H_o; there is not sufficient evidence to reject the claim that $\rho = 0$.

11. a. BAC (%)

```
 y
.26 ┤           .
.24 ┤              .
.22 ┤                  .
.20 ┤      .       .
.18 ┤. 
.18 ┤          .
.16 ┤                      .
.14 ┤
.12 ┤
    └──┬──┬──┬──┬──┬──┬──┬── x
      16 22 28 34 40 46 52
         age (years)
```

$n = 8$
$\Sigma x = 276.6$
$\Sigma y = 1.66$
$\Sigma x^2 = 10680.48$
$\Sigma y^2 = .3522$
$\Sigma xy = 57.191$

b. $n(\Sigma xy) - (\Sigma x)(\Sigma y) = 8(57.191) - (276.6)(1.66) = -1.628$
$n(\Sigma x^2) - (\Sigma x)^2 = 8(10680.48) - (276.6)^2 = 8936.28$
$n(\Sigma y^2) - (\Sigma y)^2 = 8(.3522) - (1.66)^2 = .062$
$r = [n(\Sigma xy) - (\Sigma x)(\Sigma y)]/[\sqrt{n(\Sigma x^2) - (\Sigma x)^2} \cdot \sqrt{n(\Sigma y^2) - (\Sigma y)^2}]$
$= -1.628/[\sqrt{8936.28} \cdot \sqrt{.062}]$
$= -.069$

c. $H_0: \rho = 0$
$H_1: \rho \neq 0$
$\alpha = .05$
C.R. $r < -.707$ OR C.R. $t < -t_{6,.025} = -2.447$
$\qquad r > .707$ $\qquad t > t_{6,.025} = 2.447$
calculations: calculations:
$\qquad r = -.069$ $\qquad t_r = (r - \mu_r)/s_r$
$\qquad\qquad\qquad\qquad\qquad = (-.069 - 0)/\sqrt{(1-(-.069)^2)/6}$
$\qquad\qquad\qquad\qquad\qquad = -.069/.407$
$\qquad\qquad\qquad\qquad\qquad = -.170$

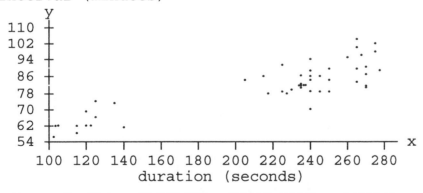

conclusion:
　　Do not reject H_0; there is not sufficient evidence to reject the claim that $\rho = 0$.
No; based on this result, the BAC level does not seem to be related to the age of the person tested.

13. a. interval (minutes)

```
 y
110 ┤
102 ┤                              .: .
 94 ┤                        .   .  . .
 86 ┤                    . .  . : .  .
 78 ┤                 . .:. . .  . .
 70 ┤      . .  .
 62 ┤ ..  . ..      .
 54 ┤.
    └──┬──┬──┬──┬──┬──┬──┬──┬──┬── x
     100 120 140 160 180 200 220 240 260 280
            duration (seconds)
```

$n = 50$
$\Sigma x = 10832$
$\Sigma y = 4033$
$\Sigma x^2 = 2513280$
$\Sigma y^2 = 332331$
$\Sigma xy = 903488$

b. $n(\Sigma xy) - (\Sigma x)(\Sigma y) = 50(903488) - (10832)(4033) = 1488944$
$n(\Sigma x^2) - (\Sigma x)^2 = 50(2513280) - (10832)^2 = 8331776$
$n(\Sigma y^2) - (\Sigma y)^2 = 50(332331) - (4033)^2 = 351461$
$r = [n(\Sigma xy) - (\Sigma x)(\Sigma y)]/[\sqrt{n(\Sigma x^2) - (\Sigma x)^2} \cdot \sqrt{n(\Sigma y^2) - (\Sigma y)^2}]$
$= 1488944/[\sqrt{8331776} \cdot \sqrt{351461}]$
$= .870$

c. $H_o: \rho = 0$
$H_1: \rho \neq 0$
$\alpha = .05$
C.R. $r < -.279$ OR C.R. $t < -t_{48,.025} = -1.960$
     $r > .279$ $t > t_{48,.025} = 1.960$
calculations: calculations:
    $r = .870$ $t_r = (r - \mu_r)/s_r$

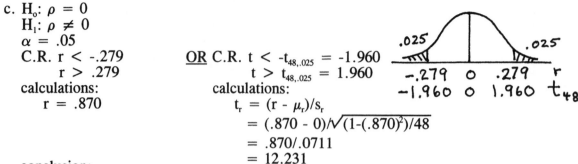

$$= (.870 - 0)/\sqrt{(1-(.870)^2)/48}$$
$$= .870/.0711$$
$$= 12.231$$

conclusion:
    Reject H_o; there is sufficient evidence to reject the claim that $\rho = 0$ and to conclude that $\rho \neq 0$ (in fact, $\rho > 0$).

Yes; there is a significant positive linear correlation, suggesting that the interval after an eruption is related to the duration of the eruption. NOTE: The longer the duration of an eruption, the more pressure has been released and the longer it will take the geyser to build back up for another eruption. In fact, the park rangers use the duration of one eruption to predict the time of the next eruption.

15. a. nicotine (mg)

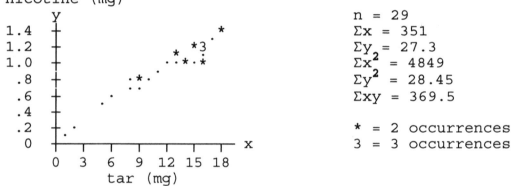

$n = 29$
$\Sigma x = 351$
$\Sigma y = 27.3$
$\Sigma x^2 = 4849$
$\Sigma y^2 = 28.45$
$\Sigma xy = 369.5$

* = 2 occurrences
3 = 3 occurrences

b. $n(\Sigma xy) - (\Sigma x)(\Sigma y) = 29(369.5) - (351)(27.3) = 1133.2$
$n(\Sigma x^2) - (\Sigma x)^2 = 29(4849) - (351)^2 = 17420$
$n(\Sigma y^2) - (\Sigma y)^2 = 29(28.45) - (27.3)^2 = 79.76$
$r = [n(\Sigma xy) - (\Sigma x)(\Sigma y)]/[\sqrt{n(\Sigma x^2) - (\Sigma x)^2} \cdot \sqrt{n(\Sigma y^2) - (\Sigma y)^2}]$
   $= 1133.2/[\sqrt{17420} \cdot \sqrt{79.76}]$
   $= .961$

c. $H_o: \rho = 0$
$H_1: \rho \neq 0$
$\alpha = .05$
C.R. $r < -.361$ OR C.R. $t < -t_{27,.025} = -2.052$
     $r > .361$ $t > t_{27,.025} = 2.052$
calculations: calculations:
    $r = .961$ $t_r = (r - \mu_r)/s_r$
$$= (.961 - 0)/\sqrt{(1-(.961)^2)/27}$$
$$= .961/.0530$$
$$= 18.148$$

conclusion:
    Reject H_o; there is sufficient evidence to reject the claim that $\rho = 0$ and to conclude that $\rho \neq 0$ (in fact, $\rho > 0$).

Yes; based on this result there is a significant positive linear correlation, suggesting that researchers might be able to reduce their laboratory expenses by measuring only one of these two variables.

17. a. production (million bu)

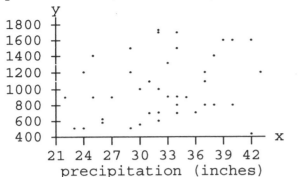

precipitation (inches)

$n = 41$
$\Sigma x = 1331.4$
$\Sigma y = 41630$
$\Sigma x^2 = 44489.00$
$\Sigma y^2 = 48268200$
$\Sigma xy = 1374872.0$

b. $n(\Sigma xy) - (\Sigma x)(\Sigma y) = 41(1374872.0) - (1331.4)(41630) = 943570.0$
$n(\Sigma x^2) - (\Sigma x)^2 = 41(44489.00) - (1331.4)^2 = 51423.04$
$n(\Sigma y^2) - (\Sigma y)^2 = 41(48268200) - (41630)^2 = 245939300$
$r = [n(\Sigma xy) - (\Sigma x)(\Sigma y)]/[\sqrt{n(\Sigma x^2) - (\Sigma x)^2} \cdot \sqrt{n(\Sigma y^2) - (\Sigma y)^2}]$
$\quad = 943570.0/[\sqrt{51423.04} \cdot \sqrt{245939300}]$
$\quad = .265$

c. $H_o: \rho = 0$
$H_1: \rho \neq 0$
$\alpha = .05$
C.R. $r < -.312$ OR C.R. $t < -t_{39,.025} = -1.960$
$\quad\quad r > .312$ $\quad\quad\quad t > t_{39,.025} = 1.960$
calculations: calculations:
$\quad r = .265$ $\quad\quad t_r = (r - \mu_r)/s_r$
$\quad\quad\quad = (.265 - 0)/\sqrt{(1-(.265)^2)/39}$
$\quad\quad\quad = .265/.154$
$\quad\quad\quad = 1.719$

conclusion:
Do not reject H_o; there is not sufficient evidence to reject the claim that $\rho = 0$.
No, there is not sufficient evidence to say there is a significant linear correlation.

19. A linear correlation coefficient very close to zero indicates <u>no</u> significant linear correlation and no tendencies can be inferred.

21. A linear correlation coefficient very close to zero indicates no significant <u>linear</u> correlation, but there may some other type of relationship between the variables.

23. a. No change. Graphically, this reverses the x and y axes and has the effect of rotating the scatter diagram; rotating the figure does not change the nature of the relationship between the points. Mathematically,
$\quad r = [n(\Sigma xy) - (\Sigma x)(\Sigma y)]/[\sqrt{n(\Sigma x^2) - (\Sigma x)^2} \cdot \sqrt{n(\Sigma y^2) - (\Sigma y)^2}]$
does not change because x and y play exactly the same roles in the equation.

b. No change in absolute value. Graphically, this spreads out (or condenses, if the constant is less than 1) the scatter diagram horizontally; changing the scale does not change the nature of the relationship between the points. Mathematically,
$\quad r = [n(\Sigma cxy) - (\Sigma cx)(\Sigma y)]/[\sqrt{n(\Sigma c^2 x^2) - (\Sigma cx)^2} \cdot \sqrt{n(\Sigma y^2) - (\Sigma y)^2}]$
$\quad = [n(\Sigma xy) - (\Sigma x)(\Sigma y)]/[\sqrt{n(\Sigma x^2) - (\Sigma x)^2} \cdot \sqrt{n(\Sigma y^2) - (\Sigma y)^2}]$
because the change multiplies the numerator by a factor of c and the denominator by a factor of $|c|$.
NOTE: If $c < 0$, the correlation changes sign and the points form a mirror image reflected about the y axis -- downhill trends are now uphill ones, and vice-versa.

c. No change. Graphically, this moves each point to the right (or to the left, if the constant is negative); shifting the figure does not change the nature of the relationship between the points. Mathematically,

$$r = [n \cdot \Sigma(x+c)y - \Sigma(x+c) \cdot (\Sigma y)]/[\sqrt{n \cdot \Sigma(x+c)^2 - (\Sigma(x+c))^2} \cdot \sqrt{n(\Sigma y^2) - (\Sigma y)^2}]$$
$$= [n(\Sigma xy) - (\Sigma x)(\Sigma y)]/[\sqrt{n(\Sigma x^2) - (\Sigma x)^2} \cdot \sqrt{n(\Sigma y^2) - (\Sigma y)^2}]$$

because the c's cancel out within the numerator and within the denominator -- as shown:

$$n \cdot \Sigma(x+c)y - \Sigma(x+c) \cdot (\Sigma y) = n \cdot (\Sigma xy + \Sigma cy) - (\Sigma x + \Sigma c) \cdot (\Sigma y)$$
$$= n(\Sigma xy) + nc(\Sigma y) - (\Sigma x)(\Sigma y) - (\Sigma c)(\Sigma y)$$
$$= n(\Sigma xy) - (\Sigma x)(\Sigma y), \quad \text{since } \Sigma c = nc$$
$$n \cdot \Sigma(x+c)^2 - [\Sigma(x+c)]^2 = n \cdot \Sigma(x^2 + 2xc + c^2) - [\Sigma x + \Sigma c]^2$$
$$= n(\Sigma x^2) + 2nc\Sigma x + n^2c^2 - (\Sigma x)^2 - 2\Sigma x\Sigma c - (\Sigma c)^2$$
$$= n(\Sigma x^2) - (\Sigma x)^2, \quad \text{since } \Sigma c = nc$$

25. a. For $\pm t_{48,.025} = \pm 1.960$, the critical values are $r = \pm 1.960/\sqrt{(\pm 1.960)^2 + 48} = \pm .272$.

   b. For $\pm t_{73,.05} = \pm 1.645$, the critical values are $r = \pm 1.645/\sqrt{(\pm 1.645)^2 + 73} = \pm .189$.

   c. For $-t_{18,.05} = -1.734$, the critical value is $r = -1.734/\sqrt{(-1.734)^2 + 18} = -.378$.

   d. For $t_{8,.05} = 1.860$, the critical value is $r = 1.860/\sqrt{(1.860)^2 + 8} = .549$.

   e. For $t_{10,.01} = 2.764$, the critical value is $r = 2.764/\sqrt{(2.764)^2 + 10} = .658$.

27.

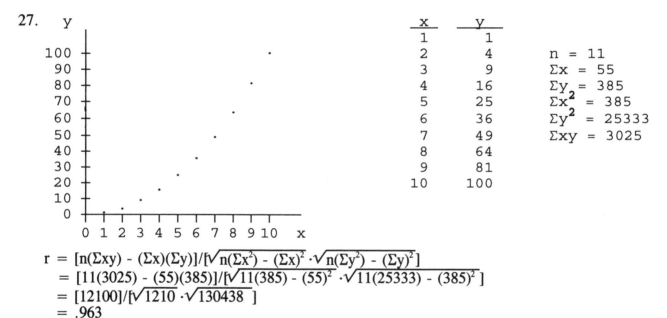

x	y
1	1
2	4
3	9
4	16
5	25
6	36
7	49
8	64
9	81
10	100

$n = 11$
$\Sigma x = 55$
$\Sigma y = 385$
$\Sigma x^2 = 385$
$\Sigma y^2 = 25333$
$\Sigma xy = 3025$

$$r = [n(\Sigma xy) - (\Sigma x)(\Sigma y)]/[\sqrt{n(\Sigma x^2) - (\Sigma x)^2} \cdot \sqrt{n(\Sigma y^2) - (\Sigma y)^2}]$$
$$= [11(3025) - (55)(385)]/[\sqrt{11(385) - (55)^2} \cdot \sqrt{11(25333) - (385)^2}]$$
$$= [12100]/[\sqrt{1210} \cdot \sqrt{130438}]$$
$$= .963$$

The $r = .963$ indicates a significant linear relationship -- at the .05 level, the critical values from Table A-6 are $\pm .602$. This section of the parabola, which does not include points on both sides of the minimum at (0,0), is very close to being a straight line.

9-3 Regression

NOTE: For exercises 1-20, the exact summary statistics (i.e., without any rounding) are given on the right. While the intermediate calculations on the left are presented rounded to various degrees of accuracy, the entire unrounded values were preserved in the calculator until the very end.

1. $\bar{x} = 3.00$
   $\bar{y} = 7.00$
   $b_1 = [n(\Sigma xy) - (\Sigma x)(\Sigma y)]/[n(\Sigma x^2) - (\Sigma x)^2]$
      $= [4(104) - (12)(28)]/[4(46) - (12)^2]$
      $= 80/40$
      $= 2.00$
   $b_0 = \bar{y} - b_1\bar{x}$
      $= 7.00 - (2.00)(3.00)$
      $= 1.00$
   $\hat{y} = b_0 + b_1x$
      $= 1.00 + 2.00x$

                                            $n = 4$
$\Sigma x = 12$
$\Sigma y = 28$
$\Sigma x^2 = 46$
$\Sigma y^2 = 236$
$\Sigma xy = 104$

3. $\bar{x} = 5.00$
   $\bar{y} = 15.00$
   $b_1 = [n(\Sigma xy) - (\Sigma x)(\Sigma y)]/[n(\Sigma x^2) - (\Sigma x)^2]$
      $= [5(489) - (25)(75)]/[5(163) - (25)^2]$
      $= 570/190$
      $= 3.00$
   $b_0 = \bar{y} - b_1\bar{x}$
      $= 15.00 - (3.00)(5.00)$
      $= 0.00$
   $\hat{y} = b_0 + b_1x$
      $= 0.00 + 3.00x$

$n = 5$
$\Sigma x = 25$
$\Sigma y = 75$
$\Sigma x^2 = 163$
$\Sigma y^2 = 1469$
$\Sigma xy = 489$

NOTE: In the exercises that follow, this manual uses the full accuracy of b_0 nand b_1 when calculating $\hat{y}$. Using only the rounded values as stated in the equation produces slightly different answers.

5. $\bar{x} = 40.9$
   $\bar{y} = 273.25$
   $b_1 = [n(\Sigma xy) - (\Sigma x)(\Sigma y)]/[n(\Sigma x^2) - (\Sigma x)^2]$
      $= 92954/8247$
      $= 11.3$
   $b_0 = \bar{y} - b_1\bar{x}$
      $= 273.25 - (11.3)(40.9)$
      $= -187.5$
   $\hat{y} = b_0 + b_1x$
      $= -187.5 + 11.3x$
   $\hat{y}_{52} = -187.5 + 11.3(52)$
      $= 398.6$ lbs

$n = 8$
$\Sigma x = 327$
$\Sigma y = 2186$
$\Sigma x^2 = 14397$
$\Sigma y^2 = 730220$
$\Sigma xy = 100972$

7. $\bar{x} = 1.825$
   $\bar{y} = 3.25$
   $b_1 = [n(\Sigma xy) - (\Sigma x)(\Sigma y)]/[n(\Sigma x^2) - (\Sigma x)^2]$
      $= 74.80/50.5456$
      $= 1.480$
   $b_0 = \bar{y} - b_1\bar{x}$
      $= 3.25 - (1.480)(1.825)$
      $= .549$

$n = 8$
$\Sigma x = 14.60$
$\Sigma y = 26$
$\Sigma x^2 = 32.9632$
$\Sigma y^2 = 104$
$\Sigma xy = 56.80$

$$\hat{y} = b_o + b_1 x$$
$$= .549 + 1.480x$$
$$\hat{y}_{.50} = .549 + 1.480(.50)$$
$$= 1.3 \text{ people}$$

9. $\bar{x} = 4.791$
$\bar{y} = 3.25$
$b_1 = [n(\Sigma xy) - (\Sigma x)(\Sigma y)]/[n(\Sigma x^2) - (\Sigma x)^2]$
$\quad = 38.06/573.1127$
$\quad = .0664$
$b_o = \bar{y} - b_1\bar{x}$
$\quad = 3.25 - (.0664)(4.781)$
$\quad = 2.93$
$\hat{y} = b_o + b_1 x$
$\quad = 2.93 + .0664x$
$\hat{y}_{8.00} = \bar{y}$ [no significant correlation]
$\quad = 3.25 \text{ people}$

$n = 8$
$\Sigma x = 38.33$
$\Sigma y = 26$
$\Sigma x^2 = 255.2877$
$\Sigma y^2 = 104$
$\Sigma xy = 129.33$

11. $\bar{x} = 34.575$
$\bar{y} = .2075$
$b_1 = [n(\Sigma xy) - (\Sigma x)(\Sigma y)]/[n(\Sigma x^2) - (\Sigma x)^2]$
$\quad = -1.628/8936.28$
$\quad = -.000182$
$b_o = \bar{y} - b_1\bar{x}$
$\quad = .2075 - (-.000182)(34.575)$
$\quad = .214$
$\hat{y} = b_o + b_1 x$
$\quad = .214 - .000182x$
$\hat{y}_{21} = \bar{y}$ [no significant correlation]
$\quad = .21\%$

$n = 8$
$\Sigma x = 276.6$
$\Sigma y = 1.66$
$\Sigma x^2 = 10680.48$
$\Sigma y^2 = .3522$
$\Sigma xy = 57.191$

13. $\bar{x} = 216.64$
$\bar{y} = 80.66$
$b_1 = [n(\Sigma xy) - (\Sigma x)(\Sigma y)]/[n(\Sigma x^2) - (\Sigma x)^2]$
$\quad = 1488944/8331776$
$\quad = .179$
$b_o = \bar{y} - b_1\bar{x}$
$\quad = 80.66 - (.179)(216.64)$
$\quad = 41.9$
$\hat{y} = b_o + b_1 x$
$\quad = 41.9 + .179x$
$\hat{y}_{210} = 41.9 + .179(210)$
$\quad = 79.5 \text{ minutes}$

$n = 50$
$\Sigma x = 10832$
$\Sigma y = 4033$
$\Sigma x^2 = 2513280$
$\Sigma y^2 = 332331$
$\Sigma xy = 903488$

15. $\bar{x} = 12.1$
$\bar{y} = .941$
$b_1 = [n(\Sigma xy) - (\Sigma x)(\Sigma y)]/[n(\Sigma x^2) - (\Sigma x)^2]$
$\quad = 1133.2/17420$
$\quad = .0651$
$b_o = \bar{y} - b_1\bar{x}$
$\quad = .941 - (.0651)(12.1)$
$\quad = .154$
$\hat{y} = b_o + b_1 x$
$\quad = .154 + .0651x$

$n = 29$
$\Sigma x = 351$
$\Sigma y = 27.3$
$\Sigma x^2 = 4849$
$\Sigma y^2 = 28.45$
$\Sigma xy = 369.5$

$\hat{y}_3 = .154 + .0651(3)$
$\quad = .3$ mg

17. $\bar{x} = 32.5$ $n = 41$
$\quad \bar{y} = 1015.4$ $\Sigma x = 1331.4$
$\quad b_1 = [n(\Sigma xy) - (\Sigma x)(\Sigma y)]/[n(\Sigma x^2) - (\Sigma x)^2]$ $\Sigma y = 41630$
$\qquad = 943570.0/51423.04$ $\Sigma x^2 = 44489.00$
$\qquad = 18.3$ $\Sigma y^2 = 48268200$
$\quad b_0 = \bar{y} - b_1\bar{x}$ $\Sigma xy = 1374872.0$
$\qquad = 1015.4 - (18.3)(32.5)$
$\qquad = 420$
$\quad \hat{y} = b_0 + b_1x$
$\qquad = 420 + 18.3x$
$\quad \hat{y}_{40.0} = \bar{y}$ [no significant correlation]
$\qquad = 1015$ million bu

19. The .05 critical values for r are taken from Table A-6.
a. CV $= \pm.632$; r $= .931$ is significant
use $\hat{y} = 4.00 + 2.00x$
$\quad \hat{y}_{3.00} = 4.00 + 2.00(3.00)$
$\qquad = 10.00$
b. CV $= \pm.220$; r $= -.033$ is not significant
use $\hat{y} = \bar{y}$
$\quad \hat{y}_{3.00} = \bar{y}$
$\qquad = 2.50$

21. The scatter diagram is given below, with the new point of (120,800) indicated by *.

weight (lbs)

	originally	now
	n = 8	n = 9
	$\Sigma x = 516.5$	$\Sigma x = 636.5$
	$\Sigma y = 2176$	$\Sigma y = 2976$
	$\Sigma x^2 = 34525.75$	$\Sigma x^2 = 48925.75$
	$\Sigma y^2 = 728520$	$\Sigma y^2 = 1368520$
	$\Sigma xy = 151879$	$\Sigma xy = 247879$

originally
$\hat{y} = b_0 + b_1x = -352 + 9.66x$

$b_1 = [n(\Sigma xy) - (\Sigma x)(\Sigma y)]/[n(\Sigma x^2) - (\Sigma x)^2]$
$\quad = 336687/35199.5$
$\quad = 9.57$
$b_0 = \bar{y} - b_1\bar{x}$
$\quad = (2976/9) - (9.57)(636.5/9)$
$\quad = -346$
$\hat{y} = b_0 + b_1x$
$\quad = -346 + 9.57x$

The new point is an outlier because it is far away from the others. It is not an influential point, because it does not significantly alter the regression line.

23. <u>original data</u>
   n=5
   $\Sigma x = 4,234,178$
   $\Sigma y = 576$
   $\Sigma x^2 = 3,595,324,583,102$
   $\Sigma y^2 = 67552$
   $\Sigma xy = 491,173,342$

   $\bar{x} = 846835.6$
   $\bar{y} = 115.2$
   $n\Sigma xy - (\Sigma x)(\Sigma y) = 16,980,182$
   $n\Sigma x^2 - (\Sigma x)^2 = 48,459,579,826$
   $b_1 = 16,980,182/48,459,579,826$
   $\quad = .0003504$
   $b_0 = \bar{y} - b_1\bar{x}$
   $\quad = 115.2 - .0003504(846835.6)$
   $\quad = -181.53$
   $\hat{y} = b_0 + b_1x$
   $\quad = -181.53 + .0003504x$

<u>original data divided by 1000</u>
   n = 5
   $\Sigma x = 4,234.178$
   $\Sigma y = 576$
   $\Sigma x^2 = 3,595,324.583102$
   $\Sigma y^2 = 67552$
   $\Sigma xy = 491,173.342$

   $\bar{x} = 846.8356$
   $\bar{y} = 115.2$
   $n\Sigma xy - (\Sigma x)(\Sigma y) = 16,980.182$
   $n\Sigma x^2 - (\Sigma x)^2 = 48,459.579826$
   $b_1 = 16,980.182/48,459.579826$
   $\quad = .3504$
   $b_0 = \bar{y} - b_1\bar{x}$
   $\quad = 115.2 - .3504(846.8356)$
   $\quad = -181.53$
   $\hat{y} = b_0 + b_1x$
   $\quad = -181.53 + .3504x$

Dividing each x by 1000 multiplies b_1, the coefficient of x in the regression equation, by 1000; multiplying the x coefficient by 1000 and dividing x by 1000 will "cancel out" and all predictions remain the same.

Dividing each y by 1000 divides both b_1 and b_0 by 1000; consistent with the new "units" for y, all predictions will also turn out divided by 1000.

25.   original data   n = 4

x	y
2.0	12.0
2.5	18.7
4.2	53.0
10.0	225.0

$\Sigma x = 18.7$
$\Sigma y = 308.7$
$\Sigma x^2 = 127.89$
$\Sigma y^2 = 53927.69$
$\Sigma xy = 2543.35$

$b_1 = [n(\Sigma xy) - (\Sigma x)(\Sigma y)]/[n(\Sigma x^2) - (\Sigma x)^2]$
$\quad = 4400.71/161.87$
$\quad = 27.2$
$b_0 = \bar{y} - b_1\bar{x}$
$\quad = (308.7/4) - (27.2)(18.7/4)$
$\quad = -49.9$
$\hat{y} = b_0 + b_1x = -49.9 + 27.2x$
$r = [n(\Sigma xy) - (\Sigma x)(\Sigma y)]/[\sqrt{n(\Sigma x^2) - (\Sigma x)^2} \cdot \sqrt{n(\Sigma y^2) - (\Sigma y)^2}]$
$\quad = 4400.71/[\sqrt{161.87} \cdot \sqrt{120415.07}]$
$\quad = .9968$

Based on the value of the associated correlations (.9968 > .9631), the above equation using the original data seems to fit the data better than the following equation using ln(x) instead of x.

NOTE: Both x and y (perhaps, especially y) seem to be growing exponentially. A wiser choice for a transformation might be to use both ln(x) for x and ln(y) for y (or, perhaps, only ln(y) for y).

using ln(x) for x

x	y
.693	12.0
.916	18.7
1.435	53.0
2.303	225.0

$n = 4$
$\Sigma x = 5.347$
$\Sigma y = 308.7$
$\Sigma x^2 = 8.681$
$\Sigma y^2 = 53927.69$
$\Sigma xy = 619.594$

$b_1 = [n(\Sigma xy) - (\Sigma x)(\Sigma y)]/[n(\Sigma x^2) - (\Sigma x)^2]$
$= 827.7220/6.134071$
$= 134.9$

$b_0 = \bar{y} - b_1\bar{x}$
$= (308.7/4) - (134.9)(5.347/4)$
$= 103.2$

$\hat{y} = b_0 + b_1 \cdot \ln(x)$ [since the "x" is really ln(x)]
$= -103.2 + 134.9 \cdot \ln(x)$

$r = [n(\Sigma xy) - (\Sigma x)(\Sigma y)]/[\sqrt{n(\Sigma x^2) - (\Sigma x)^2} \cdot \sqrt{n(\Sigma y^2) - (\Sigma y)^2}]$
$= 827.7220/[\sqrt{6.134071} \cdot \sqrt{120415.07}]$
$= .9631$

9-4 Variation and Prediction Intervals

1. The coefficient of determination is $r^2 = (.2)^2 = .04$.
The portion of the total variation explained by the regression line is $r^2 = .04 = 4\%$.

3. The coefficient of determination is $r^2 = (-.225)^2 = .051$.
The portion of the total variation explained by the regression line is $r^2 = .051 = 5.1\%$.

5. The predicted values were calculated using the regression line $\hat{y} = 2 + 3x$.

x	y	$\hat{y}$	$\bar{y}$	$\hat{y}-\bar{y}$	$(\hat{y}-\bar{y})^2$	$y-\hat{y}$	$(y-\hat{y})^2$	$y-\bar{y}$	$(y-\bar{y})^2$
1	5	5	12.2	-7.2	51.84	0	0	-7.2	51.84
2	8	8	12.2	-4.2	17.64	0	0	-4.2	17.64
3	11	11	12.2	-1.2	1.44	0	0	-1.2	1.44
5	17	17	12.2	4.8	23.04	0	0	4.8	23.04
6	20	20	12.2	7.8	60.84	0	0	7.8	60.84
17	61	61	61.0	0	154.80	0	0	0	154.80

a. The explained variation is $\Sigma(\hat{y}-\bar{y})^2 = 154.80$

b. The unexplained variation is $\Sigma(y-\hat{y})^2 = 0$

c. The total variation is $\Sigma(y-\bar{y})^2 = 154.80$

d. $r^2 = \Sigma(\hat{y}-\bar{y})^2/\Sigma(y-\bar{y})^2 = 154.80/154.80 = 1.00$

e. $s_e^2 = \Sigma(y-\hat{y})^2/(n-2) = 0/3 = 0$
$s_e = 0$

NOTE: A table such as the one in the preceding problem organizes the work and provides all the values needed to discuss variation. In such a table, the following must always be true and can be used as a checkbefore proceeding.

* $\Sigma y = \Sigma \hat{y} = \Sigma \bar{y}$
* $\Sigma(\hat{y}-\bar{y}) = \Sigma(y-\hat{y}) = \Sigma(y-\bar{y}) = 0$
* $\Sigma(y-\bar{y})^2 + \Sigma(y-\hat{y})^2 = \Sigma(y-\bar{y})^2$

7. The predicted values were calculated using the regression line $\hat{y} = .549270 + 1.47985x$.

x	y	$\hat{y}$	$\bar{y}$	$\hat{y}-\bar{y}$	$(\hat{y}-\bar{y})^2$	$y-\hat{y}$	$(y-\hat{y})^2$	$y-\bar{y}$	$(y-\bar{y})^2$
.27	2	.949	3.25	-2.301	5.295	1.051	1.105	-1.25	1.5625
1.41	3	2.636	3.25	-.614	.377	.364	.133	-.25	.0625
2.19	3	3.790	3.25	.540	.292	-.790	.624	-.25	.0625
2.83	6	4.737	3.25	1.487	2.212	1.263	1.595	2.75	7.5625
2.19	4	3.790	3.25	.540	.292	.210	.044	.75	.5625
1.81	2	3.228	3.25	-.022	.000	-1.228	1.507	-1.25	1.5625
.85	1	1.807	3.25	-1.443	2.082	-.807	.651	-2.25	5.0625
3.05	5	5.063	3.25	1.813	3.286	-.063	.004	1.75	3.0625
14.60	26	26.000	26.00	0	13.837	0	5.663	0	19.5000

a. The explained variation is $\Sigma(\hat{y}-\bar{y})^2 = 13.837$

b. The unexplained variation is $\Sigma(y-\hat{y})^2 = 5.663$

c. The total variation is $\Sigma(y-\bar{y})^2 = 19.500$

d. $r^2 = \Sigma(\hat{y}-\bar{y})^2/\Sigma(y-\bar{y})^2 = 13.837/19.500 = .7096$

e. $s_e^2 = \Sigma(y-\hat{y})^2/(n-2) = 5.663/6 = .9438$
   $s_e = .9715$

9. a. $\hat{y} = 2 + 3x$
   $\hat{y}_4 = 2 + 3(4)$
   $= 14$

   b. $\hat{y} \pm t_{n-2,\alpha/2}s_e\sqrt{1 + 1/n + n(x_o-\bar{x})^2/[n\Sigma x^2-(\Sigma x)^2]}$
   $\hat{y} \pm 0$, since $s_e = 0$
   The prediction "interval" in this case shrinks to a single point. Since $r^2 = 1.00$ (i.e., 100% of the variability in the y's can be explained by the regression), a perfect prediction can be made. For a practical example of such a situation, consider the regression line $\hat{y} = 1.399x$ for predicting the amount of money y due for purchasing x gallons of gasoline at $1.399 per gallon -- the "prediction" will be exactly correct every time because of the perfect correlation between the number of gallons purchased and the amount of money due.

11. a. $\hat{y} = .549270 + 1.47985x$
    $\hat{y}_{2.50} = .549270 + 1.47985(2.50) = 4.25$

    b. preliminary calculations
    $n = 8$
    $\Sigma x = 14.60$ $\bar{x} = 14.60/8 = 1.825$
    $\Sigma x^2 = 32.9632$ $n\Sigma x^2-(\Sigma x)^2 = 8(32.9632)-(14.60)^2 = 50.5456$
    $\hat{y} \pm t_{n-2,\alpha/2}s_e\sqrt{1 + 1/n + n(x_o-\bar{x})^2/[n\Sigma x^2-(\Sigma x)^2]}$
    $\hat{y}_{2.50} \pm t_{6,.025}(.9715)\sqrt{1 + 1/8 + 8(2.50-1.825)^2/[50.5456]}$
    $4.25 \pm (2.447)(.9715)\sqrt{1.19711}$
    4.25 ± 2.60
    $1.65 < y_{2.50} < 6.85$ [Fractional values may represent part-time occupancy.]

Exercises 13-16 refer to the chapter problem of Table 9-1. They use the following, which are calculated and/or discussed at various places in the text,

$n = 8$

$\Sigma x = 516.6$ $\hat{y} = -351.660 + 9.65979x$

$\Sigma x^2 = 34525.75$ $s_e = 66.5994$

and the values obtained below.

$\bar{x} = (\Sigma x)/n = 516.5/8 = 64.5625$

$n\Sigma x^2 - (\Sigma x)^2 = 8(34525.75) - (516.5)^2 = 9433.75$

13. $\hat{y} \pm t_{n-2,\alpha/2}s_e\sqrt{1 + 1/n + n(x_o-\bar{x})^2/[n\Sigma x^2-(\Sigma x)^2]}$

$\hat{y}_{50.0} \pm t_{6,.025}(66.5994)\sqrt{1 + 1/8 + 8(50.0-64.5625)^2/[9433.75]}$

$131.33 \pm (2.447)(66.5994)\sqrt{1.30484}$

131.33 ± 186.16

$-54.8 < y_{50.0} < 317.5$

$0 < y_{50.0} < 317.5$

15. $\hat{y} \pm t_{n-2,\alpha/2}s_e\sqrt{1 + 1/n + n(x_o-\bar{x})^2/[n\Sigma x^2-(\Sigma x)^2]}$

$\hat{y}_{49.7} \pm t_{6,.05}(66.5994)\sqrt{1 + 1/8 + 8(49.7-64.5625)^2/[9433.75]}$

$128.43 \pm (1.943)(66.5994)\sqrt{1.31232}$

128.43 ± 148.24

$-19.8 < y_{49.7} < 276.7$

$0 < y_{49.7} \ 276.7$

17. This exercise uses the following values from the chapter problem of Table 9-1, which are calculated and/or discussed at various places in the text,

$n = 8$ $\Sigma x = 516.5$ $b_o = -351.660$

                $\Sigma x^2 = 34525.75$ $b_1 = 9.65979$

and the values obtained below. $s_e = 66.5994$

$\bar{x} = (\Sigma x)/n = 516.5/8 = 64.5625$

$\Sigma x^2 - (\Sigma x)^2/n = 34525.75 - (516.5)^2/8 = 1179.21875$

a. $b_o \pm t_{n-2,\alpha/2}s_e\sqrt{1/n + \bar{x}^2/[\Sigma x^2-(\Sigma x)^2/n]}$

$-351.66 \pm t_{6,.025}(66.5994)\sqrt{1/8 + (64.5625)^2/[1179.21875]}$

$-351.66 \pm (2.447)(66.5994)\sqrt{3.65981}$

-351.66 ± 311.77

$-663.4 < \beta_o < -39.9$

b. $b_1 \pm t_{n-2,\alpha/2}s_e/\sqrt{\Sigma x^2-(\Sigma x)^2/n}$

$9.660 \pm t_{6,.025}(66.5994)/\sqrt{1179.21875}$

$9.660 \pm (2.447)(.971554)/\sqrt{1179.21875}$

9.660 ± 4.746

$4.914 < \beta_1 < 14.406$

19. a. Since $s_e^2 = \Sigma(y-\hat{y})^2/(n-2)$, then $(n-2)s_e^2 = \Sigma(y-\hat{y})^2$.
The unexplained variation is equal to $(n-2)s_e^2$.

b.
$$r^2 = (\text{explained variation})/(\text{total variation})$$
$$(\text{total variation})\cdot r^2 = (\text{explained variation})$$
$$[(\text{explained variation}) + (\text{unexplained variation})]\cdot r^2 = (\text{explained variation})$$
$$(\text{explained variation})\cdot r^2 + (\text{unexplained variation})\cdot r^2 = (\text{explained variation})$$
$$(\text{unexplained variation})\cdot r^2 = (\text{explained variation}) - (\text{explained variation})\cdot r^2$$
$$(\text{unexplained variation})\cdot r^2 = (\text{explained variation})(1 - r^2)$$
$$(\text{unexplained variation})\cdot r^2/(1 - r^2) = (\text{explained variation})$$

The explained variation is equal to $(\text{unexplained variation})\cdot r^2/(1 - r^2)$.

c. If $r^2 = .900$, then $r = \pm.949$.
Since the regression line has a negative slope (i.e., $b_1 = -2$), we choose the negative root.
The linear correlation coefficient, therefore, is $r = -.949$.

9-5 Multiple Regression

1. WEIGHT $= -285 - 1.38\cdot(AGE) - 11.2\cdot(HEADWDTH) + 28.6\cdot(NECK)$
$\hat{y} = -285 - 1.38x_1 - 11.2x_2 + 28.6x_3$

3. Yes, since the overall P-value of .002 is less than .05.

5. a. WEIGHT $= -274 + .426\cdot(LENGTH) + 12.1\cdot(CHEST)$
$\hat{y} = -274 + .426x_1 + 12.1x_2$
b. $R^2 = 92.8\% = .928$
adjusted $R^2 = 92.5\% = .925$
P-value $= .000$
c. Yes, since the overall P-value of .000 [actually, $.000^+$] is less than .05.

7. a. WEIGHT $= -235 + .403\cdot(AGE) + 5.11\cdot(NECK) - .555\cdot(LENGTH) + 9.19\cdot(CHEST)$
$\hat{y} = -235 + .403x_1 + 5.11x_2 - .555x_3 + 9.19x_4$
b. $R^2 = 94.2\% = .942$
adjusted $R^2 = 93.8\% = .938$
P-value $= .000$
c. Yes, since the overall P-value of .000 [actually, $.000^+$] is less than .05.

9. a. HHSIZE $= 3.56 + .0980\cdot(YARD)$
$\hat{y} = 3.56 + .0980x_1$
b. $R^2 = 2.1\% = .021$
adjusted $R^2 = 0.5\% = .005$
P-value $= .256$
c. No, since the overall P-value of .256 is greater than .05.

11. a. HHSIZE $= 1.15 + 1.41\cdot(PLASTIC) - .0144\cdot(PAPER)$
$\hat{y} = 1.15 + 1.41x_1 - .0144x_2$
b. $R^2 = 56.4\% = .564$
adjusted $R^2 = 54.9\% = .549$
P-value $= .000$
c. Yes, since the overall P-value of .000 [actually, $.000^+$] is less than .05.

13. a. HHSIZE $= .554 + .491\cdot(METAL) + 1.08\cdot(PLASTIC)$
$\hat{y} = .554 + .491x_1 + 1.08x_2$
b. $R^2 = 61.3\% = .613$
adjusted $R^2 = 60.0\% = .600$
P-value $= .000$
c. Yes, since the overall P-value of .000 [actually, $.000^+$] is less than .05.

15. To find the best variables for a regression to predict CORNPROD, consider the correlations. Arranged in one table, the correlations are as follows.

	CORNPROD	YEAR	PRECIP	AVTEMP	ACREHARV
CORNPROD	1.000				
YEAR	0.857	1.000			
PRECIP	0.265	0.170	1.000		
AVTEMP	-0.061	0.085	-0.152	1.000	
ACREHARV	0.800	0.580	0.119	-0.098	1.000

The variable having the highest correlation (0.857) with CORNPROD is YEAR. This positive correlation indicates that corn production has been increasing each year (probably due to more efficient farming, new fertilizers and pesticides, genetic advances, etc.), regardless of the weather or other factors. The most relevant question for predicting corn production is, therefore, "What year are you referring to?" The directions, however, specify to <u>not</u> include YEAR.

The permissible variable having the highest correlation (0.800) with CORNPROD is ACREHARV. The permissible variable having the next highest correlation (0.265) with CORNPROD is PRECIP. Since this correlation is not very high, and so much lower than the correlation for ACREHARV, including PRECIP in the regression will probably not make a significant improvement.

The best multiple regression equation with corn production as the independent variable, therefore, is probably

CORNPROD = -1399 + .217·(ACREHARV)
R^2 = 63.9% = .639
adjusted R^2 = 63.0% = .630
overall P-value = .000

The adjusted R^2 and overall P-value indicate that this is a suitable multiple regression equation for making predictions on corn production.

17. Letting $x_1 = x$ and $x_2 = x^2$, the multiple regression equation is

$\hat{y}$ = 2.1714 + 2.4357x_1 + .4643x_2
R^2 = 100.0% = 1.000
adjusted R^2 = 99.9% = .999
overall P-value = 0.000

Accurate to three decimal places, the coefficient of multiple determination is R^2 = 1.000. Since this value is extremely close to 100% (i.e. 1.000 exactly), there is almost a perfect fit between the equation and the data.

Review Exercises

1. Let x be the price and y be the consumption.

$n = 10$ $n(\Sigma xy) - (\Sigma x)(\Sigma y) = 10(4.74330) - (13.66)(3.468)$
$\Sigma x = 13.66$ $= .06012$
$\Sigma y = 3.468$ $n(\Sigma x^2) - (\Sigma x)^2 = 10(18.6694) - (13.66)^2$
$\Sigma x^2 = 18.6694$ $= .0984$
$\Sigma y^2 = 1.234888$ $n(\Sigma y^2) - (\Sigma y)^2 = 10(1.234888) - (3.468)^2$
$\Sigma xy = 4.74330$ $= .321856$
$r = [n(\Sigma xy) - (\Sigma x)(\Sigma y)]/[\sqrt{n(\Sigma x^2) - (\Sigma x)^2} \cdot \sqrt{n(\Sigma y^2) - (\Sigma y)^2}]$
$= [.06012]/[\sqrt{.0984} \cdot \sqrt{.321856}]$
$= .338$

a. H_o: $\rho = 0$
   H_1: $\rho \neq 0$
   $\alpha = .05$
   C.R. $r < -.632$ $\underline{OR}$ C.R. $t < -t_{8,.025} = -2.306$
       $r > .632$ $t > t_{8,.025} = 2.306$
   calculations: calculations:
       $r = .338$ $t_r = (r - \mu_r)/s_r$

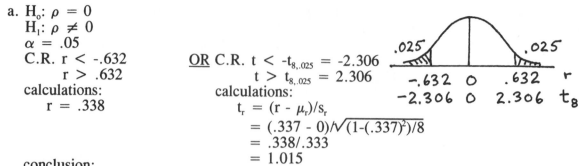

                                         $= (.337 - 0)/\sqrt{(1-(.337)^2)/8}$
                                            $= .338/.333$
                                            $= 1.015$
   conclusion:
       Do not reject H_o; there is not sufficient evidence to reject the claim that $\rho = 0$.

b. $b_1 = [n(\Sigma xy) - (\Sigma x)(\Sigma y)]/[n(\Sigma x^2) - (\Sigma x)^2]$
      $= .06012/.0984$
      $= .611$
   $b_o = \bar{y} - b_1\bar{x}$
      $= (3.468/10) - (.611)(13.66/10)$
      $= -.488$
   $\hat{y} = b_o + b_1 x$
      $= -.488 + .611x$

c. $\hat{y} = -.488 + .611x$
   $\hat{y}_{1.38} = \bar{y}$ [no significant correlation]
      $= .3468$ pints per capita per week

2. Let x be the income and y be the consumption.
      $n = 10$ $n(\Sigma xy) - (\Sigma x)(\Sigma y) = 10(1230.996) - (3548)(3.468)$
      $\Sigma x = 3548$ $= 5.496$
      $\Sigma y = 3.468$ $n(\Sigma x^2) - (\Sigma x)^2 = 10(1259524) - (3548)^2$
      $\Sigma x^2 = 1259524$ $= 6936$
      $\Sigma y^2 = 1.234888$ $n(\Sigma y^2) - (\Sigma y)^2 = 10(1.234888) - (3.468)^2$
      $\Sigma xy = 1230.996$ $= .321856$
   $r = [n(\Sigma xy) - (\Sigma x)(\Sigma y)]/[\sqrt{n(\Sigma x^2) - (\Sigma x)^2} \cdot \sqrt{n(\Sigma y^2) - (\Sigma y)^2}]$
      $= [5.496]/[\sqrt{6936} \cdot \sqrt{.321856}]$
      $= .116$

a. H_o: $\rho = 0$
   H_1: $\rho \neq 0$
   $\alpha = .05$
   C.R. $r < -.632$ $\underline{OR}$ C.R. $t < -t_{8,.025} = -2.306$
       $r > .632$ $t > t_{8,.025} = 2.306$
   calculations: calculations:
       $r = .116$ $t_r = (r - \mu_r)/s_r$
                                          $= (.116 - 0)/\sqrt{(1-(.116)^2)/8}$
                                          $= .116/.351$
   conclusion: $= .331$
       Do not reject H_o; there is not sufficient evidence to reject the claim that $\rho = 0$.

b. $b_1 = [n(\Sigma xy) - (\Sigma x)(\Sigma y)]/[n(\Sigma x^2) - (\Sigma x)^2]$
      $= 5.496/6936$
      $= .000792$
   $b_o = \bar{y} - b_1\bar{x}$
      $= (3.468/10) - (.000792)(3548/10)$
      $= .0657$
   $\hat{y} = b_o + b_1 x$
      $= .0657 + .000792x$

c. $\hat{y} = .0657 + .000792x$
$\hat{y}_{365} = \bar{y}$ [no significant correlation]
$\quad\quad = .3468$ pints per capita per week

3. Let x be the temperature and y be the consumption.
$\quad$ $n = 10$ $\quad\quad\quad$ $n(\Sigma xy) - (\Sigma x)(\Sigma y) = 10(189.038) - (526)(3.468)$
$\quad$ $\Sigma x = 526$ $\quad\quad\quad\quad\quad\quad\quad\quad = 66.212$
$\quad$ $\Sigma y = 3.468$ $\quad\quad\quad$ $n(\Sigma x^2) - (\Sigma x)^2 = 10(29926) - (526)^2$
$\quad$ $\Sigma x^2 = 29926$ $\quad\quad\quad\quad\quad\quad\quad = 22584$
$\quad$ $\Sigma y^2 = 1.234888$ $\quad\quad$ $n(\Sigma y^2) - (\Sigma y)^2 = 10(1.234888) - (3.468)^2$
$\quad$ $\Sigma xy = 189.038$ $\quad\quad\quad\quad\quad\quad\quad = .321856$
$\quad$ $r = [n(\Sigma xy) - (\Sigma x)(\Sigma y)]/[\sqrt{n(\Sigma x^2) - (\Sigma x)^2} \cdot \sqrt{n(\Sigma y^2) - (\Sigma y)^2}]$
$\quad\quad = [66.212]/[\sqrt{22584} \cdot \sqrt{.321856}]$
$\quad\quad = .777$

a. H_o: $\rho = 0$
$\quad$ H_1: $\rho \neq 0$
$\quad$ $\alpha = .05$
$\quad$ C.R. $r < -.632$ $\quad\quad$ OR C.R. $t < -t_{8,.025} = -2.306$
$\quad\quad\quad$ $r > .632$ $\quad\quad\quad\quad\quad\quad$ $t > t_{8,.025} = 2.306$
$\quad$ calculations: $\quad\quad\quad\quad$ calculations:
$\quad\quad$ $r = .777$ $\quad\quad\quad\quad\quad\quad$ $t_r = (r - \mu_r)/s_r$
$\quad\quad\quad\quad\quad\quad\quad\quad\quad\quad = (.777 - 0)/\sqrt{(1-(.777)^2)/8}$
$\quad\quad\quad\quad\quad\quad\quad\quad\quad\quad = .777/.223$
$\quad\quad\quad\quad\quad\quad\quad\quad\quad\quad = 3.486$

$\quad$ conclusion:
$\quad\quad$ Reject H_o; there is sufficient evidence to reject the claim that $\rho = 0$ and to conclude that $\rho \neq 0$ (in fact, $\rho > 0$)

b. $b_1 = [n(\Sigma xy) - (\Sigma x)(\Sigma y)]/[n(\Sigma x^2) - (\Sigma x)^2]$
$\quad\quad = 66.212/22584$
$\quad\quad = .00293$
$\quad$ $b_o = \bar{y} - b_1\bar{x}$
$\quad\quad = (3.468/10) - (.00293)(526/10)$
$\quad\quad = .193$
$\quad$ $\hat{y} = b_o + b_1 x$
$\quad\quad = .193 + .00293x$

c. $\hat{y} = .193 + .00293x$
$\quad$ $\hat{y}_{32} = .193 + .00293(32)$
$\quad\quad = .2864$ pints per capita per week

4. Using Minitab and the given notation,
$\quad$ $\hat{y} = -.053 + .747x_1 - .00220x_2 + .00303x_3$
$\quad$ $R^2 = 72.6\% = .726$
$\quad$ adjusted $R^2 = 58.9\% = .589$
$\quad$ overall P-value $= .040$

Yes; since the overall P-value of .040 is less than .05, the regression equation can be used to predict ice cream consumption. Individually, x_1 and x_2 were not significantly related to consumption; only x_3 could be used to predict consumption. This suggests that variables x_1 and x_2 might not be making a worthwhile contribution to the regression, and that the equation from exercise #3 (using x_3 alone) might actually be a better and more efficient predictive equation.

To check this, calculate the adjusted R^2 for exercise #3.

$$\text{adjusted } R^2 = 1 - \{(n\text{-}1)/[n\text{-}(k+1)]\} \cdot (1\text{-}R^2)$$
$$= 1 - \{9/[10\text{-}(2)]\} \cdot (1\text{-}R^2)$$
$$= 1 - (9/8) \cdot (1\text{-}.777^2)$$
$$= .554$$

Since the adjusted R^2 of .589 is higher than the adjusted R^2 of .554 from exercise #3, however, there is a well-defined statistical sense in which this multiple regression equation is the best of the equations considered in these exercises.

5. Let x be the minutes and y be the savings.

   $n = 9$ $n(\Sigma xy) - (\Sigma x)(\Sigma y) = 9(909.927) - (75.42)(101.2)$
   $\Sigma x = 75.42$ $= 556.839$
   $\Sigma y = 101.2$ $n(\Sigma x^2) - (\Sigma x)^2 = 9(729.3314) - (75.42)^2$
   $\Sigma x^2 = 729.3314$ $= 875.8062$
   $\Sigma y^2 = 1177.36$ $n(\Sigma y^2) - (\Sigma y)^2 = 9(1177.36) - (101.2)^2$
   $\Sigma xy = 909.927$ $= 354.8$

   a. $r = [n(\Sigma xy) - (\Sigma x)(\Sigma y)]/[\sqrt{n(\Sigma x^2) - (\Sigma x)^2} \cdot \sqrt{n(\Sigma y^2) - (\Sigma y)^2}]$
   $\quad = [556.839]/[\sqrt{875.8062} \cdot \sqrt{354.8}\ ]$
   $\quad = .999$

   $H_o: \rho = 0$
   $H_1: \rho \neq 0$
   $\alpha = .05$
   C.R. $r < -.666$ OR C.R. $t < -t_{7,.025} = -2.365$
   $\quad\quad r > .666$ $t > t_{7,.025} = 2.365$
   calculations: calculations:
   $\quad r = .999$ $t_r = (r - \mu_r)/s_r$
   $\quad\quad\quad\quad\quad\quad\quad\quad\quad\quad = (.999 - 0)/\sqrt{(1\text{-}(.999)^2)/7}$
   $\quad\quad\quad\quad\quad\quad\quad\quad\quad\quad = .999/.0175$
   $\quad\quad\quad\quad\quad\quad\quad\quad\quad\quad = 57.072$

   conclusion:
       Reject H_o; there is sufficient evidence to conclude that $\rho \neq 0$ (in fact, $\rho > 0$).

   b. $b_1 = [n(\Sigma xy) - (\Sigma x)(\Sigma y)]/[n(\Sigma x^2) - (\Sigma x)^2]$
   $\quad\quad = 556.839/875.8062$
   $\quad\quad = .636$
   $\quad b_o = \bar{y} - b_1\bar{x}$
   $\quad\quad = (101.2/9) - (.636)(75.42/9)$
   $\quad\quad = 5.92$
   $\quad \hat{y} = b_o + b_1 x$
   $\quad\quad = 5.92 + .636x$

   c. $\hat{y} = b_o + b_1 x$
   $\quad \hat{y}_{5.0} = 5.92 + (.636)(5)$
   $\quad\quad = 9.10$ index of savings

6. Let x be the periodical values and y be the savings.

   $n = 9$ $n(\Sigma xy) - (\Sigma x)(\Sigma y) = 9(11.230) - (1.04)(101.2)$
   $\Sigma x = 1.04$ $= -4.178$
   $\Sigma y = 101.2$ $n(\Sigma x^2) - (\Sigma x)^2 = 9(.1436) - (1.04)^2$
   $\Sigma x^2 = .1436$ $= .2108$
   $\Sigma y^2 = 1177.36$ $n(\Sigma y^2) - (\Sigma y)^2 = 9(1177.36) - (101.2)^2$
   $\Sigma xy = 11.230$ $= 354.8$

   a. $r = [n(\Sigma xy) - (\Sigma x)(\Sigma y)]/[\sqrt{n(\Sigma x^2) - (\Sigma x)^2} \cdot \sqrt{n(\Sigma y^2) - (\Sigma y)^2}]$
   $\quad = [-4.178]/[\sqrt{.2108} \sqrt{354.8}\ ] = -.483$

H_o: $\rho = 0$
H_1: $\rho \neq 0$
$\alpha = .05$

C.R. r < -.666   OR C.R. t < $-t_{7,.025}$ = -2.365
    r > .666   t > $t_{7,.025}$ = 2.365

calculations: calculations:
   r = -.483 $t_r = (r - \mu_r)/s_r$
                           = $(-.483 - 0)/\sqrt{(1-(-.483)^2)/7}$
                           = -.483/.331
                           = -1.460

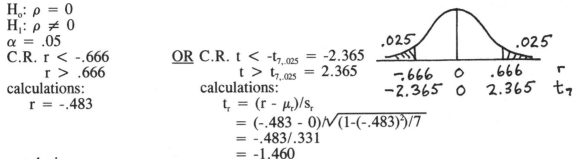

conclusion:
   Do not reject H_o; there is not sufficient evidence to reject the claim that $\rho = 0$.

b. $b_1 = [n(\Sigma xy) - (\Sigma x)(\Sigma y)]/[n(\Sigma x^2) - (\Sigma x)^2]$
    = -4.178/.2108
    = -19.82
  $b_o = \bar{y} - b_1\bar{x}$
    = (101.2/9) - (-19.82)(1.04/9)
    = 13.53
  $\hat{y} = b_o + b_1 x$
    = 13.53 - 19.82x

c. $\hat{y} = b_o + b_1 x$
  $\hat{y}_{.20} = \bar{y}$ [no significant correlation]
    = (101.2)/9
    = 11.24 index of savings

7. Refer to the notation and summary calculations of exercise #5. This exercise requires the calculation of s_e. Since r = .999 is very close to 1.000, there is very little variability unexplained by the regression line and s_e will be very close to zero. This necessitates extra care in the calculations for s_e -- STORE and RECALL with complete accuracy any intermediate values other than n or the primary summations given exactly above.

$s_e^2 = [\Sigma y^2 - b_o(\Sigma y) - b_1(\Sigma xy)]/(n-2)$
   $= [1177.36 - (5.9164)(101.2) - (.6358)(909.927)]/7 = .084540/7 = .012077$
$s_e = .1099$

$\hat{y} \pm t_{n-2,\alpha/2}s_e\sqrt{1 + 1/n + n(x_o-\bar{x})^2/[n\Sigma x^2-(\Sigma x)^2]}$
$\hat{y}_{5.0} \pm t_{7,.025}(.1099)\sqrt{1 + 1/9 + 9(5.0-8.38)^2/[875.8062]}$
$9.095 \pm (2.365)(.1099)\sqrt{1.2285}$
$9.095 \pm .288$
$8.81 < y_{5.0} < 9.38$

8. Using Minitab and the given notation,
$\hat{y} = 6.060 + .630x_1 - .815x_2$
$R^2 = .998$
adjusted $R^2 = .998$
overall P-value = .000
Yes; since the overall P-value of .000 [actually $.000^+$] is less than .05, the multiple regression equation may be used for making predictions.

Cumulative Review Exercises

1. concerns μ: $n > 30$, use z [with s for σ]
   summary statistics: $n = 50$,　$\Sigma x = 4033$,　$\Sigma x^2 = 332331$,　$\bar{x} = 80.66$,　$s = 11.98$

   a. original claim $\mu > 66$
   H_o: $\mu \le 66$
   H_1: $\mu > 66$
   $\alpha = .05$ [assumed]
   C.R. $z > z_{.05} = 1.645$
   calculations:
   $$z_{\bar{x}} = (\bar{x} - \mu)/\sigma_{\bar{x}}$$
   $$= (80.66 - 66)/(11.98/\sqrt{50}$$
   $$= 14.66/1.694$$
   $$= 8.655$$
   conclusion: reject H_o; there is sufficient evidence to conclude that $\mu > 66$.

   b. $\bar{x} \pm z_{.025} \cdot \sigma_{\bar{x}}$
   $80.66 \pm 1.96 \cdot 11.98/\sqrt{50}$
   80.66 ± 3.32
   $77.3 < \mu < 84.0$

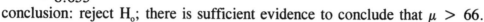

2. concerns μ_d; $n \le 30$, use t
   the $d = x_{humorous} - x_{serious}$ values are: 5 3 5 4 5 5 2 5
   summary statistics: $n = 8$,　$\Sigma d = 34$,　$\Sigma d^2 = 154$,　$\bar{d} = 4.25$,　$s_d = 1.165$
   original claim $\mu_d = 0$
   H_o: $\mu_d = 0$
   H_1: $\mu_d \ne 0$
   $\alpha = .05$ [assumed]
   C.R. $t < -t_{7,.025} = -2.365$
   　　　$t > t_{7,.025} = 2.365$
   calculations:
   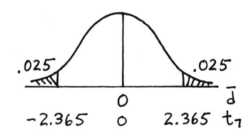
   $$t_{\bar{d}} = (\bar{d} - \mu_d)/s_{\bar{d}}$$
   $$= (4.25 - 0)/(1.165/\sqrt{8}$$
   $$= 4.25/.4118$$
   $$= 10.319$$
   conclusion: reject H_o; there is sufficient evidence to reject the claim that $\mu_d = 0$ and to
   　conclude that $\mu \ne 0$ (in fact, $\mu_d > 0$).
   Yes; based on this result, the humorous commercial seems to be better.
   No; the issue of correlation is not relevant to this situation of seeing whether one commercial is
   better. It might be relevant, however, in a different situation. If there were a significant
   positive correlation, for example, that would indicate that those who liked one type tended to
   like the other type and that those who disliked one type tended to dislike the other. If there
   were a significant correlation, a consumer's score for one type could be predicted from his
   score for the other -- independent of whether one type of commercial seemed to be better.

3. Let x and y be as given.
   $n = 12$ 　　　$n(\Sigma xy) - (\Sigma x)(\Sigma y) = 12(122836) - (1189)(1234)$
   $\Sigma x = 1189$ 　　　　　　　　　　$= 6806$
   $\Sigma y = 1234$ 　　　$n(\Sigma x^2) - (\Sigma x)^2 = 12(118599) - (1189)^2$
   $\Sigma x^2 = 118599$ 　　　　　　　　$= 9467$
   $\Sigma y^2 = 127724$ 　　　$n(\Sigma y^2) - (\Sigma y)^2 = 12(127724) - (1234)^2$
   $\Sigma xy = 122836$ 　　　　　　　　$= 6632$
   $r = [n(\Sigma xy) - (\Sigma x)(\Sigma y)]/[\sqrt{n(\Sigma x^2) - (\Sigma x)^2} \cdot \sqrt{n(\Sigma y^2) - (\Sigma y)^2}]$
   　$= [6806]/[\sqrt{9467} \sqrt{9932}] = .702$

H_o: $\rho = 0$
H_1: $\rho \neq 0$
$\alpha = .05$ [assumed]
C.R. r < -.576   OR C.R. t < $-t_{10,.025}$ = -2.228
   r > .576   t > $t_{10,.025}$ = 2.228
calculations: calculations:
   r = .702 $t_r = (r - \mu_r)/s_r$
                                $= (.702 - 0)/\sqrt{(1-(.702)^2)/10}$
                                $= .702/.225$
                                $= 3.116$

conclusion:
   Reject H_o; there is sufficient evidence to reject the claim that $\rho = 0$ and to conclude that $\rho \neq 0$ (in fact, $\rho > 0$).

Ordinarily, the conclusion would be that about $R^2 = (.702)^2 = .493 = 49.3\%$ of the variation in x can be explained in terms of y (and vice-versa). In this context that means that about 49.3% of the variation among the IQ's in one group can be explained in terms of the IQ's of their twins (i.e, in terms of heredity). In simplest terms, it seems intelligence is about ½ due to heredity and ½ due to environment.

NOTE: The study in exercise #3 contains an interesting subtlety. Within each pair, the older twin was designated x. That was an arbitrary decision to produce an objective rule, and there is no biological basis for putting all the older twins in one group and the younger twins in another. The x-y designation within pairs could just as properly have been made alphabetically, randomly, or by another rule. But the rule affects the results. If it happens to designate all the twins with the higher IQ as x, the correlation rises to r = .810. If it happens to designate the twins with the higher IQ as x in the first six pairs and as y in the last six, the correlation falls to r = .633.

One technique which does not depend upon an x-y designation at all involves comparing the variation within pairs to the overall variation in IQ's. If there is significantly less variability between twins than there is variability in the general population, then there is a significant relationship between the IQ's of twins. One possible approach is given below.

pair	x_1	x_2	$\overline{x}$	s^2
1	107	111	109.0	8.0
2	96	97	96.5	0.5
3	103	116	109.5	84.5
4	90	107	98.5	144.5
5	96	99	97.5	4.5
6	113	111	112.0	2.0
7	86	85	85.5	0.5
8	99	108	103.5	40.5
9	109	102	105.5	24.5
10	105	105	105.0	0.0
11	96	100	98.0	8.0
12	89	93	91.0	8.0

To estimate the overall s^2 use the variance for all 24 IQ scores taken as a group: $s_1^2 = 73.955$.

To estimate the s^2 between twins, use the mean of the 12 s^2 values at the left: $s_2^2 = 27.125$.

Now use the F statistic to test the claim that $s_1^2 > s_2^2$.

Chapter 10

Multinomial Experiments and Contingency Tables

10-2 Multinomial Experiments

NOTE: In multinomial problems, always verify that $\Sigma E = \Sigma O$ before proceeding. If these sums are not equal, then an error has been made and further calculations have no meaning.

1. H_o: $p_0 = p_1 = p_2 = \ldots = p_9 = .10$
    H_1: at least one of the proportions is different from .10
    $\alpha = .01$
    C.R. $\chi^2 > \chi^2_{9,.01} = 21.666$
    calculations:

    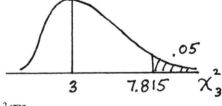

digit	O	E	$(O-E)^2/E$
0	35	8	91.125
1	0	8	8.000
2	2	8	4.500
3	1	8	6.125
4	4	8	2.000
5	24	8	32.000
6	1	8	6.125
7	4	8	2.000
8	7	8	.125
9	2	8	4.500
	80	80	156.500

    $\chi^2 = \Sigma[(O-E)^2/E]$
    $= 156.500$

    conclusion:
    Reject H_o; there is sufficient evidence to conclude that at least one of the proportions is different from .10.

    Based on these results it appears that the students were asked to report their own weights and were not actually weighed.

3. H_o: $p_{LF} = p_{RF} = p_{LR} = p_{RR} = .25$
    H_1: at least one of the proportions is different from .25
    $\alpha = .05$
    C.R. $\chi^2 > \chi^2_{3,.05} = 7.815$
    calculations:

tire	O	E	$(O-E)^2/E$
LF	11	10	.100
RF	15	10	2.500
LR	8	10	.400
RR	2	10	1.600
	40	40	4.600

    $\chi^2 = \Sigma[(O-E)^2/E]$
    $= 4.600$

    conclusion:
    Do not reject H_o; there is not sufficient evidence to conclude that at least one of the proportions is different from .25.

    While we cannot be 95% certain that there is a tendency to pick one tire more than any other, it might be worthwhile to take a larger sample to see if the trend toward picking front tires becomes significant. If so, the probability calculated in section 3-4 under the assumption of equal probabilities would not be valid.

5. H_o: $p_{Sun} = p_{Mon} = p_{Tue} = \ldots = p_{Sat} = 1/7$
   H_1: at least one of the proportions is different from 1/7
   $\alpha = .05$
   C.R. $\chi^2 > \chi^2_{6,.05} = 12.592$
   calculations:

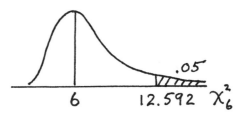

day	O	E	$(O-E)^2/E$
Sun	40	30.857	2.709
Mon	24	30.857	1.524
Tue	25	30.857	1.112
Wed	28	30.857	.265
Thu	29	30.857	.112
Fri	32	30.857	.042
Sat	38	30.857	1.653
	216	216.000	7.417

$\chi^2 = \Sigma[(O-E)^2/E]$
$= 7.417$

conclusion:
   Do not reject H_o; there is not sufficient evidence to conclude that at least one of the proportions is different from 1/7.

There is not enough evidence to reject with 95% confidence the daily drinker theory in favor of theory of the casual drinker who binges on Friday and Saturday.

7. H_o: $p_{Mon} = .30$, $p_{Tue} = .15$, $p_{Wed} = .15$, $p_{Thu} = .20$, $p_{Fri} = .20$
   H_1: at least one of the proportions is different from what is claimed
   $\alpha = .05$
   C.R. $\chi^2 > \chi^2_{4,.05} = 9.488$
   calculations:

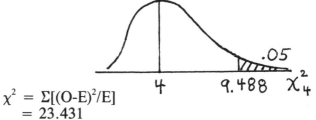

day	O	E	$(O-E)^2/E$
Mon	31	44.10	3.891
Tue	42	22.05	18.050
Wed	18	22.05	.744
Thu	25	29.40	.659
Fri	31	29.40	.087
	147	147.00	23.431

$\chi^2 = \Sigma[(O-E)^2/E]$
$= 23.431$

conclusion:
   Reject H_o; there is sufficient evidence to conclude that at least one of the proportions is different from what is claimed.

Rejection of this claim may indirectly help to correct the accident problem. It may indicate that the safety expert is wrong and that a better such person should be employed. It may indicate that the safety expert is correct but that Tuesday (which makes an unusually large contribution to $\Sigma[(O-E)^2/E]$) involves circumstances unique to the plant being studied. At the very least, rejection of the hypothesis indicates that medical staffing should not be based on the proportions claimed.

9. H_o: $p_0 = p_1 = p_2 = ... = p_9 = .10$
 H_1: at least one of the proportions is different from .10
 $\alpha = .05$
 C.R. $\chi^2 > \chi^2_{9,.05} = 16.919$
 calculations:

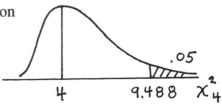

digit	O	E	$(O-E)^2/E$
0	8	10	.400
1	8	10	.400
2	12	10	.400
3	11	10	.100
4	10	10	.000
5	8	10	.400
6	9	10	.100
7	8	10	.400
8	12	10	.400
9	14	10	1.600
	100	100	4.200

$\chi^2 = \Sigma[(O-E)^2/E]$
$= 4.200$

conclusion:
 Do not reject H_o; there is not sufficient evidence to conclude that at least one of the proportions is different from .10.

11. H_o: $p_1 = .16$, $p_2 = .44$, $p_3 = .27$, $p_4 = .13$
 H_1: at least one of the proportions is different from the license proportions
 $\alpha = .05$
 C.R. $\chi^2 > \chi^2_{3,.05} = 7.815$
 calculations:

group	O	E	$(O-E)^2/E$
1: < 25	36	14.08	34.125
2: 25-44	21	38.72	8.109
3: 45-64	12	23.76	5.821
4: > 64	19	11.44	4.996
	88	88.00	53.051

$\chi^2 = \Sigma[(O-E)^2/E]$
$= 53.051$

conclusion:
 Reject H_o; there is sufficient evidence to conclude that at least one of the proportions is different from the license proportions.

Yes; the "under 25" group appears to have a disproportionate number of crashes. It would be fairer, but much more difficult, to base the E values on the proportion of miles driven and not on the proportion of licenses possessed.

13. NOTE: Usually H_o is given in terms of p_i's. When a specific distribution is named, H_o is often stated using the term "goodness of fit" without mentioning specific probabilities -- although they must typically be calculated using the named distribution in order to determine the E_i's.
 H_o: there is goodness of fit to the uniform distribution
 H_1: there is not goodness of fit to the uniform distribution
 $\alpha = .05$ [assumed]
 C.R. $\chi^2 > \chi^2_{4,.05} = 9.488$
 calculations:

interval	O	E	$(O-E)^2/E$
55- 46	8	10	.400
65- 74	5	10	2.500
75- 84	18	10	6.400
85- 94	13	10	.900
95-104	6	10	1.600
	50	50	11.800

$\chi^2 = \Sigma[(O-E)^2/E]$
$= 11.800$

conclusion:
 Reject H_o; there is sufficient evidence to conclude that there is not goodness of fit to the uniform distribution.

15. H_o: $p_{poor} = p_{fair} = p_{good} = p_{exce} = .25$
    H_1: at least one of the proportions is different from .25
    $\alpha = .05$
    C.R. $\chi^2 > \chi^2_{3,.05} = 7.815$
    calculations:

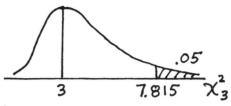

type	O	E	$(O-E)^2/E$
poor	4	15	8.067
fair	13	15	.267
good	29	15	13.067
exce	14	15	.067
	60	60	21.467

$\chi^2 = \Sigma[(O-E)^2/E]$
$= 21.467$

conclusion:
Reject H_o; there is sufficient evidence to conclude that at least one of the proportions is different from .25.

NOTE: Before giving movie critic James Harrington two thumbs down on the basis of the above test, one should determine exactly how the list of movies in the appendix was obtained and whether it is a fair test of the critic's claim. Is it possible, for instance, that the given list was obtained from a list of movies being shown at some point in time? Since poor movies have shorter runs, they would be under represented in the sample. Or is it possible that the critic as talking of "all films produced" and not merely those that make it onto some list. As a recognized critic with inside connections, he might be aware of films so poor that they never make it to the box office or onto lists that are compiled.

17. Because $4.168 < 4.400 < 14.684$ [i.e., $\chi^2_{9,.90} < 4.400 < \chi^2_{9,.10}$],
    it must be $.10 < $ P-value $< .90$ [and the P-value is very close to .90].

19. NOTE: Both outcomes having the same expected frequency is equivalent to $p_1 = p_2 = .5$.
    a. H_o: $p_1 = p_2 = .5$
    H_1: at least one of the proportions is different from .5
    $\alpha = .05$
    C.R. $\chi^2 > \chi^2_{1,.05} = 3.841$
    calculations

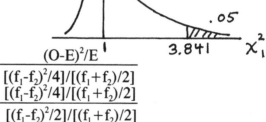

type	O	E	O-E	$(O-E)^2$	$(O-E)^2/E$
A	f_1	$(f_1+f_2)/2$	$(f_1-f_2)/2$	$(f_1-f_2)^2/4$	$[(f_1-f_2)^2/4]/[(f_1+f_2)/2]$
B	f_2	$(f_1+f_2)/2$	$(f_2-f_1)/2$	$*(f_1-f_2)^2/4$	$[(f_1-f_2)^2/4]/[(f_1+f_2)/2]$
	f_1+f_2	f_1+f_2			$[(f_1-f_2)^2/2]/[(f_1+f_2)/2]$

*NOTE: $(f_2-f_1)^2 = (f_1-f_2)^2$
$\chi^2 = \Sigma[(O-E)^2/E]$
$= [(f_1-f_2)^2/2]/[(f_1+f_2)/2]$
$= (f_1-f_2)^2/(f_1+f_2)$

b. H_o: $p = .5$
    H_1: $p \neq .5$
    $\alpha = .05$
    C.R. $z < -z_{.025} = -1.960$
    $z > z_{.025} = 1.960$
    calculations:

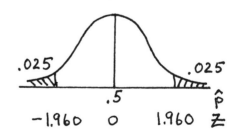

$z_{\hat{p}} = (\hat{p} - \mu_{\hat{p}})/\sigma_{\hat{p}}$
$= [f_1/(f_1+f_2) - .5]/\sqrt{(.5)(.5)/(f_1+f_2)}$
$= [.5(f_1-f_2)/(f_1+f_2)]/\sqrt{(.5)(.5)/(f_1+f_2)}$
$= [(f_1-f_2)/(f_1+f_2)]/\sqrt{1/(f_1+f_2)}$
$= (f_1-f_2)/\sqrt{f_1+f_2}$

Note that $z^2 = \chi^2$ since $[(f_1-f_2)/\sqrt{f_1+f_2}]^2 = (f_1-f_2)^2/(f_1+f_2)$ and $(\pm1.960)^2 = 3.841$.

21. NOTE: Usually the hypothesized p_i's are given and the formula $E_i = np_i$ is used to find the individual expected values. Here, the individual expected values are given. To state a null hypothesis, either use the formula "in reverse" to solve for the hypothesized p_i's or use the term "goodness of fit" without mentioning specific probabilities.

H_o: there is goodness of fit to the Poisson distribution
H_1: there is not goodness of fit to the Poisson distribution
$\alpha = .05$
C.R. $\chi^2 > \chi^2_{2,.05} = 5.991$
calculations:

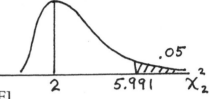

homicides	O	E	$(O-E)^2/E$
0	268	265.6	.022
1	79	84.4	.345
2+	18	15.0*	.600
	365	365.0	.967

$\chi^2 = \Sigma[(O-E)^2/E]$
$= .967$

conclusion:
   Do not reject H_o; there is not sufficient evidence to conclude that there is not goodness of fit to the Poisson distribution.

* Combine the last three cells so that $E_i \geq 5$ for each i. Make $E_{2+} = 15.0$ so that $\Sigma E_i = 365$. This assumes the $\Sigma E_i = .1$ for $i \geq 5$ not given in the original table.

10-3 Contingency Tables: Independence and Homogeneity

NOTE: For each row and each column it must be true that $\Sigma O = \Sigma E$. After the marginal row and column totals are calculated, both the row totals and the column totals must sum to produce the same grand total. If either of the preceding is not true, then an error has been made and further calculations have no meaning. In addition, the following are true for all χ^2 contingency table analyses in this manual.
* The E values for each cell are given in parentheses below the O values.
* The addends used to calculate the χ^2 test statistic follow the physical arrangement of the cells in the original contingency table. This practice makes it easier to monitor the large number of intermediate steps involved and helps to prevent errors caused by missing or double-counting cells.
* The accompanying chi-square illustration follows the "usual" shape as pictured with Table A-6, even though that shape is not correct for df=1 or df=2.

1. H_o: there is homogeneity of proportions across gender of interviewer
   H_1: there is not homogeneity of proportions
   $\alpha = .01$
   C.R. $\chi^2 > \chi^2_{1,.01} = 6.635$
   calculations:

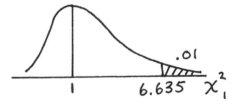

		INTERVIEWER		
		M	F	
FEMALE	A	512	336	848
RESPONSE		(565.3)	(282.7)	
	D	288	64	352
		(234.7)	(117.3)	
		800	400	1200

$\chi^2 = \Sigma[(O-E)^2/E]$
$= 5.031 + 10.063$
$\quad 12.121 + 24.242$
$= 51.458$

conclusion:
   Reject H_o; there is sufficient evidence to conclude that there is not homogeneity of proportions across gender of interviewer -- i.e., the proportion of agree/disagree responses vary according to the gender of the interviewer.

3. H_o: drug treatment and oral reaction are independent
   H_1: drug treatment and oral reaction are related
   $\alpha = .05$
   C.R. $\chi^2 > \chi^2_{1,.05} = 3.841$
   calculations:

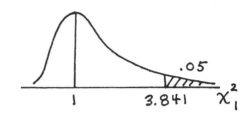

TREATMENT

		D	P	
RCTN	S	43 (38.87)	35 (39.13)	78
	N	109 (113.13)	118 (113.87)	227
		152	153	305

$\chi^2 = \Sigma[(O-E)^2/E]$
$\quad = .4383 + .4355$
$\quad\quad .1506 + .1496$
$\quad = 1.174$

conclusion:

Do not reject H_o; there is not sufficient evidence to conclude that the drug treatment and the oral reaction are related.

A person thinking about using Nicorette might still want to be concerned about mouth soreness. The direction of the data is toward those using the real drug being more likely to experience soreness. While we cannot be 95% sure that there is soreness associated with Nicorette, neither can we be sure that there is not such soreness.

5. NOTE: We assume the sample sizes of 200 men and 300 women were predetermined, making this exercise "homogeneity of proportions analysis" rather than "independence of variables."
   H_o: there is homogeneity of proportions across gender of respondent
   H_1: there is not homogeneity of proportions
   $\alpha = .05$
   C.R. $\chi^2 > \chi^2_{2,.05} = 5.991$
   calculations:

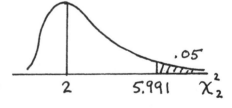

CONFIDENCE

		GREAT	SOME	LITTLE	
RESPDNT	M	115 (116)	56 (60)	29 (24)	200
	F	175 (174)	94 (90)	31 (36)	300
		290	150	60	500

$\chi^2 = \Sigma[(O-E)^2/E]$
$\quad = .009 + .267 + 1.042$
$\quad\quad .006 + .178 + .694$
$\quad = 2.195$

conclusion:

Do not reject H_o; there is not sufficient evidence to conclude that there is not homogeneity of proportions across gender of respondent -- i.e., the proportion of persons in the given confidence categories does not vary significantly according to the gender of the respondent.

7. H_o: type of crime and criminal/victim connection are independent
   H_1: type of crime and criminal/victim connection are related
   $\alpha = .05$
   C.R. $\chi^2 > \chi^2_{2,.05} = 5.991$
   calculations:

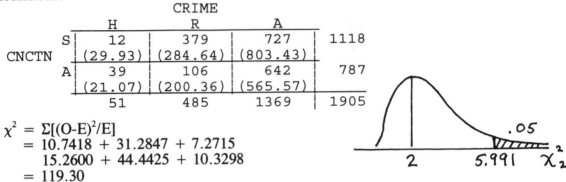

		CRIME		
		H	R	A
CNCTN	S	12 (29.93)	379 (284.64)	727 (803.43)	1118
	A	39 (21.07)	106 (200.36)	642 (565.57)	787
		51	485	1369	1905

$\chi^2 = \Sigma[(O-E)^2/E]$
$= 10.7418 + 31.2847 + 7.2715$
$\quad 15.2600 + 44.4425 + 10.3298$
$= 119.30$

conclusion:
   Reject H_o; there is sufficient evidence to conclude that the type of crime and the criminal/victim connection are related.

9. H_o: sentence and plea are independent
   H_1: sentence and plea are related
   $\alpha = .05$
   C.R. $\chi^2 > \chi^2_{1,.05} = 3.841$
   calculations:

		PLEA	
		G	NG
SNTCE	P	392 (418.48)	58 (31.52)	450
	NP	564 (537.52)	14 (40.48)	578
		956	72	1028

$\chi^2 = \Sigma[(O-E)^2/E]$
$= 1.6759 + 22.2518$
$\quad 1.3047 + 17.3241$
$= 42.557$

conclusion:
   Reject H_o; there is sufficient evidence to conclude that a person's sentence and his original plea are related.

Yes; assuming that those who are really guilty will indeed be convicted with a trial, these results suggest that a guilty plea should be encouraged. But the study reported only those who plead not guilty and were convicted in trials. Suppose there were also guilty 50 persons who plead not guilty and were acquitted. Including them in the no prison category makes the table

		PLEA	
		G	NG
SNTCE	P	392 (399.07)	58 (50.93)	450
	NP	564 (556.93)	64 (71.07)	628
		956	112	1078

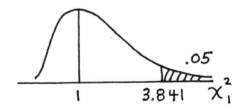

$\chi^2 = \Sigma[(O-E)^2/E]$
$= .125 + .982$
$\quad .090 + .704$
$= 1.901$

11. H_o: smoking and age are independent
    H_1: smoking and age are related
    $\alpha = .05$
    C.R. $\chi^2 > \chi^2_{3,.05} = 7.815$
    calculations:

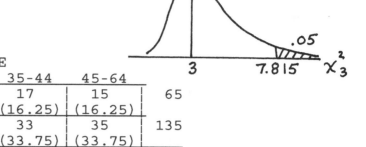

		20-24	25-34	35-44	45-64	
SMOKE	S	18	15	17	15	65
		(16.25)	(16.25)	(16.25)	(16.25)	
	N	32	35	33	35	135
		(33.75)	(33.75)	(33.75)	(33.75)	
		50	50	50	50	200

AGE

$\chi^2 = \Sigma[(O-E)^2/E]$
$= .1885 + .0963 + .0346 + .0962$
$.0907 + .0463 + .0167 + .0463$
$= .615$

conclusion:

Do not reject H_o; there is not sufficient evidence to conclude that smoking and age are related.

At present, there are not a higher percentage of smokers in one age group than in another. Targeting cigarette advertising to younger smokers, however, will reach persons who will be around longer and (if they continue smoking) will ultimately raise the percentages in all age groups.

13. H_o: alcohol use and type of crime are independent
    H_1: alcohol use and type of crime are related
    $\alpha = .05$ [assumed]
    C.R. $\chi^2 > \chi^2_{5,.05} = 11.071$
    calculations:

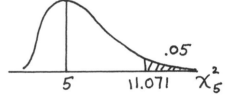

CRIME

		arson	rape	violence	stealing	coining	fraud	
USE	Y	50	88	155	379	18	63	753
		(49.11)	(79.21)	(139.93)	(358.55)	(16.90)	(109.31)	
	N	43	62	110	300	14	144	673
		(43.89)	(70.79)	(125.07)	(320.45)	(15.10)	(97.69)	
		93	150	265	679	32	207	1426

$\chi^2 = \Sigma[(O-E)^2/E]$
$= .016 + 0.976 + 1.622 + 1.167 + .072 + 19.617$
$.018 + 1.092 + 1.815 + 1.306 + .080 + 21.949$
$= 49.731$

conclusion:

Reject H_o; there is sufficient evidence to conclude that alcohol use and type of crime are related.

Fraud seems to be different from the other crimes in that it is more like to be committed by someone who abstains from alcohol.

15. H_o: gender and smoking are independent
H_1: gender and smoking are related
$\alpha = .05$ [assumed]
C.R. $\chi^2 > \chi^2_{1,.05} = 3.841$
calculations:

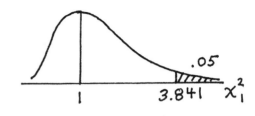

SMOKING

		Y	N	
	M	15 (14.25)	42 (42.75)	57
GENDER				
	F	10 (41.24)	33 (32.25)	43
		25	75	100

$$\chi^2 = \Sigma[(O-E)^2/E]$$
$$= .039 + .013$$
$$.052 + .017$$
$$= .122$$

conclusion:
Do not reject H_o; there is not sufficient evidence to conclude that gender and smoking are related.

17. H_o: having an accident and using a cellular phone are independent
H_1: having an accident and using a cellular phone are related
$\alpha = .05$
C.R. $\chi^2 > \chi^2_{1,.05} = 3.841$
calculations:

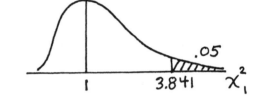

ACCIDENT

		Y	N	
	Y	23 (27.76)	282 (277.24)	305
PHONE				
	N	46 (41.24)	407 (411.76)	453
		69	689	758

$$\chi^2 = \Sigma[(\,|\,O\text{-}E\,|\,-.5)^2/E]$$
$$= .6548 + .0656$$
$$.4409 + .0442$$
$$= 1.205$$

conclusion:
Do not reject H_o; there is not sufficient evidence to conclude that having an accident and using a cellular phone are related.

Without the correction for continuity [see the solution for exercise #9 for the details] the calculated test statistic is 1.505. Since $(\,|\,O\text{-}E\,|\,-.5)^2 < (O\text{-}E)^2$ whenever $|\,O\text{-}E\,| > .25$, Yates' correction generally lowers the calculated test statistic.

Review Exercises

1. H_o: there is homogeneity of proportions across cities
   H_1: there is not homogeneity of proportions
   $\alpha = .05$
   C.R. $\chi^2 > \chi^2_{2,.05} = 5.991$
   calculations:

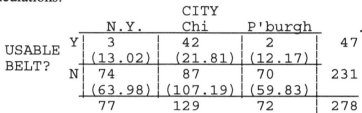

		CITY			
		N.Y.	Chi	P'burgh	
USABLE BELT?	Y	3	42	2	47
		(13.02)	(21.81)	(12.17)	
	N	74	87	70	231
		(63.98)	(107.19)	(59.83)	
		77	129	72	278

$$\chi^2 = \Sigma[(O\text{-}E)^2/E]$$
$$= 7.709 + 18.692 + 8.501$$
$$1.569 + 3.803 + 1.730$$
$$= 42.004$$

conclusion:
   Reject H_o; there is sufficient evidence to conclude that there is not homogeneity of proportions across cities.

2. H_o: $p_{Mon} = p_{Tue} = p_{Wed} = ... = p_{Sun} = 1/7$
   H_1: at least one of the proportions is different from 1/7
   $\alpha = .05$
   C.R. $\chi^2 > \chi^2_{6,.05} = 12.592$
   calculations:

day	O	E	$(O\text{-}E)^2/E$
Mon	74	66.286	.8978
Tue	60	66.286	.5961
Wed	66	66.286	.0012
Thu	71	66.286	.3353
Fri	51	66.286	3.5249
Sat	66	66.286	.0012
Sun	76	66.286	1.4236
	464	464.000	6.7801

$$\chi^2 = \Sigma[(O\text{-}E)^2/E]$$
$$= 6.780$$

conclusion:
   Do not reject H_o; there is not sufficient evidence to conclude that at least one of the proportions is different from 1/7.

The data are very close [calculated $\chi^2 = 6.78$, $E(\chi^2) = 6.00$ if H_o is true] to precisely the amount of random fluctuation we expect if there are no differences among the days, and they do not support the theory that more gunfire deaths occur on weekends when more people are at home.

3. H_o: $p_{Jan} = p_{Feb} = p_{Mar} = \ldots = p_{Dec} = 1/12$
   H_1: at least one of the proportions is different from 1/12
   $\alpha = .05$
   C.R. $\chi^2 > \chi^2_{11,.05} = 19.675$
   calculations:

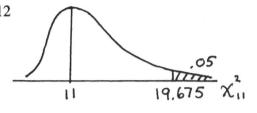

month	O	E	$(O-E)^2/E$
Jan	8	10	.400
Feb	12	10	.400
Mar	9	10	.100
Apr	15	10	2.500
May	6	10	1.600
Jun	12	10	.400
Jul	4	10	3.600
Aug	7	10	.900
Sep	11	10	.100
Oct	11	10	.100
Nov	5	10	2.500
Dec	20	10	10.000
	120	120	22.600

$\chi^2 = \Sigma[(O-E)^2/E]$
$= 22.600$

conclusion:
   Reject H_o; there is sufficient evidence to conclude that at least one of the proportions is different from 1/12.

   No; the fact that some months were offered as answers more than other months says nothing about how often the offered answers were correct and whether or not the subjects have ESP.

4. H_o: there is homogeneity of proportions across airlines
   H_1: there is not homogeneity of proportions
   $\alpha = .05$
   C.R. $\chi^2 > \chi^2_{2,.05} = 5.991$
   calculations:

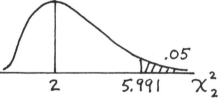

		AIRLINE			
		USAir	Americ	Delta	
ON TIME?	Y	80	77	76	233
		(77.67)	(77.67)	(77.67)	
	N	20	23	24	67
		(22.23)	(23.33)	(22.33)	
		100	100	100	300

$\chi^2 = \Sigma[(O-E)^2/E]$
$= .070 + .006 + .036$
$\quad .244 + .020 + .124$
$= .500$

conclusion:
   Do not reject H_o; there is not sufficient evidence to conclude that there is not homogeneity of proportions.

5. H_o: headache occurrence and drug usage are independent
H_1: headache occurrence and drug usage are related
$\alpha = .05$
C.R. $\chi^2 > \chi^2_{2,.05} = 5.991$
calculations:

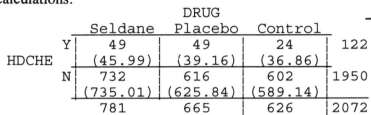

		DRUG			
		Seldane	Placebo	Control	
HDCHE	Y	49	49	24	122
		(45.99)	(39.16)	(36.86)	
	N	732	616	602	1950
		(735.01)	(625.84)	(589.14)	
		781	665	626	2072

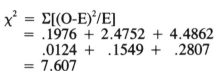

$\chi^2 = \Sigma[(O-E)^2/E]$
$= .1976 + 2.4752 + 4.4862$
$.0124 + .1549 + .2807$
$= 7.607$

conclusion:
Reject H_o; there is sufficient evidence to conclude that headache occurrence and drug usage are related.

Careful examination of the table indicates that while Seldane users have more headaches than those taking no medication (6.3% vs. 3.8%), they have fewer headaches than those taking a placebo (6.3% vs. 7.4%). This suggests that Seldane usage is associated with but not necessarily responsible for the headaches.

Cumulative Review Exercises

1. n = 4 $n(\Sigma xy) - (\Sigma x)(\Sigma y) = 4(26210) - (303)(344) = 608$
   $\Sigma x = 303$ $n(\Sigma x^2) - (\Sigma x)^2 = 4(23105) - (303)^2 = 611$
   $\Sigma y = 344$ $n(\Sigma y^2) - (\Sigma y)^2 = 4(29742) - (344)^2 = 632$
   $\Sigma x^2 = 23105$
   $\Sigma y^2 = 29742$
   $\Sigma xy = 26210$
   $r = [n(\Sigma xy) - (\Sigma x)(\Sigma y)]/[\sqrt{n(\Sigma x^2) - (\Sigma x)^2} \cdot \sqrt{n(\Sigma y^2) - (\Sigma y)^2}]$
   $= 608/[\sqrt{611} \cdot \sqrt{632}] = .978$
   original claim: $\rho \neq 0$

   H_o: $\rho = 0$
   H_1: $\rho \neq 0$
   $\alpha = .05$ [assumed]
   C.R. $r < -.950$ OR C.R. $t < -t_{2,.025} = -4.303$
   $r > .950$ $t > t_{2,.025} = 4.303$
   calculations: calculations:
   $r = .978$ $t_r = (r - \mu_r)/s_r$
            $= (.978 - 0)/\sqrt{(1-(.978)^2)/2}$
            $.978/.146$
            $= 6.696$
   conclusion:
   Reject H_o; there is sufficient evidence to conclude that $\rho \neq 0$ (in fact, $\rho > 0$).

2. original claim: $\mu_d > 0$ [$n \leq 30$ and σ_d unknown, use t]
   d = y - x: 11 9 12 9
      n = 4
      $\Sigma d = 41$ $\bar{d} = 10.25$
      $\Sigma d^2 = 427$ $s_d = 1.50$
   H_o: $\mu_d \leq 0$
   H_1: $\mu_d > 0$
   $\alpha = .05$ [assumed]
   C.R. $t > t_{3,.05} = 2.353$
   calculations:

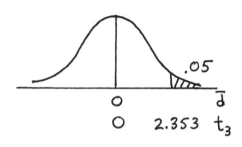

      $t_{\bar{d}} = (\bar{d} - \mu_{\bar{d}})/s_{\bar{d}}$
          $= (10.25 - 0)/(1.50/\sqrt{4})$
          $= 10.25/.750$
          $= 13.667$
   conclusion:
      Reject H_o; there is sufficient evidence to conclude that $\mu_d > 0$.

3. H_o: gender and selection are independent
   H_1: gender and selection are related
   $\alpha = .05$ [assumed]
   C.R. $\chi^2 > \chi^2_{3,.05} = 7.815$
   calculations:

		SELECTION				
		A	B	C	D	
	Y	66	80	82	75	303
GENDER		(66.97)	(79.15)	(82.42)	(74.46)	
	N	77	89	94	84	344
		(76.03)	(89.85)	(93.58)	(84.54)	
		143	169	176	159	647

   $\chi^2 = \Sigma[(O-E)^2/E]$
       $= .014 + .009 + .002 + .004$
         $.012 + .008 + .002 + .003$
       $= .055$
   conclusion:
      Do not reject H_o; there is not sufficient evidence to reject the claim that gender and selection are independent.

4. summary statistics

    males: n = 4 Σx = 303 Σx^2 = 23105 $\bar{x}$ = 75.75 s^2 = 50.917

    females: n = 4 Σx = 344 Σx^2 = 29742 $\bar{x}$ = 86.00 s^2 = 52.667

Let the females be group 1.

original claim: $\mu_1-\mu_2 = 0$ [small samples and σ unknown, first test H_o: $\sigma_1^2 = \sigma_2^2$]

    H_o: $\sigma_1^2 = \sigma_2^2$

    H_1: $\sigma_1^2 \neq \sigma_2^2$

    α = .05 [assumed]

    C.R. F < $F_{3,.975}^3$ = [.0648]

        F > $F_{3,.025}^3$ = 15.439

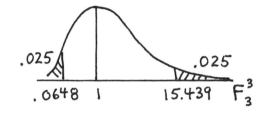

    calculations:

        F = s_1^2/s_2^2

          = 52.667/50.917

          = 1.0344

    conclusion:

        Do not reject H_o; there is not sufficient evidence to conclude that $\sigma_1^2 \neq \sigma_2^2$.

Now proceed using s_p^2 for both s_1^2 and s_2^2

    $\bar{x}_1-\bar{x}_2$ = 86.00 - 75.75 = 10.25

    s_p^2 = $(df_1 \cdot s_1^2 + df_2 \cdot s_2^2)/(df_1 + df_2)$

        = $(3 \cdot 52.667 + 3 \cdot 50.917)/(3 + 3)$ = 310.75/6 = 51.792

H_o: $\mu_1-\mu_2 = 0$

H_1: $\mu_1-\mu_2 \neq 0$

α = .05 [assumed]

C.R. t < $-t_{6,.025}$ = -2.447

    t > $t_{6,.025}$ = 2.447

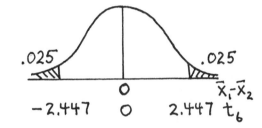

calculations:

    $t_{\bar{x}_1-\bar{x}_2}$ = $(\bar{x}_1-\bar{x}_2 - \mu_{\bar{x}_1-\bar{x}_2})/s_{\bar{x}_1-\bar{x}_2}$

        = $(10.25 - 0)/\sqrt{51.792/4 + 51.792/4}$

        = 10.25/5.089

        = 2.014

conclusion:

    Do not reject H_o; there is not sufficient evidence to reject the claim that $\mu_1-\mu_2 = 0$.

Chapter 11

Analysis of Variance

11-2 One-Way ANOVA

1. a. $F = .49$
   b. C.V. is $F = 3.8853$ [C.R. is $F > F^2_{12,.05} = 3.8853$]
   c. P-value $= .627$
   d. The decision can be made using either the F statistic or the P-value.
      Using F: since $.49 < 3.8853$, fail to reject $H_o: \mu_1 = \mu_2 = \mu_3$.
      Using P-value: since $.627 > .05$, fail to reject $H_o: \mu_1 = \mu_2 = \mu_3$.
      The two methods will always agree with each other.
   e. The differences among consumer responses to television commercials from the three companies are well within the limits expected due to chance variation. There is no evidence to contradict the hypothesis that the mean response is the same for all three companies.

3. a. $k = 3$, $\Sigma \overline{x}_i = 294.32$, $\Sigma \overline{x}_i^2 = 28875.1$
   $$\overline{\overline{x}} = \Sigma \overline{x}_i / k$$
   $$= 294.32/3$$
   $$= 98.107$$
   $$s_{\overline{x}}^2 = \Sigma (\overline{x}_i - \overline{\overline{x}})^2 / (k-1)$$
   $$= [(97.94-98.107)^2 + (98.58-98.107)^2 + (97.8-98.107)^2]/2$$
   $$= .3459/2$$
   $$= .1729$$
   $$s_{\overline{x}}^2 = [k \cdot \Sigma \overline{x}_i^2 - (\Sigma \overline{x}_i)^2]/[k(k-1)]$$
   $$= [3 \cdot 28875.1 - (294.32)^2]/[3(2)]$$
   $$= 1.0376/6$$
   $$= .1729$$
   NOTE: Either of the above formulas can be used to calculate $s_{\overline{x}}^2$. While the latter (short-cut) formula was generally preferred in the earlier chapters, we choose to use the former (definition) formula here. Since the ANOVA calculations involve so many steps, using the definition formula helps to keep the calculations more organized and reinforces the concepts involved -- especially when the n_i values are not equal.
   $$ns_{\overline{x}}^2 = 5 \cdot .1729 = .8647$$
   b. $s_p^2 = \Sigma df_i s_i^2 / \Sigma df_i = [4(.568)^2 + 4(.701)^2 + 4(.752)^2]/12 = .4598$ [NOTE: The manual used the rounded s's provided. Using the original raw data yields a slightly different value.]
   c. $F = ns_{\overline{x}}^2 / s_p^2 = .8647/.4598 = 1.8804$
   d. C.V. is $F = 3.8853$ [C.R. is $F > F^2_{12,.05} = 3.8853$]
   e. Using F: since $1.8804 < 3.8853$, fail to reject $H_o: \mu_{18-20} = \mu_{21-29} = \mu_{30+}$.
   NOTE: Exercise 3 was worked in complete detail, showing even work done on the calculator without having to be written down. Subsequent exercises are worked showing intermediate steps, but without writing down detail for routine work done on the calculator. While the manual typically shows only three decimal places for the intermediate steps, all decimal places were carried in the calculator. DO NOT ROUND OFF INTERMEDIATE ANSWERS. SAVE calculated values that will be used again. See your instructor or class assistant if you need help using your calculator accurately and efficiently.

NOTE: This section is calculation-oriented. Do not get so involved with the formulas that you miss concepts. This manual arranges the calculations to promote both computational efficiency and understanding of the underlying principles. The following notation is used in this section.

k = the number of groups

n_i = the number of scores in group i (where i = 1,2,...,k)

$\bar{x}_i$ = the mean of group i

s_i^2 = the variance of group i

$\bar{\bar{x}}$ = $\Sigma n_i\bar{x}_i/\Sigma n_i$ = the (weighted) mean of the group means
= the overall mean of all the scores in all the groups
= $\Sigma\bar{x}_i/k$ = simplified form when each group has equal size n

s_B^2 = the variance between the groups
= $\Sigma n_i(\bar{x}_i-\bar{\bar{x}})^2/(k-1)$
= $n\Sigma(\bar{x}_i-\bar{\bar{x}})^2/(k-1)$ = $ns_{\bar{x}}^2$ = simplified form when each group has equal size n

s_p^2 = the variance within the groups
= $\Sigma df_i s_i^2/\Sigma df_i$ = the (weighted) mean of the group variances
= the formula for s_p^2 from the two-sample problem generalized to k samples
= $\Sigma s_i^2/k$ = simplified form when each group has equal size n

numerator df = k-1

denominator df = Σdf_i
= k(n-1) = simplified form when each group has equal size n

F = s_B^2/s_p^2 = (variance between groups)/(variance within groups)

5. Since each group has equal size n, the simplified forms can be used.
The following preliminary values are identified and/or calculated.

	1	4	7
n	10	10	10
Σx	1493	1445	1387
Σx^2	225,541	236,287	210,189
$\bar{x}$	149.3	144.5	138.7
s^2	292.900	3053.833	1979.122

k = 3
n = 10
$s_{\bar{x}}^2 = \Sigma(\bar{x}_i-\bar{\bar{x}})^2/(k-1)$
= 56.347/2 = 28.173

$\bar{\bar{x}} = \Sigma\bar{x}_i/k$
= 144.167
$s_p^2 = \Sigma s_i^2/k$
= 5325.856/3 = 1775.285

$H_o: \mu_1 = \mu_4 = \mu_7$
$H_1:$ at least one mean is different
α = .05
C.R. $F > F_{27,.05}^2$ = 3.3541
calculations:
$F = ns_{\bar{x}}^2/s_p^2$
= 10(28.173)/1775.185
= .1587

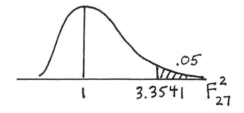

conclusion:
Do not reject H_o; there is not sufficient evidence to conclude that at least one mean is different.

No; no zone seems to have homes with higher (or lower) selling prices.

NOTE: The selling prices (whose overall average is $144,167) can be analyzed in terms of 1000's of dollars instead of dollars. See exercise #17 (c) for an explanation of why this is so.

7. Since each group has equal size n, the simplified forms can be used. The following preliminary values are identified and/or calculated.

	1	4	7
n	10	10	10
Σx	8.65	5.71	21.39
Σx^2	9.4507	5.0457	78.9271
$\bar{x}$	.865	.571	2.139
s^2	.219	.198	3.686

$k = 3$
$n = 10$
$s_{\bar{x}}^2 = \Sigma(\bar{x}_i - \bar{\bar{x}})^2/(k-1)$
     $= 1.389/2 = .695$

$\bar{\bar{x}} = \Sigma \bar{x}_i/k$
     $= 1.192$
$s_p^2 = \Sigma s_i^2/k$
     $= 4.103/3 = 1.368$

$H_o: \mu_1 = \mu_4 = \mu_7$
$H_1:$ at least one mean is different
$\alpha = .05$
C.R. $F > F_{27,.05}^2 = 3.3541$
calculations:
  $F = ns_{\bar{x}}^2/s_p^2$
    $= 10(.695)/1.368$
    $= 5.0793$

conclusion:
  Reject H_o; there is sufficient evidence to conclude that at least one mean is different.

Yes, one zone (viz., zone 7) does appear to have larger lots.

NOTE: While the exercise requests a specific deduction on the basis of rejecting H_o, the only conclusion possible from the ANOVA test is the one stated -- that "at least one mean is different." Considering the ordering of the sample means, there are three distinct possibilities: $\mu_7 > \mu_1 > \mu_4$, $\mu_7 > \mu_1 = \mu_4$, $\mu_7 = \mu_1 > \mu_4$. There are further statistical procedures that can be employed to determine which of the preceding cases should be concluded, but they are beyond the scope of this text.

NOTE ALSO: One of the assumptions of the ANOVA test is that all the groups have the same variance (and that common variance is estimated by s_p^2 -- the average of the group variances). In this exercise, one of the sample variances is almost 20 times larger than the others, and it is questionable whether the population variances are truly equal. This means that perhaps the ANOVA test is not appropriate and that a professional statistician should be consulted about a more advanced procedure.

9. The following preliminary values are identified and/or calculated.

	1	2	3	4	5	total
n	11	11	10	9	7	48
Σx	36.7	39.6	32.5	27.6	25.5	161.9
Σx^2	124.49	144.68	106.73	85.22	94.65	555.77
$\bar{x}$	3.3364	3.6000	3.2500	3.0667	3.6429	3.3729
s^2	.2045	.2120	.1228	.0725	.2929	

$$\bar{\bar{x}} = \Sigma n_i \bar{x}_i / \Sigma n_i$$
$$= [11(3.3364) + 11(3.6000) + 10(3.2500) + 9(3.0667) + 7(3.6429)]/48$$
$$= 161.9/48$$
$$= 3.3729$$

NOTE: This must always agree with the $\bar{x}$ in the "total" column.

$$\Sigma n_i(\bar{x}_i - \bar{\bar{x}})^2 = 11(3.3364 - 3.3729)^2 + 11(3.6000 - 3.3729)^2 + 10(3.2500 - 3.3729)^2$$
$$+ 9(3.0667 - 3.3729)^2 + 7(3.6429 - 3.3729)^2$$
$$= 2.087$$

$$\Sigma df_i s_i^2 = 10(.2045) + 10(.2120) + 9(.1228) + 8(.0725) + 6(.2929)$$
$$= 7.608$$

$$s_B^2 = \Sigma n_i(\bar{x}_i - \bar{\bar{x}})^2/(k-1)$$
$$= 2.087/4$$
$$= .5217$$

$$s_p^2 = \Sigma df_i s_i^2/\Sigma df_i$$
$$= 7.606/43$$
$$= .1769$$

$H_o: \mu_1 = \mu_2 = \mu_3 = \mu_4 = \mu_5$
$H_1:$ at least one mean is different
$\alpha = .05$
C.R. $F > F_{43,.05}^4 = 2.6060$
calculations:

$$F = s_B^2/s_p^2$$
$$= .5217/.1769$$
$$= 2.9491$$

.05

2.6060 F_{43}^4

conclusion:

Reject H_o; there is sufficient evidence to conclude that at least one mean is different.

NOTE: An ANOVA table may be completed as follows:

	source	SS	df	MS	F
$F = MS_{Trt}/MS_{Err}$	Trt	2.087	4	.522	2.9491
$= .522/.177$	Error	7.608	43	.177	
$= 2.9491$	Total	9.694	47		

(1) Enter $SS_{Trt} = \Sigma n_i(\bar{x}_i - \bar{\bar{x}})^2$ and $SS_{Err} = \Sigma df_i s_i^2$ values from the preliminary calculations.
(2) Enter $df_{Trt} = k-1$ and $df_{Err} = \Sigma df_i = \Sigma(n_i-1) = \Sigma n_i - k$.
(3) Add the SS and df columns to find SS_{Tot} and df_{Tot}. [The df_{Tot} must equal $\Sigma n_i - 1$.]
(4) Calculate $MS_{Trt} = SS_{Trt}/df_{Trt}$ and $MS_{Err} = SS_{Err}/df_{Err}$.
(5) Calculate $F = MS_{Trt}/MS_{Err}$.

As a final check, calculate s^2 (i.e., the variance of all the scores in one large group) two different ways as indicated below. If these answers agree, the problem is probably correct.

* from the "total" column in the table for the preliminary calculations:
$$s^2 = [n\Sigma x^2 - (\Sigma x)^2]/[n(n-1)]$$
$$= [48(555.77)-(161.9)^2]/[48(47)] = 465.35/2256 = .206$$

* from the "total" row of the ANOVA table
$$s^2 = SS_{Tot}/df_{Tot}$$
$$= 9.694/47 = .206$$

11. Since each group has equal size n, the simplified forms can be used.
The following preliminary values are identified and/or calculated.

	Apr	May	Jun	Jul	Aug	Sep
n	12	12	12	12	12	12
$\bar{x}$	1978.25	6061.33	8401.58	9844.58	6196.75	2074.17
s^2	958852	3529138	8810033	23599775	3770743	1177073

$k = 6$ $\qquad\qquad\qquad \bar{\bar{x}} = \Sigma\bar{x}_i/k$

$n = 12$ $\qquad\qquad\qquad\quad = 5759.44$

$s_{\bar{x}}^2 = \Sigma(\bar{x}_i - \bar{\bar{x}})^2/(k-1)$ $\qquad s_p^2 = \Sigma s_i^2/k$

$\qquad = 51832409/5 = 10366482$ $\qquad = 41845614/6 = 6904269$

H_o: $\mu_{Apr} = \mu_{May} = \mu_{Jun} = \mu_{Jul} = \mu_{Aug} = \mu_{Sep}$
H_1: at least one mean is different
$\alpha = .05$ [assumed]
C.R. $F > F_{66,.05}^4 = 2.3683$

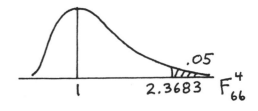

calculations:
$\quad F = ns_{\bar{x}}^2/s_p^2$
$\qquad = 12(10366482)/6974269$
$\qquad = 17.8367$
conclusion:
   Reject H_o; there is sufficient evidence to conclude that at least one mean is different.

NOTE: The ANOVA assumption that the groups have equal population variances may not be met. If that is the case, the above test is not appropriate.

13. The following preliminary values are identified and/or calculated.

	R	O	Y	Br	Bl	G	total
n	21	8	26	33	5	7	100
Σx	19.104	7.401	23.849	30.123	4.507	6.846	91.470
Σx^2	17.934278	6.862429	21.904543	27.546793	4.075729	6.018734	83.802506
$\bar{x}$	.90971	.92513	.91272	.91282	.90140	.92657	.91470
s^2	.000755	.002226	.001144	.001562	.003280	.001499	

$\bar{\bar{x}} = \Sigma n_i \bar{x}_i / \Sigma n_i$
$= [21(.90971) + 8(.92513) + 26(.91272) + 33(.91282) + 5(.90140) + 7(.92657)]/100$
$= 91.470/100$
$= .91470$

NOTE: This must always agree with the $\bar{x}$ in the "total" column.

$\Sigma n_i(\bar{x}_i - \bar{\bar{x}})^2 = 21(.90971 - .91470)^2 + 8(.92513 - .91470)^2 + 26(.91272 - .91470)^2$
$\qquad\qquad + 33(.91282 - .91470)^2 + 5(.90140 - .91470)^2 + 7(.92657 - .91470)^2$
$\qquad = .00355$

$\Sigma df_i s_i^2 = 20(.000755) + 7(.002226) + 25(.001144)$
$\qquad\qquad + 32(.001562) + 4(.003280) + 6(.001499)$
$\qquad = .13135$

$s_B^2 = \Sigma n_i(\bar{x}_i - \bar{\bar{x}})^2/(k-1)$
$\quad = .00355/5$
$\quad = .00071$

$s_p^2 = \Sigma df_i s_i^2 / \Sigma df_i$
$\quad = .13135/94$
$\quad = .001398$

H_o: $\mu_R = \mu_O = \mu_Y = \mu_{Br} = \mu_{Bl} = \mu_G$
H_1: at least one mean is different
$\alpha = .05$
C.R. $F > F_{94,.05}^5 = 2.2899$
calculations:
$\quad F = s_B^2/s_p^2$
$\qquad = .00071/.001398$
$\qquad = .5081$

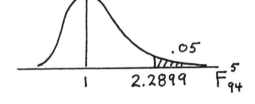

conclusion:
Do not reject H_o; there is not sufficient evidence to conclude that at least one mean is different.

No. If the intent is to make the different colors have the same mean, there is no evidence that this is not being accomplished. Corrective action is not required.

NOTE: An ANOVA table may be completed as follows:

	source	SS	df	MS	F
$F = MS_{Trt}/MS_{Err}$	Trt	.00355	5	.00071	.51
$= .00071/.00140$	Error	.13135	94	.00140	
$= .51$	Total	.13490	99		

(1) Enter $SS_{Trt} = \Sigma n_i(\bar{x}_i - \bar{\bar{x}})^2$ and $SS_{Err} = \Sigma df_i s_i^2$ values from the preliminary calculations.
(2) Enter $df_{Trt} = k-1$ and $df_{Err} = \Sigma df_i = \Sigma(n_i - 1) = \Sigma n_i - k$.
(3) Add the SS and df columns to find SS_{Tot} and df_{Tot}. [The df_{Tot} must equal $\Sigma n_i - 1$.]
(4) Calculate $MS_{Trt} = SS_{Trt}/df_{Trt}$ and $MS_{Err} = SS_{Err}/df_{Err}$.
(5) Calculate $F = MS_{Trt}/MS_{Err}$.

As a final check, calculate s^2 (i.e., the variance of all the scores in one large group) two different ways as indicated below. If these answers agree, the problem is probably correct.

* from the "total" column in the table for the preliminary calculations:
$\quad s^2 = [n\Sigma x^2 - (\Sigma x)^2]/[n(n-1)]$
$\qquad = [100(83.802506) - (91.470)^2]/[100(99)] = 13.4897/9900 = .001363$

* from the "total" row of the ANOVA table
$\quad s^2 = SS_{Tot}/df_{Tot}$
$\qquad = .13490/99 = .001363$

15. Adding $2°$ to each temperature in the 18-20 age group will raise $\bar{x}_1$ by $2°$ to 99.940, but adding a constant to each value will not affect the spread of the values -- i.e., $s_1 = .568$ does not change. Re-working exercise #3 produces the following.

a. $\bar{\bar{x}} = \Sigma\bar{x}_i/k$
$= 296.32/3$
$= 98.773$

$s_{\bar{x}}^2 = \Sigma(\bar{x}_i-\bar{\bar{x}})^2/(k-1)$
$= [(99.94-98.773)^2 + (98.58-98.773)^2 + (97.8-98.773)^2]/2$
$= 2.346/2$
$= 1.1729$

$ns_{\bar{x}}^2 = 5 \cdot 1.1729 = 5.8647$

b. $s_p^2 = \Sigma df_i s_i^2/\Sigma df_i$
$= [4(.568)^2 + 4(.701)^2 + 4(.752)^2]/12$
$= .4598$

NOTE: The manual used the rounded s's provided. Using the original raw data yields a slightly different value.

c. $F = ns_{\bar{x}}^2/s_p^2$
$= 5.8647/.4598$
$= 12.7548$

d. C.V. is $F = 3.8853$ [C.R. is $F > F_{12,.05}^2 = 3.8853$]

e. Using F: since $12.7458 > 3.8853$, reject $H_o: \mu_{18-20} = \mu_{21-29} = \mu_{30+}$ and conclude that at least one of the means is different.

17. The calculated test ratio is $F = s_B^2/s_p^2 = [\Sigma n_i(\bar{x}_i-\bar{\bar{x}})^2/(k-1)]/[\Sigma df_i s_i^2/\Sigma df_i]$. Because manipulating the scores does not change k or the n_i's, all the df_i's and the critical F value remain the same.

a. No change. Because adding the same constant c to each score increases all the means by c, each $\bar{x}_i-\bar{\bar{x}}$ difference remains unchanged and s_B^2 is not changed. Because adding the same constant to each score does not affect the spread of the scores, each s_i^2 remains unchanged and s_p^2 is not changed. Since both the numerator and the denominator are unchanged, the calculated F ratio remains the same.

b. No change. Because multiplying each score by the same constant c multiplies all the means by c, each $\bar{x}_i-\bar{\bar{x}}$ difference changes by a factor of c and s_B^2 is changed by a factor of c^2. Because multiplying each score by the same constant c changes standard deviations by a factor of c and variances by a factor of c^2, each s_i^2 and the s_p^2 are changed by a factor of c^2. Since both the numerator and the denominator are changed by a factor of c^2, the calculated F ratio remains the same.

c. No change. Changing the order of the samples will change the order of the addends in the numerator $\Sigma n_i(\bar{x}_i-\bar{\bar{x}})^2/(k-1)$ and the denominator $\Sigma df_i s_i^2/\Sigma df_i$ of the F ratio. Since changing the order of the addends does not affect the sum, both the numerator and the denominator are unchanged and the calculated F ratio remains the same.

19. The following preliminary values are identified and/or calculated.

	1	4
n	10	10
Σx	1493	1445
Σx^2	225,541	236,287
$\bar{x}$	149.3	144.5
s^2	292.900	3053.833

a. $n_1 \le 30$ and $n_4 \le 30$, use t (assume $\sigma_1 = \sigma_4$ and use s_p^2)

$\bar{x}_1 - \bar{x}_4 = 149.3 - 144.5 = 4.8$

$s_p^2 = [9(292.000) + 9(2053.833)]/(9 + 9) = 1673$

H_o: $\mu_1 - \mu_4 = 0$

H_1: $\mu_1 - \mu_4 \ne 0$

$\alpha = .05$

C.R. $t < -t_{18,.025} = -2.101$

$\quad\quad t > t_{18,.025} = 2.101$

calculations:

$\quad t_{\bar{x}_1 - \bar{x}_4} = (\bar{x}_1 - \bar{x}_4 - \mu_{\bar{x}_1 - \bar{x}_4})/s_{\bar{x}_1 - \bar{x}_4}$

$\quad\quad = (4.8 - 0)/\sqrt{1673/10 + 1673/10}$

$\quad\quad = 4.8/18.294$

$\quad\quad = .262$

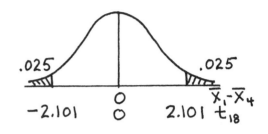

conclusion:

Do not reject H_o; there is not sufficient evidence to conclude that $\mu_1 - \mu_2 \ne 0$.

b. The following preliminary values are identified and/or calculated.

$k = 2$ $\bar{\bar{x}} = \Sigma \bar{x}_i/k$

$n = 10$ $= 146.9$

$s_{\bar{x}}^2 = \Sigma(\bar{x}_i - \bar{\bar{x}})^2/(k-1)$ $s_p^2 = \Sigma s_i^2/k$

$\quad = 11.52/1 = 11.52$ $= 3346.733/2 = 1673.3365$

H_o: $\mu_1 = \mu_2$

H_1: at least one mean is different

$\alpha = .05$

C.R. F $> F_{18,.05}^1 = 4.4139$

calculations:

$\quad F = n s_{\bar{x}}^2/s_p^2$

$\quad\quad = 10(11.52)/1673.3665$

$\quad\quad = .0688$

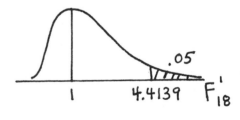

conclusion:

Do not reject H_o; there is not sufficient evidence to conclude that at least one mean is different.

c. critical value: $t^2 = (\pm 2.101)^2 = 4.414 = F$

calculated statistic: $t^2 = (.262)^2 = .069 = F$

11-3 Two-Way ANOVA

NOTE: The formulas and principles in this section are logical extensions of the previous sections.

$SS_{Row} = \Sigma n_i(\bar{x}_i - \bar{\bar{x}})^2$ for $i=1,2,3...$ [for each row]
$SS_{Col} = \Sigma n_j(\bar{x}_j - \bar{\bar{x}})^2$ for $j=1,2,3...$ [for each column]
$SS_{Tot} = \Sigma(x - \bar{\bar{x}})^2$ [for all the x's]
When there is only one observation per cell...
  the unexplained variation is
    $SS_{Err} = SS_{Tot} - SS_{Row} - SS_{Col}$
  and there is not enough data to measure interaction.
When there is more than one observation per cell...
  the unexplained variation (i.e., the failure of items in the same cell to respond the same) is
    $SS_{Err} = \Sigma(x - \bar{x}_{ij})^2 = \Sigma df_{ij} s_{ij}^2$ [for each cell -- i,e., for each i,j (row,col) combination]
  and the interaction sum of squares is
    $SS_{Int} = SS_{Tot} - SS_{Row} - SS_{Col} - SS_{Err}$.

Since the data will be analyzed from statistical software packages, however, the above formulas need not be used by hand.

1. a. $MS_{Int} = 1263$
   b. $MS_E = 695$
   c. $MS_{Star} = 350$
   d. $MS_{MPAA} = 14$

3. H_o: $\mu_P = \mu_F = \mu_G = \mu_E$
   H_1: at least one mean is different
   $\alpha = .05$ [assumed]
   C.R. $F > F^3_{8,.05} = 4.0662$
   calculations:
     $F = MS_{Star}/MS_E$
       $= 350/695$
       $= .5036$

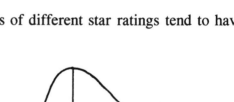

   conclusion:
     Do not reject H_o; there is not sufficient evidence to conclude that at least one mean is different.

   The evidence does not support a conclusion that movies of different star ratings tend to have different lengths.

5. H_o: $\mu_{G/PG/PG-13} = \mu_R$
   H_1: the means are different
   $\alpha = .05$
   C.R. $F > F^1_{3,.05} = 10.128$
   calculations:
     $F = MS_{MPAA}/MS_E$
       $= 0/266$
       $= 0$
   conclusion:
     Do not reject H_o; there is not sufficient evidence to conclude that the means are different.

7. H_o: $\mu_1 = \mu_2 = \mu_3 = \mu_4$
   H_1: at least one mean is different
   $\alpha = .05$
   C.R. $F > F^3_{6,.05} = 4.7571$
   calculations:
   $\qquad$ $F = MS_{Operator}/MS_E$
   $\qquad\quad = 19.86/8.03$
   $\qquad\quad = 2.47$
   conclusion:
   $\qquad$ Do not reject H_o; there is not sufficient evidence to conclude that at least one operator mean is different.

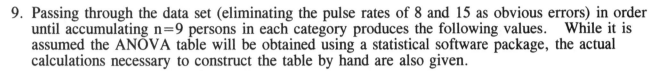

9. Passing through the data set (eliminating the pulse rates of 8 and 15 as obvious errors) in order until accumulating $n=9$ persons in each category produces the following values.   While it is assumed the ANOVA table will be obtained using a statistical software package, the actual calculations necessary to construct the table by hand are also given.

GENDER

	male	female		SUMMARY STATISTICS		
yes	63 52 60	73 67 72	$\Sigma x = 561$	$\Sigma x = 687$		
	66 60 60	82 90 60	$\Sigma x^2 = 35183$	$\Sigma x^2 = 53195$		
	68 69 63	88 75 80	$\bar{x} = 62.33$	$\bar{x} = 76.33$	$\bar{x} = 69.333$	
SMOKE			$s^2 = 26.75$	$s^2 = 94.25$		
no	69 67 66	97 88 83	$\Sigma x = 594$	$\Sigma x = 680$		
	55 47 55	77 60 78	$\Sigma x^2 = 40322$	$\Sigma x^2 = 52644$		
	75 76 84	70 67 60	$\bar{x} = 66.00$	$\bar{x} = 75.56$	$\bar{x} = 70.778$	
			$s^2 = 139.75$	$s^2 = 158.28$		
			$\bar{x} = 64.167$	$\bar{x} = 75.944$	$\bar{\bar{x}} = 70.056$	
					$s^2 = 133.25$	

$SS_{Smoke} = \Sigma n_i(\bar{x}_i - \bar{\bar{x}})^2 = 18(60.333-70.056)^2 + 18(70.778-70.056)^2 = 18.778$
$SS_{Gender} = \Sigma n_j(\bar{x}_j - \bar{\bar{x}})^2 = 18(64.167-70.056)^2 + 18(75.944-70.056)^2 = 1248.444$
$SS_{Tot} = \Sigma(x-\bar{\bar{x}})^2 = df \cdot s^2 = 35(133.25) = 4663.889$
$SS_{Err} = \Sigma(x-\bar{x}_{ij})^2 = \Sigma df_{ij}s^2_{ij}\ 8(26.75) + 8(94.25) + 8(139.75) + 8(158.28) = 3352.222$
$SS_{Int} = SS_{Tot} - SS_{Smoke} - SS_{Gender} - SS_{Err}$
$\qquad = 4663.889 - 18.778 - 1248.444 - 3352.222 = 44.444$

The resulting ANOVA tables is used for the test of hypothesis in exercises #9-11.

Source	df	SS	MS	F
Smoke	1	18.778	18.778	.1793
Gender	1	1248.444	1248.444	11.9175
Interaction	1	44.444	44.444	.4243
Error	32	3352.222	104.757	
Total	35	4663.889		

H_o: there is no smoking/gender interaction
H_1: there is smoking/gender interaction
$\alpha = .05$ [assumed]
C.R. $F > F^1_{32,.05} = 4.1709$
calculations:
$\qquad$ $F = MS_{Int}/MS_{Err}$
$\qquad\quad = 44.444/104.757$
$\qquad\quad = .4243$
conclusion:
$\qquad$ Do not reject H_o; there is not sufficient evidence to conclude that there is a smoking/gender interaction.

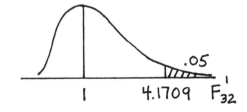

11. H_o: $\mu_{smoker} = \mu_{non\text{-}smoker}$
    H_1: the means are different
    $\alpha = .05$ [assumed]
    C.R. $F > F^1_{32,.05} = 4.1709$
    calculations:
    $\quad F = MS_{Smoke}/MS_{Err}$
    $\quad\quad = 18.778/104.757$
    $\quad\quad = .1793$
    conclusion:

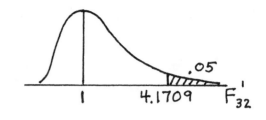

    Do not reject H_o; there is not sufficient evidence to conclude that the pulse rates for smokers and non-smokers are different.

13. Following the directions using a software package produces an ANOVA table identical to the one used for exercises #1-4. The analysis of variance technique tests for differences between means by comparing various variances. Since adding a constant to each score does not affect any variability, no values in the ANOVA summary table are changed at all.

15. The group means for the n=5 movies in each of the 6 cells of Table 11-4 are as follows.

		STAR RATING			
		fair	good	excellent	
MPAA RATING	G/PG/PG-13	102.4	99.4	134.6	112.1
	R	115.2	115.2	112.2	114.2
		108.8	107.7	123.4	

  a. Adding 100 minutes to each G/PG/PG-13 film rated excellent yields the desired effect.

		STAR RATING			
		fair	good	excellent	
MPAA RATING	G/PG/PG-13	102.4	99.4	234.6	145.4
	R	115.2	115.2	112.2	114.2
		108.8	107.7	173.4	

  b. Adding 100 minutes to each G/PG/PG-13 film yields the desired effect.

		STAR RATING			
		fair	good	excellent	
MPAA RATING	G/PG/PG-13	202.4	199.4	234.6	212.1
	R	115.2	115.2	112.2	114.2
		158.8	157.7	173.4	

  c. Adding 100 minutes to each film rated excellent yields the desired effect.

		STAR RATING			
		fair	good	excellent	
MPAA RATING	G/PG/PG-13	102.4	99.4	234.6	145.4
	R	115.2	115.2	212.2	147.5
		108.8	107.7	223.4	

Review Exercises

1. H_o: $\mu_A = \mu_B = \mu_C$
   H_1: at least one mean is different
   $\alpha = .05$
   C.R. $F > F^2_{14,.05} = 3.7389$
   calculations:
   $\quad F = MS_{Trt}/MS_{Err}$
   $\quad\quad = .0088286/.0000816$
   $\quad\quad = 46.90$
   conclusion:
   $\quad$ Reject H_o; there is sufficient evidence to conclude that at least one mean is different.

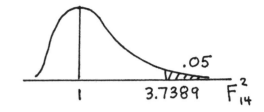

2. Since each group has equal size n, the simplified forms can be used.
   The following preliminary values are identified and/or calculated.

	oceanside	oceanfront	bayside	bayfront
n	6	6	6	6
Σx	2294	2956	1555	3210
Σx^2	944,422	1,494,108	435,465	1,772,790
$\bar{x}$	382.333	492.667	259.167	535.000
s^2	13469.867	7557.067	6492.167	11088.000

   $k = 4$ $\qquad\qquad\qquad$ $\bar{\bar{x}} = \Sigma\bar{x}_i/k$
   $n = 6$ $\qquad\qquad\qquad\quad = 417.292$
   $s^2_{\bar{x}} = \Sigma(\bar{x}_i-\bar{\bar{x}})^2/(k-1)$ $\qquad$ $s^2_p = \Sigma s^2_i/k$
   $\quad = 45762.243/3 = 15254.081$ $\qquad = 38607.100/4 = 9651.775$

   H_o: $\mu_{OS} = \mu_{OF} = \mu_{BS} = \mu_{BF}$
   H_1: at least one mean is different
   $\alpha = .05$
   C.R. $F > F^3_{20,.05} = 3.0984$
   calculations:
   $\quad F = ns^2_{\bar{x}}/s^2_p$
   $\quad\quad = 6(15254.081)/9651.775$
   $\quad\quad = 9.4827$
   conclusion:
   $\quad$ Reject H_o; there is sufficient evidence to conclude that at least one mean is different.

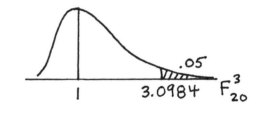

3. H_o: there is no engine-size/transmission-type interaction
   H_1: there is engine-size/transmission-type interaction
   $\alpha = .05$
   C.R. $F > F^2_{6,.05} = 5.1433$
   calculations:
   $\quad F = MS_{Int}/MS_{Err}$
   $\quad\quad = .6/11.3$
   $\quad\quad = .0531$
   conclusion:
   $\quad$ Do not reject H_o; there is not sufficient evidence to conclude that there is engine-size/transmission-type interaction.

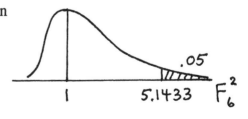

4. H_o: $\mu_{1.5} = \mu_{2.2} = \mu_{2.5}$
H_1: at least one mean is different
$\alpha = .05$
C.R. $F > F_{6,.05}^2 = 5.1433$
calculations:
   $F = MS_{Size}/MS_{Err}$
   $= 21.6/11.3$
   $= 1.9115$
conclusion:
   Do not reject H_o; there is not sufficient evidence to conclude that at least one mean is different.

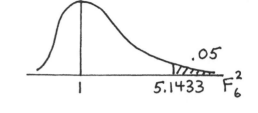

5. H_o: $\mu_A = \mu_M$
H_1: the means are different
$\alpha = .05$
C.R. $F > F_{6,.05}^1 = 5.9874$
calculations:
   $F = MS_{Trans}/MS_{Err}$
   $= 40.3/11.3$
   $= 3.5664$
conclusion:
   Do not reject H_o; there is not sufficient evidence to conclude that the means are different.

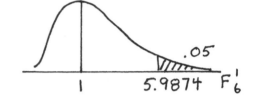

Cumulative Review Exercises

1. The ordered values and summary statistics for each agency are as follows.
   BAC: 40 52 55 63 68 72 75 77 $\Sigma x = 502$ $\Sigma x^2 = 32660$ $\bar{x} = 62.75$ $s^2 = 165.6429$
   SFA: 59 66 69 70 73 75 82 84 $\Sigma x = 578$ $\Sigma x^2 = 42232$ $\bar{x} = 72.25$ $s^2 = 67.3571$
   DFA: 42 53 57 61 69 73 73 74 $\Sigma x = 502$ $\Sigma x^2 = 32438$ $\bar{x} = 62.75$ $s^2 = 133.9286$

   a. For each agency, n=8: $P_{25} = (x_2+x_3)/2$ $P_{50} = (x_4+x_5)/2$ $P_{25} = (x_6+x_7)/2$
      The five number summaries x_1, P_{25}, P_{50}, P_{75}, x_n are as follows.
      BAC: 40 53.5 65.5 73.5 77   SFA: 59 67.5 71.5 78.5 84   DFA: 42 55.0 65.0 73.0 74

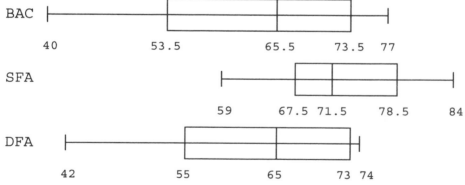

   While there is considerable overlap of scores, the boxplots seem to indicate that the Solomon & Ford agency scores tend to be higher and more consistent.

   b. BAC: $\bar{x} = 62.75$ $s^2 = 165.6429$ $s = 12.87$
      SFA: $\bar{x} = 72.25$ $s^2 = 67.3571$ $s = 8.21$
      DFA: $\bar{x} = 62.75$ $s^2 = 133.9286$ $s = 11.57$

c. Let Barnum agency scores be group 1.

original claim: $\mu_1 - \mu_2 = 0$ [small samples and σ unknown, first test H_o: $\sigma_1^2 = \sigma_2^2$]

H_o: $\sigma_1^2 = \sigma_2^2$
H_1: $\sigma_1^2 \neq \sigma_2^2$
$\alpha = .05$ [assumed]
C.R. $F < F_{7,.975}^{7} = [.2002]$
    $F > F_{7,.025}^{7} = 4.9949$

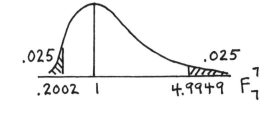
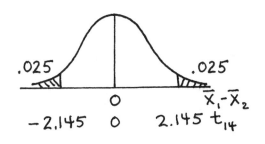

calculations:
    $F = s_1^2/s_2^2$
       $= 165.6429/67.3571$
       $= 2.4592$

conclusion:
    Do not reject H_o; there is not sufficient evidence to conclude that $\sigma_1^2 \neq \sigma_2^2$.

Now proceed using s_p^2 for both s_1^2 and s_2^2

$\bar{x}_1 - \bar{x}_2 = 62.75 - 72.25 = -9.50$
$s_p^2 = (df_1 \cdot s_1^2 + df_2 \cdot s_2^2)/(df_1 + df_2)$
     $= (7 \cdot 165.6429 + 7 \cdot 67.3571)/(7 + 7) = 1631/14 = 116.5$

H_o: $\mu_1 - \mu_2 = 0$
H_1: $\mu_1 - \mu_2 \neq 0$
$\alpha = .05$
C.R. $t < -t_{14,.025} = -2.145$
    $t > t_{14,.025} = 2.145$

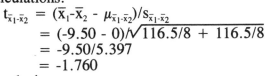

calculations:
$t_{\bar{x}_1 - \bar{x}_2} = (\bar{x}_1 - \bar{x}_2 - \mu_{\bar{x}_1 - \bar{x}_2})/s_{\bar{x}_1 - \bar{x}_2}$
    $= (-9.50 - 0)/\sqrt{116.5/8 + 116.5/8}$
    $= -9.50/5.397$
    $= -1.760$

conclusion:
    Do not reject H_o; there is not sufficient evidence to reject the claim that $\mu_1 - \mu_2 = 0$.

d. For each agency it is true that $n \leq 30$ and σ unknown, use t

BAC	SFA	DFA
$\bar{x} \pm t_{7,.025} \cdot s/\sqrt{n}$	$\bar{x} \pm t_{7,.025} \cdot s/\sqrt{n}$	$\bar{x} \pm t_{7,.025} \cdot s/\sqrt{n}$
$62.75 \pm 2.365 \cdot 12.87/\sqrt{8}$	$72.25 \pm 2.365 \cdot 8.207/\sqrt{8}$	$62.75 \pm 2.365 \cdot 11.57/\sqrt{8}$
62.75 ± 10.76	72.25 ± 6.86	62.75 ± 9.68
$51.99 < \mu < 73.51$	$65.39 < \mu < 79.11$	$53.07 < \mu < 72.43$

Since the intervals overlap, it is possible that the agencies could have a common mean (say, for example, $\mu = 70$). The interval for Solomon & Ford, however, was slightly higher and narrower than the other two.

e. Since each group has equal size n, the simplified forms can be used.

$k = 3$ $\quad\quad\quad\quad\quad\quad\quad$ $\bar{\bar{x}} = \Sigma\bar{x}_i/k$

$n = 8$ $\quad\quad\quad\quad\quad\quad\quad\quad$ $= 65.917$

$s_{\bar{x}}^2 = \Sigma(\bar{x}_i-\bar{\bar{x}})^2/(k-1)$ $\quad\quad$ $s_p^2 = \Sigma s_i^2/k$

$\quad\quad = 60.167/2 = 30.083$ $\quad\quad\quad = 366.9286/3 = 122.31$

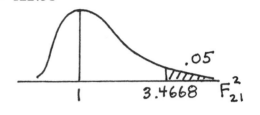

H_o: $\mu_{BAC} = \mu_{SFA} = \mu_{DFA}$
H_1: at least one mean is different
$\alpha = .05$
C.R. $F > F_{21,.05}^2 = 3.4668$
calculations:

$F = ns_{\bar{x}}^2/s_p^2$

$\quad = 8(30.083)/122.31$

$\quad = 1.9677$

conclusion:

Do not reject H_o; there is not sufficient evidence to conclude that at least one mean is different.

On the basis of these results, I cannot be 95% confident that there is any difference in the mean agency scores. Since one of the three must be selected, however, I would choose Solomon & Ford because their scores have both the highest mean and the lowest standard deviation.

2. a. let x = the weight of a newborn baby

$P(x > 8.00)$

$\quad = P(z > .42)$

$\quad = .5000 - .1618$

$\quad = .3372$

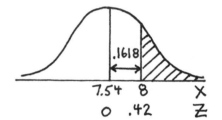

b. The $\bar{x}$'s are normally distributed because the original distribution is normal.

$\mu_{\bar{x}} = \mu = 7.54$

$\sigma_{\bar{x}} = \sigma/\sqrt{n} = 1.09/\sqrt{16} = .2725$

$P(\bar{x} > 8.00)$

$\quad = P(z > 1.69)$

$\quad = .5000 - .4545$

$\quad = .0455$

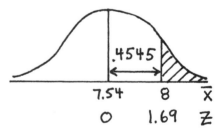

c. let B = a baby weighs more than 7.54 lbs

$P(B) = .5000$, for each birth

$P(B_1 \text{ and } B_2 \text{ and } B_3) = P(B_1) \cdot P(B_2) \cdot P(B_3)$

$\quad\quad\quad\quad\quad\quad\quad = (.5000) \cdot (.5000) \cdot (.5000)$

$\quad\quad\quad\quad\quad\quad\quad = .125$

Chapter 12

Statistical Process Control

12-2 Control Charts for Variation and Mean

NOTE: In this section, k = number of sample subgroups
        n = number of observations per sample subgroup

1. There are $k \cdot n = (25 \text{ years}) \cdot (6 \text{ observations/year}) = 150$ observations on the run chart.
   $\bar{\bar{x}} = \Sigma\bar{x}/k = 1811.9/25 = 72.48$

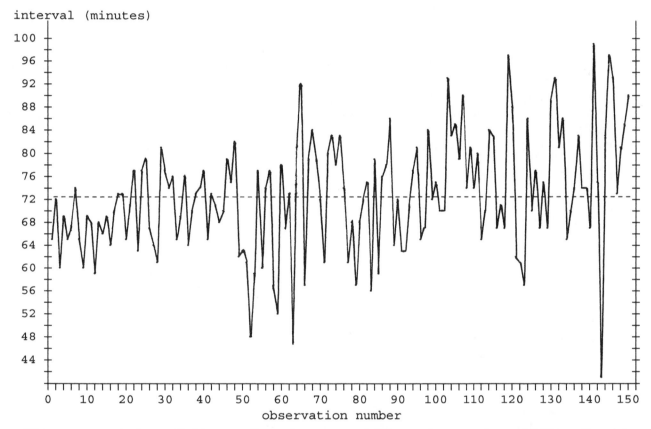

There appears to be a slight upward trend accompanied by an increase in variability. Tourists will have to wait longer to see an eruption, and the times for eruptions will be predicted with less accuracy.

NOTE: Although the line $\bar{\bar{x}} = 72.48$ is not specified in the text as a necessary part of the run chart, such a line is included in the Minitab examples and is included in this manual.

3. $\bar{\bar{x}} = \Sigma\bar{x}/k = 1811.9/25 = 72.48$
$\bar{R} = \Sigma R/k = 564/25 = 22.56$
$LCL = \bar{\bar{x}} - A_2\bar{R} = 72.476 - (.483)(22.56) = 72.476 - 10.896 = 61.58$
$UCL = \bar{\bar{x}} + A_2\bar{R} = 72.476 + (.483)(22.56) = 72.476 + 10.896 = 83.37$

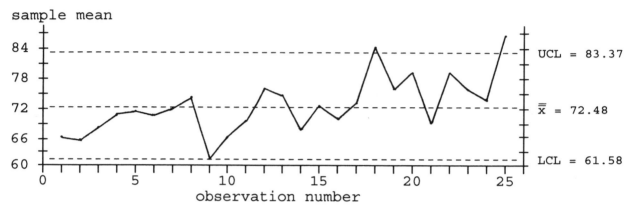

The process mean is not within statistical control. To be out of control, the process must meet one of three out-of-control criteria. This process actually meets two of the three out-of-control criteria. (1) There is a pattern to the $\bar{x}$ values -- in this case, an upward drift. (2) There is a point lying beyond the upper or lower control limits -- in this case, there are two points above the UCL. The practical implication for tourists is that the progressively longer intervals indicate that the time between eruptions is not stable -- but that it is growing longer.

5. $\bar{R} = \Sigma R/k = 7.292/20 = .365$ $\qquad$ $LCL = D_3\bar{R} = 0.000(.3646) = 0.000$
$\qquad\qquad\qquad\qquad\qquad\qquad\qquad$ $UCL = D_4\bar{R} = 2.114(.3646) = 0.771$

sample range

```
1.200 ┬
1.000 ┬
 .800 ┬- - - - - - - - - - - - - - - - - - - - - - - - - -  UCL = .771
 .600 ┬
 .400 ┬- - - - - - - - - - - - - - - - - - - - - - - - - -  R̄ = .365
 .200 ┬
 .000 ┬- - - - - - - - - - - - - - - - - - - - - - - - - -  LCL = .000
      └┬┬┬┬┬┬┬┬┬┬┬┬┬┬┬┬┬┬┬┬┬
       0       5       10      15      20
             observation number
```

The process variation is not within statistical control. To be out of control, the process must meet one of three out-of-control criteria. This process actually meets all three out-of-control criteria. (1) There is a pattern to the R values -- in this case, an upward drift. (2) There is a point lying beyond the upper or lower control limits -- in this case, there are 3 points above the UCL. (3) The run rule of eight applies -- in this case, the first 13 points are all below the center line.

7. There are $k \cdot n$ = (24 days)·(6 observations/day) = 144 observations on the run chart.
$\bar{\bar{x}} = \Sigma\bar{x}/k = 487.35/24 = 20.306$

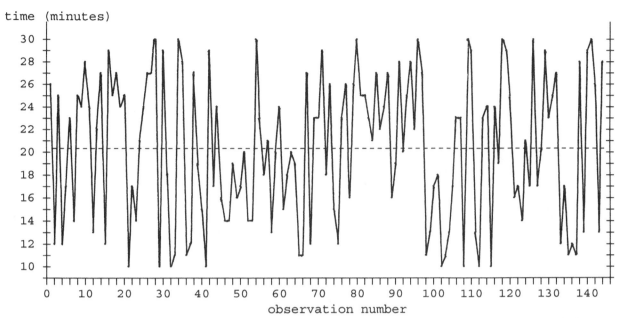

No; there does not appear to be a pattern suggesting that the process is not within statistical control.

NOTE: Although the line $\bar{\bar{x}} = 20.306$ is not specified in the text as a necessary part of the run chart, such a line is included in the Minitab examples and is included in this manual.

9. $\bar{\bar{x}} = \Sigma\bar{x}/k = 487.35/24 = 20.306$
$\bar{R} = \Sigma R/k = 359/24 = 14.958$
LCL $= \bar{\bar{x}} - A_2\bar{R} = 20.306 - (.483)(14.958) = 20.306 - 7.225 = 13.08$
UCL $= \bar{\bar{x}} + A_2\bar{R} = 20.306 + (.483)(14.958) = 20.306 + 7.225 = 27.53$

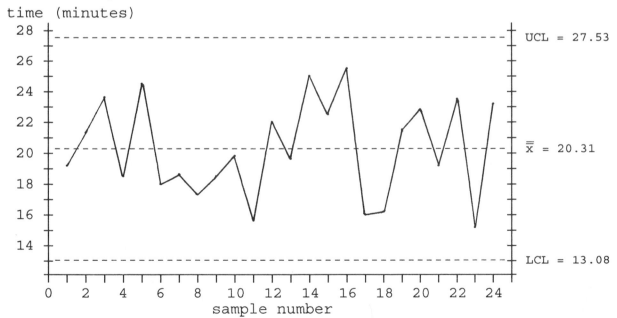

The process is working well. If there were a downward trend, then the delivery times would be being improved and no correction should be made.

11. $\bar{R} = \Sigma R/k = 10.50/20 = .525$ $LCL = D_3\bar{R} = 0(.525) = 0$
    $UCL = D_4\bar{R} = 2.114(.525) = 1.110$

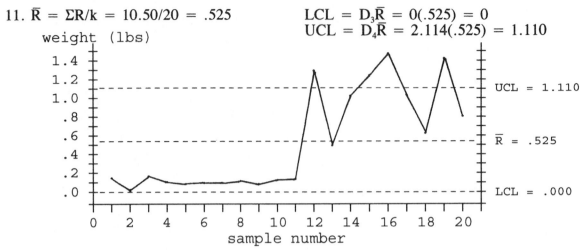

The process variation is not within statistical control. To be out of control, the process must meet one of three out-of-control criteria. This process actually meets all three out-of-control criteria. (1) There is a pattern to the R values -- in this case, an upward shift following sample #11. (2) There is a point lying beyond the upper or lower control limits -- in this case, there are 4 points above the UCL. (3) The run rule of eight applies -- in this case, the first 11 points are all below the center line.

13. $\bar{s} = \Sigma s/k = 501.2/25 = 20.048$ $LCL = B_3\bar{s} = .118(20.048) = 2.366$
    $UCL = B_4\bar{s} = 1.882(20.048) = 37.730$

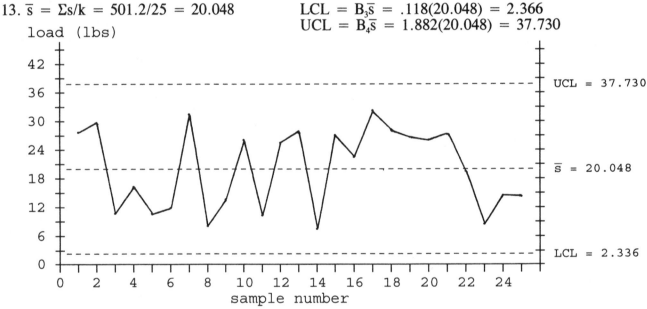

This is very similar to the R chart given in this section, and both charts indicate the process variation is within statistical control.

12-3 Control Charts for Attributes

1. This process is within statistical control. Since the first third of the sample means are generally less than the overall mean, the middle third are generally more than the overall mean, and the final third are generally less than the overall mean, however, one may wish to check future analyses to see whether such a patter tends to repeat itself.

3. This process is out of statistical control. There is an upward trend, and there is a point above the upper control limit.

5. $\bar{p} = (\Sigma x)/(\Sigma n) = (30+29+...+23)/(13)(100,000) = 332/1,300,000 = .000255$
$\sqrt{\bar{p}\cdot\bar{q}/n} = \sqrt{(.000255)(.999745)/100000} = .0000505$
LCL $= \bar{p} - 3\sqrt{\bar{p}\cdot\bar{q}/n} = .000255 - 3(.0000505) = .000255 - .000152 = .000104$
UCL $= \bar{p} + 3\sqrt{\bar{p}\cdot\bar{q}/n} = .000255 + 3(.0000505) = .000255 + .000152 = .000407$
NOTE: The 13 sample proportions are: .00030 .00029 .00029 .00027 ... .00023

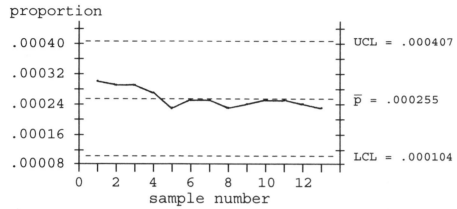

There are three out-of-control criteria, and meeting any one of them means that the process is not within statistical control. This process meets two of those criteria. (1) There is a pattern -- in this case, a downward trend. (3) The run of 8 rule applies -- in this case, the last 9 points are all below the center line. By definition, therefore, process is statistically unstable. In this instance, however, that is good and means that the age 0-4 death rate for infectious diseases is decreasing.

7. $\bar{p} = (\Sigma x)/(\Sigma n) = (29+33+...+24)/(20)(1000) = 572/20000 = .0286$
$\sqrt{\bar{p}\cdot\bar{q}/n} = \sqrt{(.0286)(.9714)/1000} = .00527$
LCL $= \bar{p} - 3\sqrt{\bar{p}\cdot\bar{q}/n} = .0286 - 3(.00527) = .0286 - .0158 = .0128$
UCL $= \bar{p} + 3\sqrt{\bar{p}\cdot\bar{q}/n} = .0286 + 3(.00527) = .0286 + .0158 = .0444$
NOTE: The 20 sample proportions are: .029 .033 .024 .029 ... .024

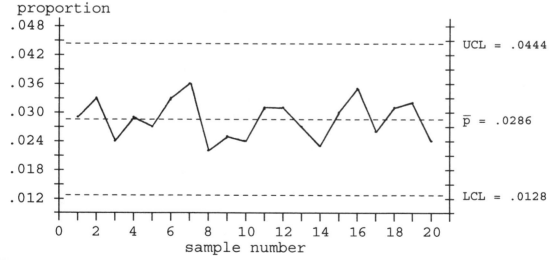

The process is within statistical control. Viewing the data as a statistical process, they do not suggest a problem that should be corrected. Viewing the data as crime statistics, they indicate that there has been no reduction in the crime rate -- and that could be considered a problem that should be corrected. On the positive side, however, the data indicate that there has been no increase in the crime rate either.

9. $\bar{p} = (\Sigma x)/(\Sigma n) = (25+24+...+31)/(13)(100,000) = 375/1,300,000 = .000288$
$n\cdot\bar{p} = (100,000)(.000288) = 28.846 \ [= \bar{x}]$
$\sqrt{n\cdot\bar{p}\cdot\bar{q}} = \sqrt{(100000)(.000288)(.999712)} = 5.370$
$LCL = n\cdot\bar{p} - 3\sqrt{n\cdot\bar{p}\cdot\bar{q}} = 28.846 - 3(5.370) = 28.846 - 16.110 = 12.736$
$UCL = n\cdot\bar{p} + 3\sqrt{n\cdot\bar{p}\cdot\bar{q}} = 28.846 + 3(5.370) = 28.846 + 16.110 = 44.956$

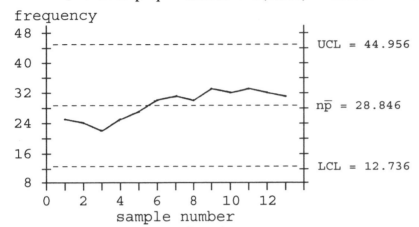

The charts are identical except for the labels. Since the labels on the np chart are n = 100,000 times the labels on the chart for p, the center line is 100,000(.000288) = 28.486 and the upper control limit is 100,000(.000450) = 44.956 and the lower control limit is 100,000(.000127) = 12.736. The placement of the points does not change.

10. $\bar{p} = (\Sigma x)/(\Sigma n) = .05$
In both parts (a) and (b), the center line occurs at .05.

a. n = 100
$\sqrt{\bar{p}\cdot\bar{q}/n} = \sqrt{(.05)(.95)/100} = .0218$
$LCL = \bar{p} - 3\sqrt{\bar{p}\cdot\bar{q}/n} = .05 - 3(.0218) = .05 - .0654 = 0 \ \text{[since it cannot be negative]}$
$UCL = \bar{p} + 3\sqrt{\bar{p}\cdot\bar{q}/n} = .05 + 3(.0218) = .05 + .0654 = .1154$

b. n = 300
$\sqrt{\bar{p}\cdot\bar{q}/n} = \sqrt{(.05)(.95)/300} = .0126$
$LCL = \bar{p} - 3\sqrt{\bar{p}\cdot\bar{q}/n} = .05 - 3(.0126) = .05 - .0378 = .0123$
$UCL = \bar{p} + 3\sqrt{\bar{p}\cdot\bar{q}/n} = .05 + 3(.0126) = .05 + .0378 = .0878$

c. The lower and upper control limits are closer to the center line in part (b). This has the advantage of being better able (i.e., on the basis of less deviance from the long run average) to detect when the process is out of statistical control, but it has the disadvantage of requiring the examination of a larger sample size. In addition, the chart in part (a) has LCL = 0.000 and cannot determine if the observed proportion is significantly less than .05. The chart in part (b) would be better able to detect a shift from 5% to 10% because the larger sample size would cause less fluctuation about the 5% or 10% long run averages and make the shift more noticeable.

Review Exercises

1. There are k · n = (20 hours) · (5 observations per hour) = 100 observations on the run chart. $\bar{\bar{x}} = \Sigma\bar{x}/k = 99.860/20 = 4.993$

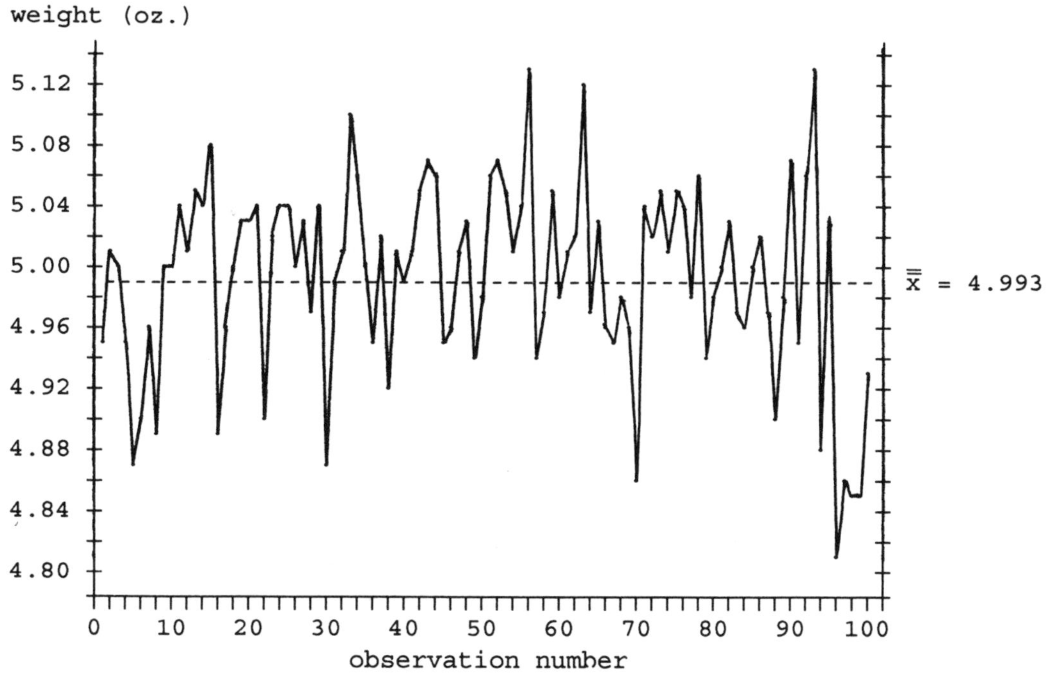

No; there does not appear to be a pattern suggesting that the process is not within statistical control.

2. $\bar{R} = \Sigma R/k = 2.48/20 = .124$ $LCL = D_3\bar{R} = 0(.124) = 0$
   $UCL = D_4\bar{R} = 2.114(.124) = .262$

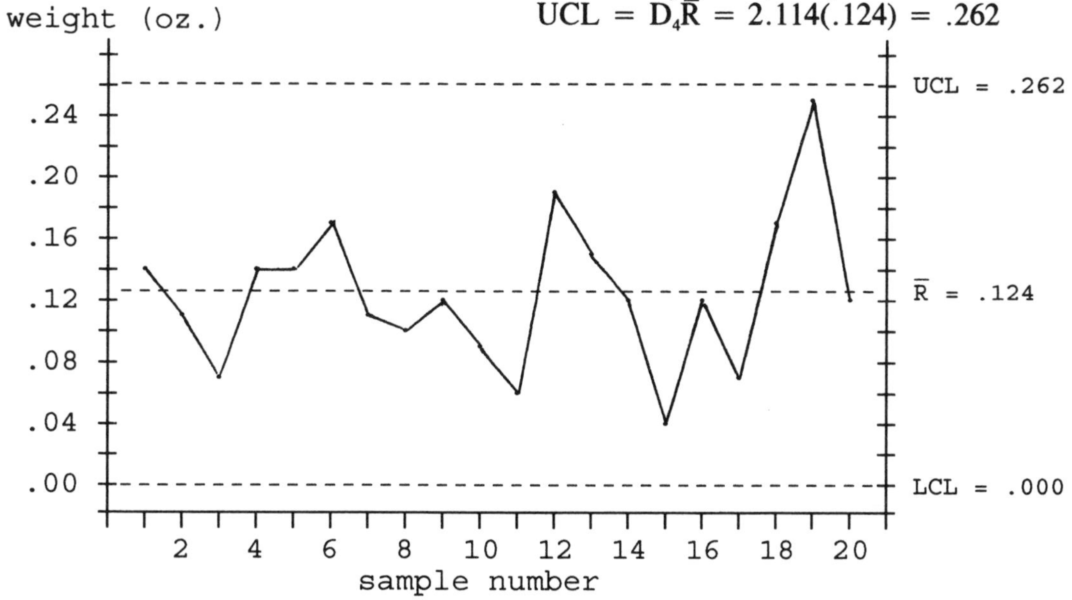

The process variation is within statistical control.

3. $\bar{\bar{x}} = \Sigma\bar{x}/k = 99.860/20 = 4.993$
   $\bar{R} = \Sigma R/k = 2.48/20 = .124$
   $LCL = \bar{\bar{x}} - A_2\bar{R} = 1.993 - (.577)(.124) = 4.993 - .072 = 4.921$
   $UCL = \bar{\bar{x}} + A_2\bar{R} = 1.993 + (.577)(.124) = 4.993 + .072 = 5.065$

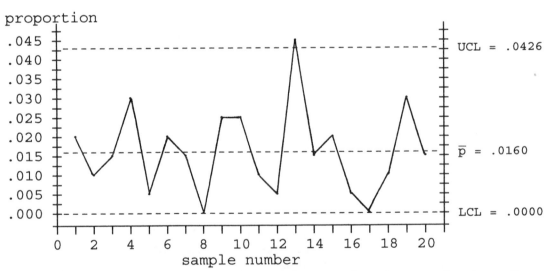

The process mean is not within statistical control because there is a point below the lower control limit. The process requires correction.

4. $\bar{p} = (\Sigma x)/(\Sigma n) = (4+2+...+3)/(20)(200) = 64/4000 = .0160$
   $\sqrt{\bar{p}\cdot\bar{q}/n} = \sqrt{(.0160)(.9840)/200} = .00887$
   $LCL = \bar{p} - 3\sqrt{\bar{p}\cdot\bar{q}/n} = .0160 - 3(.00887) = .0160 - .0266 = 0$ [since it cannot be negative]
   $UCL = \bar{p} + 3\sqrt{\bar{p}\cdot\bar{q}/n} = .0160 + 3(.00887) = .0160 + .0266 = .0426$
   NOTE: The 20 sample proportions are: .020 .010 .015 .030 ... .015

The process may be considered out of control because there is one point above the upper control limit. Since it is only one isolated and unusual point, however, the company may choose to consider the process in control. It may be worthwhile to see if there were any unusual circumstances (substitute workers, pre-holiday workday, high humidity, etc.) on the particular day when the proportion exceeded the UCL.

Cumulative Review Exercises

1. a. $\bar{p} = (\Sigma x)/(\Sigma n) = (10+8+...+11)/(20)(400) = 150/8000 = .01875$
   $\sqrt{\bar{p}\cdot\bar{q}/n} = \sqrt{(.01875)(.98125)/400} = .006782$
   $LCL = \bar{p} - 3\sqrt{\bar{p}\cdot\bar{q}/n} = .01875 - 3(.006782) = .01875 - .02035 = 0$ [since LCL $\geq$ 0]
   $UCL = \bar{p} + 3\sqrt{\bar{p}\cdot\bar{q}/n} = .01875 + 3(.006782) = .01875 + .02035 = .0391$
   NOTE: The 20 sample proportions are: .020 .010 .015 .030 ... .015

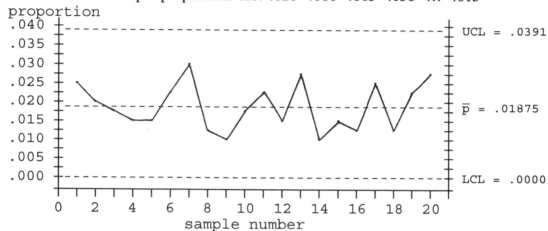

The process is within statistical control, and so the data can be treated as coming from a binomial population with a fixed mean and variance.

   b. Using all the data combined, $\hat{p} = x/n = 150/8000$
   $\hat{p} \pm z_{.025} \cdot \sqrt{\hat{p}\hat{q}/n}$
   $.01875 \pm 1.960 \cdot \sqrt{(.01875)(.98125)/8000}$
   $.01875 \pm .00297$
   $.0158 < p < .0217$

   c. original claim $p \leq .01$
   $H_o: p \leq .01$
   $H_1: p > .01$
   $\alpha = .05$
   C.R. $z > 1.645$
   calculations:
   $z_{\hat{p}} = (\hat{p} - \mu_{\hat{p}})/\sigma_{\hat{p}}$
   $= (.01875 - .01)/\sqrt{(.01875)(.98125)/8000}$
   $= .00875/.001517$
   $= 5.770$
   conclusion:
   Reject H_o; there is sufficient evidence to reject the claim that $p \leq .01$ and to conclude that $p > .01$.

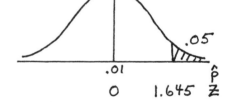

2. a. let A_i = the ith point is above the center line
   $P(A_i) = .5$ for each value of i
   $P(A) = P(\text{all 8 above}) = P(A_1) \cdot P(A_2) \cdot ... \cdot P(A_8) = (.5)^8 = .00391$
   b. let B_i = the ith point is below the center line
   $P(B_i) = .5$ for each value of i
   $P(B) = P(\text{all 8 below}) = P(B_1) \cdot P(B_2) \cdot ... \cdot P(B_8) = (.5)^8 = .00391$
   c. Notice that the events in parts (a) and (b) are mutually exclusive.
   $P(A \text{ or } B) = P(A) + P(B) - P(A \text{ and } B)$
   $= .00391 + .00391 - 0$
   $= .00781$

Chapter 13

Nonparametric Statistics

13-2 Sign Test

NOTE for $n \leq 25$: Table A-7 gives only x_L, the <u>lower</u> critical value for the sign test. Accordingly, the text lets x be the <u>smaller</u> of the number of +'s or the number of -'s and warns the user to use common sense to avoid concluding the reverse of what the data indicates.

An alternative approach maintains the natural agreement between the alternative hypothesis and the critical region and is consistent with the logic and notation of parametric tests. Let x <u>always</u> be the number of +'s. The problem's symmetry means that the upper critical value is $x_U = n-x_L$.

Since this alternative approach builds directly on established patterns and provides insight into rationale of the sign test, its C.R. and calculations and picture [notice that if H_o is true then $\mu_x = n/2$] for each exercise are given to the right of the method using only lower critical values.

NOTE for $n > 25$: The correction for continuity is a conservative adjustment intending to make less likely a false rejection of H_o by shifting the x value .5 units toward the middle. When x is the smaller of the number of +'s or the number of -'s, this always involves replacing x with $x+.5$. In the alternative approach, x is replaced with either $x+.5$ or $x-.5$ according to which one shifts the value toward the middle -- i.e., with $x+.5$ when $x < \mu_x = (n/2)$,
and with $x-.5$ when $x > \mu_x = (n/2)$.

1. Let After be group 1.
   claim: median difference = 0

pair	A	B	C	D	E	F	G	H	I	J
A-B	+	0	-	+	+	+	-	+	+	+

   n = 9 +'s and -'s

   H_o: median difference = 0
   H_1: median difference $\neq$ 0
   $\alpha = .05$
   C.R. $x \leq x_{L,9,.025} = 1$ <u>OR</u>   C.R. $x \leq x_{L,9,.025} = 1$
   $\qquad\qquad\qquad\qquad\qquad\qquad\qquad\qquad x \geq x_{U,9,.025} = 9\text{-}1 = 8$

   calculations: calculations:
   $\quad x = 2$ $\quad x = 7$
   conclusion:
   Do not reject H_o; there is not sufficient evidence to reject the claim that median difference = 0 (i.e., there is not enough evidence to reject the claim that the course has no effect).

NOTE: Compare the above results to those obtained in exercise #5 of section 8-2, where a parametric test was used on the same data. Although the text does not so indicate, several exercises in this chapter are re-tests of data that was previously analyzed parametrically; comparisons of the two results are usually informative and provide statistical insights.

3. Let favoring Covariant be a success.
   claim: p > .5
       7 +'s   3 -'s
       n = 10 +'s or -'s

   H_o: p ≤ .5
   H_1: p > .5
   α = .05
   C.R. x ≤ $x_{L,10,.05}$ = 1   <u>OR</u>   C.R. x ≥ $x_{U,10,.05}$ = 10-1 = 9

   calculations: calculations:
      x = 3   x = 7
   conclusion:
   Do not reject H_o; there is not sufficient evidence to conclude that p > .5.
   No; the sample data do not provide a reasonable basis for making the claim that most dentists favor Covariant toothpaste.

5. Let the males be group 1.
   claim: p_1 = p_2
       7 +'s   18 -'s   no 0's
       n = 25 +'s or -'s

   H_o: p_1 = p_2
   H_1: p_1 ≠ p_2
   α = .01
   C.R. x ≤ $x_{L,25,.005}$ = 5   <u>OR</u>   C.R. x ≤ $x_{L,25,.005}$ = 5
                                       x ≥ $x_{U,25,.005}$ = 25-5 = 20

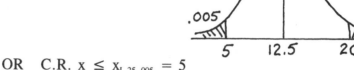

   calculations: calculations:
      x = 7   x = 7
   conclusion:
   Do not reject H_o; there is not sufficient evidence to reject the claim that p_1 = p_2.
   No; there is not sufficient evidence to charge gender bias.

7. Let weights above .9085 be +'s.
   claim: median > .9085
       18 +'s   15 -'s   no 0's
       n = 33 +'s or -'s
   Since n > 25, use z with
      μ_x = n/2 = 33/2 = 16.5
      σ_x = $\sqrt{n}$/2 = $\sqrt{33}$/2 = 2.872
   H_o: median ≤ .9085
   H_1: median > .9085
   α = .05 [assumed]
   C.R. z < -$z_{.05}$ = -1.645   <u>OR</u>   C.R. z > $z_{.05}$ = 1.645

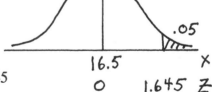

   calculations: calculations:
      x = 15   x = 18
      z_x = [(x+.5)-μ_x]/σ_x z_x = [(x-.5)-μ_x]/σ_x
        = [15.5 - 16.5]/2.872   = [17.5 - 16.5]/2.872
        = -1/2.872   = 1/2.872
        = -.348   = .348
   conclusion:
   Do not reject H_o; there is not sufficient evidence to conclude that median > .9085.
   Yes; based on this results (including the fact that there were more +'s than -'s), the package appears to be labeled correctly.
   NOTE: Compare these results to those of the parametric test in section 7-3, exercise #24.

9. Let weights above 5.670 be +'s.
   claim: median = 5.670

   10 +'s   38 -'s   2 0's
   n = 48 +'s or -'s

   Since n > 25, use z with
   $\mu_x = n/2 = 48/2 = 24$
   $\sigma_x = \sqrt{n}/2 = \sqrt{48}/2 = 3.464$
   H_0: median = 5.670
   H_1: median ≠ 5.670
   $\alpha = .05$ [assumed]
   C.R. $z < -z_{.025} = -1.960$
        $z > z_{.025} = 1.960$

   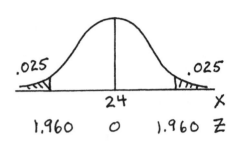

   calculations:
   x = 10
   $z_x = [(x+.5)-\mu_x]/\sigma_x$
        = [10.5 - 24]/3.464
        = -13.5/3.464
        = -3.897

   conclusion:
   Reject H_0; there is sufficient evidence to reject the claim that median = 5.670 and to conclude that median ≠ 5.670 (in fact, median < 5.670).

   This does not necessarily mean that the manufacturing specifications are not being met, but probably reflects wear from being in circulation.

   NOTE: Compare these results to those of the parametric test in section 7-2, exercise #22.

11. Let the first trial be group 1.
    claim: median difference = 0

    12 +'s   2 -'s   1 0's
    n = 14 +'s or -'s

    H_0: median difference = 0
    H_1: median difference ≠ 0
    $\alpha = .01$
    C.R. $x \le x_{L,14,.005} = 1$ OR   C.R. $x \le x_{L,14,.005} = 1$
                                              $x \ge x_{U,14,.005} = 14-1 = 13$

    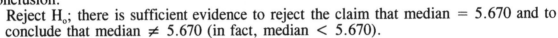

    calculations: calculations:
    x = 2   x = 12

    conclusion:
    Do not reject H_0; there is not sufficient evidence to reject the claim that median difference = 0.

    NOTE: Compare these results to those of the parametric test in section 8-2, exercise #11.

13. The n=15 scores arranged in order are: 1 2 3 6 7 8 8 9 10 11 14 17 23 25 30.
    From Table A-7, k = 3.
    The 95% confidence interval is x_{k+1} < median < x_{n-k}
                                    x_4 < median < x_{12}
                                    6 < median < 17

    NOTE: From the list of ordered scores and the fact that k = 3, it is apparent that one would reject a hypothesized median of 6 or less or of 17 or more. As in the parametric cases, the $1-\alpha$ confidence interval is composed of precisely those values not rejected in a two-tailed test at the α level.

15. Use the binomial table for $p=.5$ and reject for lower-tailed x values summing to less than .03.
    For $1 \leq n \leq 5$, $P(x=0) > .03$
    For $6 \leq n \leq 8$, $P(x \leq 0) \leq .03$ but $P(x \leq 1) > .03$
    For $9 \leq n \leq 11$, $P(x \leq 1) \leq .03$ but $P(x \leq 2) > .03$
    For $12 \leq n \leq 13$, $P(x \leq 2) \leq .03$ but $P(x \leq 3) > .03$
    For $14 \leq n \leq 15$, $P(x \leq 3) \leq .03$ but $P(x \leq 4) > .03$
    In summary, the desired critical values are as follows.

```
n:  1 2 3 4 5 6 7 8 9 10 11 12 13 14 15
x:  * * * * * 0 0 0 1  1  1  2  2  3  3
```

13-3 Wilcoxon Signed-Ranks Test for Two Dependent Samples

NOTE: Table A-8 gives only T_L, the <u>lower</u> critical value for the signed-ranks test. Accordingly, the text lets T be the <u>smaller</u> of the sum of positive ranks or the sum of the negative ranks and warns the user to use common sense to avoid concluding the reverse of what the data indicates.

An alternative approach maintains the natural agreement between the alternative hypothesis and the critical region and is consistent with the logic and notation of parametric tests. Let T <u>always</u> be the sum of the positive ranks. By symmetry, the upper critical value is $T_U = \Sigma R - T_L$.

Since this alternative approach builds directly on established patterns and provides insight into rationale of the signed-ranks test, its C.R. and calculations and picture [notice that if H_o is true then $\mu_T = \Sigma R/2$] for each exercise are given to the right of the method using only lower critical values.

1. Let the After scores be group 1.
    claim: the populations have the same distribution

subj	A	B	C	D	E	F	G	H	I	J
A-B	20	0	-10	40	30	10	-30	20	20	10
R	5	-	-2	9	7.5	2	-7.5	5	5	2

n = 9 non-zero ranks
$\Sigma R- = 9.5$
$\Sigma R+ = 35.5$
$\Sigma R = 45.0$ [check: $\Sigma R = n(n+1)/2 = 9(10)/2 = 45$]

H_o: the populations have the same distribution
H_1: the populations have different distributions
$\alpha = .01$
C.R. $T \leq T_{L,9,.005} = 2$ <u>OR</u> C.R. $T \leq T_{L,9,.005} = 2$
$T \geq T_{U,8,.005} = 45-2 = 43$

calculations:
$T = 9.5$

calculations:
$T = 35.5$

conclusion:
    Do not reject H_o; there is not sufficient evidence to reject the claim that the populations have the same distribution.

NOTE: This manual uses a minus sign preceding ranks associated with negative differences. While the ranks themselves are not negative, the use of the minus sign helps to organize the information.

3. Let the 8 am temperatures be group 1.
   claim: the populations have the same distribution

8 am	97.0	98.0	96.4	98.2	98.8	98.6	97.8	98.7	97.8	96.4	96.9
12 am	97.7	98.8	98.0	98.7	98.0	98.5	98.3	98.7	99.1	98.2	99.2
B-A	-.7	-.8	-1.6	-.5	.8	.1	-.5	0	-1.3	-1.8	-2.3
R	-4	-5.5	-8	-2.5	5.5	1	-2.5	-	-7	-9	-10

$n = 10$ non-zero ranks
$\Sigma R- = 48.5$
$\Sigma R+ = 6.5$
$\Sigma R = 55.0$ [check: $\Sigma R = n(n+1)/2 = 10(11)/2 = 55$]

H_o: the populations have the same distribution
H_1: the populations have different distributions
$\alpha = .05$

C.R. $T \leq T_{L,10,.025} = 8$ OR   C.R. $T \leq T_{L,10,.025} = 8$
                                    $T \geq T_{U,10,.025} = 55-8 = 47$

calculations: calculations:
   $T = 6.5$ $T = 6.5$

conclusion:
   Reject H_o; there is sufficient evidence to reject the claim that the populations have the same distribution and to conclude that they have different distributions (in fact, the 8 am temperatures are lower).

NOTE: This manual follows the text and the directions to the exercises of this section by using "the populations have the same distribution" as the null hypothesis. To be more precise, the signed-rank test doesn't test "distributions" but tests the "location" (i.e., central tendency -- as opposed to variation) of distributions. The test discerns whether one group taken as a whole tends to have higher or lower scores than another group taken as a whole. The test does not discern whether one group is more variable than another. This distinction is reflected in the wording of the conclusion when rejecting H_o. Notice also that the signed-rank test measures overall differences between the times and **not** whether the two times give the same results for individuals. If the 12 am temperatures, for example, were higher for half the women and the lower for the other half of the women by the same amounts, then $\Sigma R-$ would equal $\Sigma R+$ (so we could not reject H_o), but the distributions would be very different.

5. Let Before be group 1.
   claim: the populations have the same distribution

pair	A	B	C	D	E	F	G	H
H-S	-.2	4.1	1.6	1.8	3.2	2.0	2.9	9.6
R	-1	7	2	3	6	4	5	8

$n = 8$ non-zero ranks
$\Sigma R- = 1$
$\Sigma R+ = 35$
$\Sigma R = 36$ [check: $\Sigma R = n(n+1)/2 = 8(9)/2 = 36$]

H_o: the populations have the same distribution
H_1: the populations have different distributions
$\alpha = .01$

C.R. $T \leq T_{L,8,.005} = 0$ OR   C.R. $T \leq T_{L,8,.005} = 0$
                                    $T \geq T_{U,8,.005} = 36-0 = 36$

calculations: calculations:
   $T = 1$ $T = 35$

conclusion:
   Do not reject H_o; there is not sufficient evidence to reject the claim that the populations have the same distribution.

NOTE: Compare these results to those of the parametric test in section 8-2, exercise #9.

7. Let Before be group 1.
   claim: the populations have the same distribution

pair	A	B	C	D	E	F	G	H	I	J	K	L	M
B-A	21	33	2	16	-1	12	4	8	16	0	18	18	15
R	11	12	2	7.5	-1	5	3	4	7.5	-	9.5	9.5	6

   n = 12 non-zero ranks
   $\Sigma R-$ = 1
   $\Sigma R+$ = 77
   ΣR = 78   [check: $\Sigma R = n(n+1)/2 = 12(13)/2 = 78$]

H_o: the populations have the same distribution
H_1: the populations have different distributions
$\alpha = .05$ [assumed]

C.R. $T \leq T_{L,12,.025} = 14$ OR   C.R. $T \leq T_{L,12,.025} = 14$
$\qquad\qquad\qquad\qquad\qquad\qquad\qquad\qquad\qquad T \geq T_{U,12,.025} = 78-14 = 64$

calculations: calculations:
   T = 1   T = 77

conclusion:
   Reject H_o; there is sufficient evidence to reject the claim that the populations have the same distribution and to conclude that they have different distributions (in fact, the Before group has higher scores).

9. Since T is defined as the lower of $\Sigma R-$ and $\Sigma R+$, its smallest possible value is 0. Its largest value occurs when $\Sigma R- = \Sigma R+ = (1/2)(\Sigma R) = (1/2)[n(n+1)/2] = n(n+1)/4$.
   a. assuming no zeros, 0 and 10(11)/4 = 27.5
   b. assuming no zeros, 0 and 50(51)/4 = 637.5
   c. for n = 100 non-zero ranks, use the z approximation with
   $\mu_T = n(n+1)/4$
   $\quad = 100(101)/4 = 2525$
   $\sigma_T = \sqrt{[n(n+1)(2n+1)/24]}$
   $\quad = \sqrt{[100(101)(201)/24]} = 290.84$
   For $\alpha = .05$ in a two-tailed test, the critical z in 1.960.
   The critical T (i.e., the smaller of $\Sigma R-$ and $\Sigma R+$) is found by solving as follows.
   $z = (T - \mu_T)/\sigma_T < -1.96$
   $(T - 2525)/290.84 < -1.96$
   $\qquad T - 2525 < -570.05$
   $\qquad\qquad T < 1954.95$
   With the assumption of no ties in ranks, the C.R. is $T \leq 1954$.

   NOTE: In general, however, assigning the average of the tied ranks would introduce (1/2)'s -- but not (1/3)'s, (1/4)'s, etc., since the average of n consecutive integers is always a either a whole number (if n is odd) or a half number (if n is even). The proper critical value for T would then be 1954.5.

13-4   Wilcoxon Rank-Sum Test for Two Independent Samples

NOTE: As in the previous section, the manual follows the wording in the text and tests the hypothesis that "the populations have the same distribution" with the understanding that the test detects only differences in location and not differences in variability. In addition, always letting $R = \Sigma R_1$ guarantees agreement and consistency between the directions of H_1 and the C.R. as in the previous chapters.

1. Below are the scores (in order) for each group. The group listed first is considered group 1.

Red	R	Bro	R
.870	6	.856	1
.872	8	.858	2
.874	9	.860	3
.882	12	.866	4
.888	13	.867	5
.891	15	.871	7
.897	16.5	.875	10
.898	18.5	.876	11
.908	25	.889	14
.908	25	.897	16.5
.908	25	.898	18.5
.911	29	.900	20
.912	30	.902	21.5
.913	31	.902	21.5
.920	35.5	.904	23
.924	39.5	.909	27.5
.924	39.5	.909	27.5
.933	45	.914	32.5
.936	46.5	.914	32.5
.952	48	.919	34
.983	52	.920	35.5
	569.0	.921	37
		.923	38
		.928	41
		.930	42.5
		.930	42.5
		.932	44
		.936	46.5
		.955	49
		.965	50
		.976	51
		.988	53
		1.033	54
			916.0

$$n_1 = 21 \qquad \Sigma R_1 = 569$$
$$n_2 = 33 \qquad \Sigma R_2 = 916$$

$$n = \Sigma n = 54 \qquad \Sigma R = 1485$$

check: $\Sigma R = n(n+1)/2$
$$= 54(55)/2$$
$$= 1485$$

$$R = \Sigma R_1 = 569$$

$$\mu_R = n_1(n+1)/2$$
$$= 21(55)/2$$
$$= 577.5$$

$$\sigma^2_R = n_1 n_2(n+1)/12$$
$$= (21)(33)(55)/12$$
$$= 3176.25$$

H_o: the populations have the same distribution
H_1: the populations have different distributions
$\alpha = .05$
C.R. $z < -z_{.025} = -1.96$
   $z > z_{.025} = 1.96$
calculations:
$$z_R = (R - \mu_R)/\sigma_R$$
$$= (569 - 577.5)/\sqrt{3176.25}$$
$$= -8.5/56.358$$
$$= -.151$$

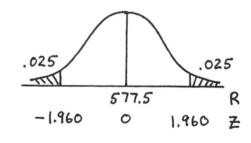

conclusion:
   Do not reject H_o; there is not sufficient evidence to reject the claim that the populations have the same distribution.

3. Below are the scores (in order) for each group. The group listed first is considered group 1.

Lab 1	R	Lab 2	R
2.9	2.5	2.7	1
2.9	2.5	3.2	9
3.0	4	3.3	10
3.1	6.5	3.4	11.5
3.1	6.5	3.4	11.5
3.1	6.5	3.6	13
3.1	6.5	3.8	16.5
3.7	14.5	3.8	16.5
3.7	14.5	4.0	19
3.9	18	4.1	20
4.2	21	4.3	22
	103.0		150.0

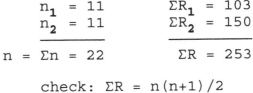

$n_1 = 11$ $\Sigma R_1 = 103$
$n_2 = 11$ $\Sigma R_2 = 150$

$n = \Sigma n = 22$ $\Sigma R = 253$

check: $\Sigma R = n(n+1)/2$
      $= 22(23)/2$
      $= 253$

$R = \Sigma R_1 = 103$

$\mu_R = n_1(n+1)/2$
    $= 11(23)/2$
    $= 126.5$

$\sigma_R^2 = n_1 n_2 (n+1)/12$
    $= (11)(11)(23)/12$
    $= 231.917$

H_o: the populations have the same distribution
H_1: the populations have different distributions
$\alpha = .05$
C.R. $z < -z_{.025} = -1.96$
      $z > z_{.025} = 1.96$
calculations:
  $z_R = (R - \mu_R)/\sigma_R$
      $= (103 - 126.5)/\sqrt{231.917}$
      $= -23.5/15.229$
      $= -1.543$

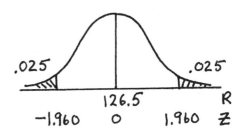

conclusion:
  Do not reject H_o; there is not sufficient evidence to reject the claim that the populations have the same distribution.
Yes; the two labs have obtained results that agree.

5. Below are the scores (in order) for each group. The group listed first is considered group 1.

Beer	R	Liqr	R
.129	1	.182	9
.146	2	.185	10
.148	3	.190	12.5
.152	4	.205	15
.154	5	.220	17
.155	6	.224	18
.164	7	.225	19.5
.165	8	.226	20.5
.187	11	.227	21
.190	12.5	.234	22
.203	14	.241	23
.212	16	.247	24
	89.5	.253	25
		.257	26
			261.5

$n_1 = 12$ $\qquad$ $\Sigma R_1 = 89.5$
$n_2 = 14$ $\qquad$ $\Sigma R_2 = 261.5$

$n = \Sigma n = 26$ $\qquad$ $\Sigma R = 351.0$

check: $\Sigma R = n(n+1)/2$
$\qquad\qquad = 26(27)/2$
$\qquad\qquad = 351$

$R = \Sigma R_1 = 89.5$

$\mu_R = n_1(n+1)/2$
$\qquad = 12(27)/2$
$\qquad = 162$

$\sigma_R^2 = n_1 n_2(n+1)/12$
$\qquad = (12)(14)(27)/12$
$\qquad = 378$

H_o: the populations have the same distribution
H_1: the populations have different distributions
$\alpha = .05$
C.R. $z < -z_{.025} = -1.96$
$\qquad z > z_{.025} = 1.96$
calculations:
$\quad z_R = (R - \mu_R)/\sigma_R$
$\qquad = (89.5 - 162)/\sqrt{378}$
$\qquad = -72.5/19.442$
$\qquad = -3.729$

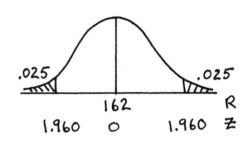

conclusion:
Reject H_o; there is sufficient evidence to reject the claim that the populations have the same distribution and to conclude that they have different distributions (in fact, population 1 has lower BAC levels).

Based on these results, it appears that the liquor drinkers have a higher BAC level and are (presumably) more dangerous.

7. Below are the scores (in order) for each group. The group listed first is considered group 1.

E to D	R	D to E	R
7.10	1	26.63	11
16.32	2	26.68	12
20.60	3	27.24	14
21.06	4	27.62	15
21.13	5	29.34	21
21.96	6	29.49	22
24.23	7	30.20	24
24.64	8	30.26	25
25.49	9	32.34	29
26.43	10	32.54	30
26.69	13	33.53	34
27.85	16	33.62	35
28.02	17	34.02	36
28.71	18	35.32	37
28.89	19	35.91	38
28.90	20	42.91	41
30.02	23		424
30.29	26		
30.72	27		
31.73	28		
32.83	31		
32.86	32		
33.31	33		
38.81	39		
39.29	40		
	437		

$n_1 = 25 \qquad \Sigma R_1 = 437$

$n_2 = 16 \qquad \Sigma R_2 = 424$

$n = \Sigma n = 41 \qquad \Sigma R = 861$

check: $\Sigma R = n(n+1)/2$
$= 41(42)/2$
$= 861$

$R = \Sigma R_1 = 437$

$\mu_R = n_1(n+1)/2$
$= 25(42)/2$
$= 525$

$\sigma_R^2 = n_1 n_2 (n+1)/12$
$= (25)(16)(42)/12$
$= 1400$

H_0: the populations have the same distribution
H_1: the populations have different distributions
$\alpha = .05$
C.R. $z < -z_{.025} = -1.96$
$\qquad z > z_{.025} = 1.96$
calculations:
$z_R = (R - \mu_R)/\sigma_R$
$\qquad = (437 - 525)/\sqrt{1400}$
$\qquad = -88/37.417$
$\qquad = -2.352$

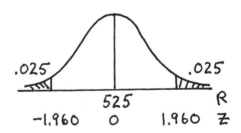

conclusion:
Reject H_0; there is sufficient evidence to reject the claim that the populations have the same distribution and to conclude that they have different distributions (in fact, population 1 has lower scores).

9. The denominators of the Mann-Whitney and Wilcoxon statistics are the same.
The numerator of the Mann-Whitney statistic, assuming $R = \Sigma R_1$, is
$U - n_1 n_2/2 = [n_1 n_2 + n_1(n_1 +1)/2 - R] - n_1 n_2/2$
$\qquad = n_1 n_2/2 + n_1(n_1 +1)/2 - R$
$\qquad = (n_1/2)(n_2 + n_1 + 1) - R$
$\qquad = -[R - n_1(n_1 + n_2 + 1)/2]$
$\qquad = -[\text{numerator of Wilcoxon statistic}]$

13-5 Kruskal-Wallis Test

NOTE: As in the previous sections, the manual follows the wording in the text and tests the hypothesis that "the populations have the same distribution" with the understanding that the test detects only differences in location and not differences in variability.

1. Below are the scores (in order) for each group. The group listed first is group 1, etc.

18-20	R	21-29	R	30+	R
97.1	2	97.9	6	97.4	1
97.7	5	98.2	8.5	97.5	3.5
98.0	7	98.2	8.5	97.5	3.5
98.4	10	99.0	14	98.6	12.5
98.5	11	99.6	15	98.6	12.5
	35		52.0		33.0

$n_1 = 5$ $R_1 = 35$
$n_2 = 5$ $R_2 = 52$
$n_3 = 5$ $R_3 = 33$

$n = \Sigma n = 15$ $\Sigma R = 120$

check:
$\Sigma R = n(n+1)/2$
$= 15(16)/2$
$= 120$

H_o: the populations have the same distribution
H_1: the populations have different distributions
$\alpha = .05$
C.R. $H > \chi^2_{2,.05} = 5.991$
calculations:

$H = [12/n(n+1)] \cdot [\Sigma(R_i^2/n_i)] - 3(n+1)$
$= [12/15(16)] \cdot [(35)^2/5 + (52)^2/5 + (33)^2/5] - 3(16)$
$= [.05] \cdot [1003.6] - 48$
$= 2.180$

conclusion:

Do not reject H_o; there is not sufficient evidence to conclude that the populations have different distributions.

NOTE: Compare these results to those of the parametric test in section 11-2, exercise #3.

3. Below are the scores (in order) for each group. The group listed first is group 1, etc.

o-sid	R	o-frt	R	b-sid	R	b-frt	R
235	4	399	11.5	190	1	389	9
279	6	435	13	199	2	489	16.5
369	8	446	14	219	3	489	16.5
395	10	499	18	239	5	549	21
469	15	538	19	309	7	599	22
547	20	639	23	399	11.5	695	24
	63		98.5		29.5		109.0

$n_1 = 6$ $R_1 = 63.0$
$n_2 = 6$ $R_2 = 98.5$
$n_3 = 6$ $R_3 = 29.5$
$n_4 = 6$ $R_4 = 109.0$

$n = \Sigma n = 24$ $\Sigma R = 300.0$

check:
$\Sigma R = n(n+1)/2$
$= 24(25)/2$
$= 300$

H_o: the populations have the same distribution
H_1: the populations have different distributions
$\alpha = .05$
C.R. $H > \chi^2_{3,.05} = 7.815$
calculations:

$H = [12/n(n+1)] \cdot [\Sigma(R_i^2/n_i)] - 3(n+1)$
$= [12/24(25)] \cdot [(63)^2/6 + (98.5)^2/6$
$\quad + (29.5)^2/6 + (109)^2/6] - 3(25)$
$= [.02] \cdot [4409.75] - 75$
$= 13.075$

conclusion:

Reject H_o; there is sufficient evidence to reject the claim that the populations have the same distribution and to conclude that they have different distributions.

NOTE: Compare these results to those of the parametric test in chapter 11, review exercise #2.

5. Below are the weights in thousandths of a gram (in order) for each group. The group listed first is group 1, etc.

red	R	ora	R	yel	R	bro	R	blu	R	gre	R
870	9.5	861	5	868	8	856	2	838	1	890	26
872	12	897	31	876	16.5	858	3	870	9.5	902	39.5
874	13	898	34	877	18	860	4	875	14.5	902	39.5
882	21	903	42	879	19.5	866	6	956	87	911	52
888	24	920	61	879	19.5	867	7	968	90.5	930	72
891	27	942	82	886	22.5	871	11		202.5	949	83.5
897	31	971	92	886	22.5	875	14.5			1002	98
898	34	1009	99	892	28	876	16.5				410.5
908	46		446	893	29	889	25				
908	46			900	36.5	897	31				
908	46			906	44	898	34				
911	52			910	50	900	36.5				
912	54			911	52	902	39.5				
913	55			917	58	902	39.5				
920	61			921	63.5	904	43				
924	67			924	67	909	48.5				
924	67			926	69	909	48.5				
933	75			934	76	914	56.5				
936	77.5			939	79	914	56.5				
952	85			940	80	919	59				
983	95			941	81	920	61				
	998.0			949	83.5	921	63.5				
				960	88	923	65				
				968	90.5	928	70				
				978	94	930	72				
				989	97	930	72				
					1392.5	932	74				
						936	77.5				
						955	86				
						965	89				
						976	93				
						988	96				
						1033	100				
							1600.5				

$$n_1 = 21 \quad R_1 = 998.0$$
$$n_2 = 8 \quad R_2 = 446.0$$
$$n_3 = 26 \quad R_3 = 1392.5$$
$$n_4 = 33 \quad R_4 = 1600.5$$
$$n_5 = 5 \quad R_5 = 202.5$$
$$n_6 = 7 \quad R_6 = 410.5$$

$$n = \Sigma n = 100 \qquad \Sigma R = 5050.0$$

check:
$$\Sigma R = n(n+1)/2$$
$$= 100(101)/2$$
$$= 5050$$

H_o: the populations have the same distribution
H_1: the populations have different distributions
$\alpha = .05$ [assumed]
C.R. $H > \chi^2_{5,.05} = 11.071$

calculations:

$$H = [12/n(n+1)] \cdot [\Sigma(R_i^2/n_i)] - 3(n+1)$$
$$= [12/100(101)] \cdot [(998)^2/21 + (446)^2/8 + (1392.5)^2/26$$
$$+ (1600.5)^2/33 + (202.5)^2/5 + (410.5)^2/7] - 3(101)$$
$$= [.00119] \cdot [256770.7] - 303$$
$$= 2.074$$

conclusion:
   Do not reject H_o; there is not sufficient evidence to reject the claim that the populations have the same distribution.

No; the results do not indicate a problem that requires corrective action.

NOTE: Compare these results to those of the parametric test in section 11-2, exercise #13.

7. Below are the scores (in order) for each group. The group listed first is group 1, etc.

0-2.5	R	3-3.5	R	4	R
91	6	82	2	72	1
92	7.5	90	4.5	88	3
92	7.5	93	9.5	90	4.5
96	16	93	9.5	103	27
96	16	94	12	104	29.5
98	19.5	94	12	106	34.5
99	21	94	12	120	46
100	23	95	14	121	47
100	23	96	16	125	50
101	25	97	18	144	55
105	32.5	98	19.5	159	57
108	37.5	100	23	160	58
108	37.5	102	26	168	59
110	39	104	29.5	193	60
114	42	104	29.5		531.5
123	48.5	104	29.5		
155	56	105	32.5		
	457.5	106	34.5		
		107	36		
		111	40.5		
		111	40.5		
		115	43		
		117	44		
		119	45		
		123	48.5		
		129	51		
		133	52		
		134	53		
		139	64		
			841.0		

$n_1 = 17$ $R_1 = 457.5$
$n_2 = 29$ $R_2 = 841.0$
$n_3 = 14$ $R_3 = 531.5$

$n = \Sigma n = 60$ $\Sigma R = 1830.0$

check:
$\Sigma R = n(n+1)/2$
$= 60(61)/2$
$= 1830$

H_o: the populations have the same distribution
H_1: the populations have different distributions
$\alpha = .05$
C.R. $H > \chi^2_{2,.05} = 5.991$
calculations:

$H = [12/n(n+1)] \cdot [\Sigma(R_i^2/n_i)] - 3(n+1)$
$= [12/60(61)] \cdot [(457.5)^2/8 + (841)^2/10 + (531.5)^2/7] - 3(61)$
$= [.00328] \cdot [56879.2] - 183$
$= 3.489$

conclusion:
Do not reject H_o; there is not sufficient evidence to reject the claim that the populations have the same distribution.

9. $H = [12/n(n+1)] \cdot [\Sigma(R_i^2/n_i)] - 3(n+1)$, which depends only on the rank of each score and the number of scores within each group

a. Adding or subtracting a constant to each score does not change the number of scores within each group. Since adding or subtracting a constant to each score does not affect the order of the scores, their ranks and the calculated H statistic are not affected.

b. Multiplying or dividing each score by a positive constant does not change the number of scores within each group. Since multiplying or dividing each score by a positive constant does not affect the order of the scores, their ranks and the calculated H statistic are not affected.

11. NOTE: Be careful when counting the number of tied ranks; in addition to the easily recognized ".5's," there are 3 8's, 5 12's and 3 21's, etc. The following table organizes the calculations.

rank	t	$T = t^3 - t$
8	3	24
12	5	120
15.5	2	6
18.5	2	6
21	3	24
24	3	24
30.5	4	60
34.5	4	60
37.5	2	6
40	3	24
43.5	2	6
45.5	2	6
		366

correction factor:
$$1 - \Sigma T/(n^3-n) = 1 - 366/(48^3-48)$$
$$= 1 - 366/110544$$
$$= 1 - .00331$$
$$= .99669$$

The original calculated test statistic is $H = 14.431$.
The corrected calculated test statistic is $H = 14.431/.99669 = 14.479$.
No; the corrected value of H does not differ substantially from the original one.

13-6 Rank Correlation

NOTE: This manual calculates $d = R_x - R_y$, thus preserving the sign of d. This convention means Σd must equal 0 and provides a check for the assigning and differencing of the ranks. In addition, it must always be true that $\Sigma R_x = \Sigma R_y = n(n+1)/2$.

1. In each case the $n = 5$ pairs are pairs of ranks, called R_x and R_y below to emphasize that fact.

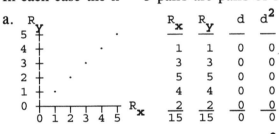

a.

R_x	R_y	d	d^2
1	1	0	0
3	3	0	0
5	5	0	0
4	4	0	0
2	2	0	0
15	15	0	0

$r_s = 1 - [6(\Sigma d^2)]/[n(n^2-1)]$
$= 1 - [6(0)]/[5(24)]$
$= 1 - 0$
$= 1$

Yes; there appears to be a perfect positive correlation between R_x and R_y.

b.

R_x	R_y	d	d^2
1	5	-4	16
2	4	-2	4
3	3	0	0
4	2	2	4
5	1	4	16
15	15	0	40

$r_s = 1 - [6(\Sigma d^2)]/[n(n^2-1)]$
$= 1 - [6(40)]/[5(24)]$
$= 1 - 2$
$= -1$

Yes; there appears to be a perfect negative correlation between R_x and R_y.

c.

R_x	R_y	d	d^2
1	2	-1	1
2	5	-3	9
3	3	0	0
4	1	3	9
5	4	1	1
15	15	0	20

$r_s = 1 - [6(\Sigma d^2)]/[n(n^2-1)]$
$= 1 - [6(20)]/[5(24)]$
$= 1 - 1$
$= 0$

No; there does not appear to be any correlation between R_x and R_y.

3. The following table summarizes the calculations.

R_x	R_y	d	d^2
2	2	0	0
6	7	-1	1
3	7	-3	9
5	4	1	1
7	5	2	4
10	8	2	4
9	9	0	0
8	10	-2	4
4	3	1	1
1	1	0	0
55	55	0	24

$$r_s = 1 - [6(\Sigma d^2)]/[n(n^2-1)]$$
$$= 1 - [6(24)]/[10(99)]$$
$$= 1 - .145$$
$$= .855$$

H_o: $\rho_s = 0$
H_1: $\rho_s \neq 0$
$\alpha = .05$
C.R. $r_s < -.648$
     $r_s > .648$
calculations:
   $r_s = .855$
conclusion:

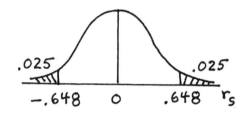

   Reject H_o; there is sufficient evidence to reject the claim that $\rho_s = 0$ and to conclude that $\rho_s \neq 0$ (in fact $\rho > 0$).

Yes; it does appear that salary increases as stress increases.

IMPORTANT NOTE: The rank correlation is correctly calculated using the ranks in the Pearson product moment correlation formula of chapter 9 to produce

$$r_s = [\Sigma R_x R_y - (\Sigma R_x)(\Sigma R_y)]/[\sqrt{\Sigma R_x^2 - (\Sigma R_x)^2} \cdot \sqrt{\Sigma R_y^2 - (\Sigma R_y)^2}]$$

Since $\Sigma R_x = \Sigma R_y = 1+2+...+n = n(n+1)/2$ [always]
  and $\Sigma R_x^2 = \Sigma R_y^2 = 1^2+2^2+...+n^2 = n(n+1)(2n+1)/6$ [when there are ties in the ranks],
it can be shown by algebra that the above formula can be shortened to

$$r_s = 1 - [6(\Sigma d^2)]/[n(n^2-1)]$$ when there are no ties in the ranks.

As the presence of ties typically does not make a difference in the first 3 decimals of r_s, this manual uses the shortened formula exclusively and notes when use of the longer formula gives a slightly different result.

5. The following table summarizes the calculations.

R_x	R_y	d	d^2
2	5	-3	9
7	2	5	25
6	3	3	9
4	8	-4	16
5	10	-5	25
8	9	-1	1
9	1	8	64
10	7	3	9
3	6	-3	9
1	4	-3	9
55	55	0	176

$$r_s = 1 - [6(\Sigma d^2)]/[n(n^2-1)]$$
$$= 1 - [6(176)]/[10(99)]$$
$$= 1 - 1.067$$
$$= -.067$$

H_o: $\rho_s = 0$
H_1: $\rho_s \neq 0$
$\alpha = .05$
C.R. $r_s < -.648$
$\quad\quad r_s > .648$
calculations:
$\quad r_s = -.067$
conclusion:
$\quad$ Do not reject H_o; there is not sufficient evidence to conclude that $\rho_s \neq 0$.

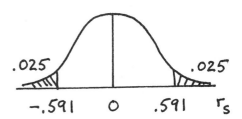

No; there does not appear to be a relationship between the stress level of jobs and their physical demands.

7. The following table summarizes the calculations.

x	R_x	y	R_y	d	d^2
107	10	111	10.5	-0.5	0.25
96	5	97	3	2	4
103	8	116	12	-4	16
90	3	107	8	-5	25
96	5	99	4	1	1
113	12	111	10.5	1.5	2.25
86	1	85	1	0	0
99	7	108	9	-2	4
109	11	102	6	5	25
105	9	105	7	2	4
96	5	100	5	0	0
89	2	93	2	0	0
	78		78.0	0.0	81.50

$$r_s = 1 - [6(\Sigma d^2)]/[n(n^2-1)]$$
$$= 1 - [6(81.5)]/[12(143)]$$
$$= 1 - .285$$
$$= .715$$

H_o: $\rho_s = 0$
H_1: $\rho_s \neq 0$
$\alpha = .05$
C.R. $r_s < -.591$
$\quad\quad r_s > .591$
calculations:
$\quad r_s = .715$
conclusion:
$\quad$ Reject H_o; there is sufficient evidence to conclude that $\rho_s \neq 0$ (in fact, $\rho_s > 0$).

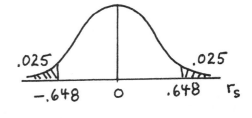

9. The following table summarizes the calculations.

x	R_x	y	R_y	d	d^2
53.0	2	80	2	0	0
67.5	3	344	5	-2	4
72.0	5.5	416	8	-2.5	6.25
72.0	5.5	348	6	-0.5	0.25
73.5	8	262	3	5	25
68.5	4	360	7	-3	9
73.0	7	332	4	3	9
37.0	1	34	1	0	0
	36.0		36.0	0.0	53.50

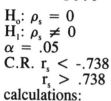

$$r_s = 1 - [6(\Sigma d^2)]/[n(n^2-1)]$$
$$= 1 - [6(53.5)]/[8(63)]$$
$$= 1 - .637$$
$$= .363$$

H_o: $\rho_s = 0$
H_1: $\rho_s \neq 0$
$\alpha = .05$
C.R. $r_s < -.738$
    $r_s > .738$
calculations:
  $r_s = .363$
conclusion:

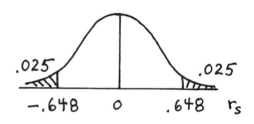

Do not reject H_o; there is not sufficient evidence to reject the claim that $\rho_s = 0$.

11. The following table summarizes the calculations.

x	R_x	y	R_y	d	d^2
29	5	31	8.5	-3.5	12.25
35	9	27	2	7	49
28	3.5	29	6.5	-3	9
44	10	25	1	9	81
25	2	31	8.5	-6.5	42.25
34	8	29	6.5	1.5	2.25
30	6	28	4	2	4
33	7	28	4	3	9
28	3.5	28	4	-0.5	0.25
24	1	33	10	-9	81
	55.0		55.0	0.0	290.00

$$r_s = 1 - [6(\Sigma d^2)]/[n(n^2-1)]$$
$$= 1 - [6(290)]/[10(99)]$$
$$= 1 - 1.758$$
$$= -.758$$

H_o: $\rho_s = 0$
H_1: $\rho_s \neq 0$
$\alpha = .05$
C.R. $r_s < -.648$
    $r_s > .648$
calculations:
  $r_s = -.758$
conclusion:

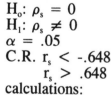

Reject H_o; there is sufficient evidence to reject the claim that $\rho_s = 0$ and to conclude that $\rho_s \neq 0$ (in fact, $\rho_s < 0$).

Yes; based on these results, you can expect to pay more for gas if you buy a heavier car. Since the calculations are based entirely on ranks, and the ranks are the same whether the weights are given in pounds or 100-pounds, such a modification would not change the results at all.

NOTE: Compare these results to those of the parametric test in section 9-2, exercise #6.

13. H_o: $\rho_s = 0$
H_1: $\rho_s \neq 0$
$\alpha = .05$
C.R. $r_s < -z_{.025}/\sqrt{n-1} = -.280$
$r_s > z_{.025}/\sqrt{n-1} = .280$

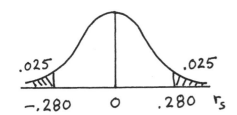

.025   .025

−.280   0   .280 r_s

calculations:
$r_s = 1 - [6(\Sigma d^2)]/[n(n^2-1)]$
$= 1 - [6(4453)]/[50(2499)]$
$= 1 - .214 = .786$

conclusion:
Reject H_o; there is sufficient evidence to reject that claim that $\rho_s = 0$ and to conclude that $\rho_s \neq 0$ (in fact, $\rho_s > 0$).

Yes; the interval after an eruption appears to be related to the duration of that eruption.

NOTE: Compare these results to those of the parametric test in section 9-2, exercise #13.

x	R_x	y	R_y	d	d^2
86	34.5	240	28	6.5	42.25
86	34.5	237	23	11.5	132.25
62	6.5	122	8	-1.5	2.25
104	50	267	40.5	9.5	90.25
62	6.5	113	4	2.5	6.25
95	45	258	37	8	64
79	18	232	19	-1	1
62	6.5	105	3	3.5	12.25
94	44	276	50	-6	36
79	18	248	34	-16	256
86	34.5	243	32	2.5	6.25
85	32	241	30	2	4
86	34.5	214	14	20.5	420.25
58	3	114	5	-2	4
89	40	272	47	-7	49
79	18	227	17	1	1
83	28	237	23	5	25
82	24.5	238	25.5	-1	1
84	30.5	203	13	17.5	306.25
82	24.5	270	44	-19.5	380.25
78	14.4	218	15	-0.5	0.25
91	43	226	16	27	729
89	40	250	36	4	16
79	18	245	33	-15	225
57	2	120	6.5	-4.5	20.25
100	48	267	40.5	7.5	56.25
62	6.5	103	2	4.5	20.25
87	37	270	44	-7	49
70	11	241	30	-19	361
88	38	239	27	11	121
82	24.5	233	20	4.5	20.25
83	28	238	25.5	2.5	6.25
56	1	102	1	0	0
81	21.5	271	46	-24.5	600.25
74	13	127	10	3	9
102	49	275	48.5	0.5	0.25
61	4	140	12	-8	64
83	28	264	38	-10	100
73	12	134	11	1	1
97	46	268	42	4	16
67	9	124	9	0	0
90	42	270	44	-2	4
84	30.5	249	35	-4.5	20.25
82	24.5	237	23	1.5	2.25
81	21.5	235	21	0.5	0.25
78	14.5	228	18	-3.5	12.25
89	40	265	39	1	1
69	10	120	6.5	3.5	12.25
98	47	275	48.5	-1.5	2.25
79	18	241	30	-12	144
	1275.0		1275.0	0.0	4453.00

15. The following table summarizes the calculations.

x	R_x	y	R_y	d	d^2
16	23.5	1.2	24	-0.5	0.25
16	23.5	1.2	24	-0.5	0.25
16	23.5	1.0	15	8.5	72.25
9	8	.8	8.5	-0.5	0.25
1	1	.1	1	0	0
8	5.5	.8	8.5	-3	9
10	10	.8	8.5	1.5	2.25
16	23.5	1.0	15	8.5	72.25
14	16.5	1.0	15	1.5	2.25
13	14	1.0	15	-1	1
13	14	1.1	20	-6	36
15	19	1.2	24	-5	25
16	23.5	1.2	24	-0.5	0.25
9	8	.7	5.5	2.5	6.25
11	11	.9	11	0	0
2	2	.2	2	0	0
18	28.5	1.4	28.5	0	0
15	19	1.2	24	-5	25
13	14	1.1	20	-6	36
15	19	1.0	15	4	16
17	27	1.3	27	0	0
9	8	.8	8.5	-0.5	0.25
12	12	1.0	15	-3	9
14	16.5	1.0	15	1.5	2.25
5	3	.5	3	0	0
6	4	.6	4	0	0
8	5.5	.7	5.5	0	0
18	28.5	1.4	28.5	0	0
16	23.5	1.1	20	3.5	12.25
	435.0		435.0	0.0	328.00

H_0: $\rho_s = 0$
H_1: $\rho_s \neq 0$
$\alpha = .05$
C.R. $r_s < -.370$
　　　$r_s > .370$
calculations:
　$r_s = 1 - [6(\Sigma d^2)]/[n(n^2-1)]$
　　　$= 1 - [6(328)]/[29(840)]$
　　　$= 1 - .081 = .919$

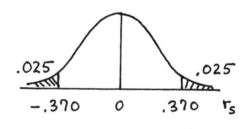

conclusion:
　Reject H_0; there is sufficient evidence to reject the claim that $\rho_s = 0$ and to conclude that $\rho_s \neq 0$ (in fact, $\rho_s > 0$).

Yes; based on this result, there does appear to be a correlation between cigarette tar and nicotine. Yes; researchers can reduce their laboratory expenses by measuring only one of these two variables.

NOTE: Using the exact formula that is always valid, instead of the shortened version that is only an approximation whenever there are ties in the ranks, yields $r_s = .918$. Compare these results to those of the parametric test in section 9-2, exercise #15.

17. H_o: $\rho_s = 0$
H_1: $\rho_s \neq 0$
$\alpha = .05$
C.R. $r_s < -z_{.025}/\sqrt{n-1} = -.310$
$r_s > z_{.025}/\sqrt{n-1} = .310$

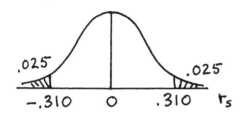

calculations:

$r_s = 1 - [6(\Sigma d^2)]/[n(n^2-1)]$
$= 1 - [6(8432.5)]/[41(1680)]$
$= 1 - .735 = .265$

conclusion:

Do not reject H_o; there is not sufficient evidence to reject that claim that $\rho_s = 0$.

No; there does not appear to be a significant correlation between precipitation and corn production. NOTE: While this may defy expectation, there may be some steadily increasing factors that influence the increasing production more than precipitation (e.g., amounts of government subsidies, advances in mechanical and biological technology, etc.). If some of these other factors could be taken into consideration, the effect of precipitation might be significant. Compare these results to those of the parametric test in section 9-2, exercise #17.

x	R_x	y	R_y	d	d^2
21.9	10	456	2	8	64
42.3	40	421	1	39	1521
29.8	13	653	9	4	16
26.0	8	573	6	2	4
34.3	27	546	5	22	484
22.7	2	499	3	-1	1
24.2	4	504	4	0	0
31.6	18.5	611	7	11.5	132.25
25.6	7	646	8	-1	1
37.9	35	789	15	20	400
33.9	25	773	14	11	121
37.4	33	753	12	21	441
31.3	16	745	11	5	25
27.0	9	852	17	-8	64
31.5	17	755	13	4	16
39.9	37	815	16	21	441
25.3	6	902	20	-14	196
30.4	14	986	24	-10	100
32.7	22	909	21	1	1
35.0	29	945	22	7	49
33.8	24	866	18	6	36
29.4	12	1178	28	-16	256
37.1	32	1230	30	2	4
42.9	41	1207	29	12	144
32.2	20	968	23	-3	9
30.5	15	1118	26	-11	121
23.7	3	1174	27	-24	576
37.0	31	1092	25	6	36
34.1	26	1478	35	-9	81
34.4	28	1664	39	-11	121
29.3	11	1463	34	-23	529
32.4	21	1731	41	-20	400
41.8	39	1578	37	2	4
36.5	30	744	10	20	400
37.5	34	1445	32	2	4
31.6	18.5	1707	40	-21.5	462.25
40.3	38	1627	38	0	0
33.3	23	1320	31	-8	64
21.6	1	899	19	-18	324
24.7	5	1446	33	-28	784
39.4	36	1562	36	0	0
	861.0		861	0.0	8432.50

19. a. $t_{6,.025} = 2.365$; $r_s^2 = (2.447)^2/[(2.447)^2 + 6] = .499$, $r_s = \pm.707$

b. $t_{13,.025} = 2.160$; $r_s^2 = (2.160)^2/[(2.160)^2 + 13] = .264$, $r_s = \pm.514$

c. $t_{28,.025} = 2.048$; $r_s^2 = (2.048)^2/[(2.048)^2 + 28] = .130$, $r_s = \pm.361$

d. $t_{28,.005} = 2.763$; $r_s^2 = (2.763)^2/[(2.763)^2 + 28] = .214$, $r_s = \pm.463$

e. $t_{6,.005} = 3.707$; $r_s^2 = (3.707)^2/[(3.707)^2 + 6] = .696$, $r_s = \pm.834$

13-7 Runs Test for Randomness

NOTE: In each exercise, the item that appears first in the sequence is considered to be of the first type and its count is designated by n_1.

1. $n_1 = 10$ (# of A's)
    $n_2 = 5$ (# of B's)
    $G = 2$ (# of runs)
    CV: 3,12 (from Table A-10)

3. $n_1 = 12$ (# of O's)
    $n_2 = 4$ (# of E's)
    $G = 4$ (# of runs)
    CV: 3,10 (from Table A-10)

5. Since $n_1 = 12$ and $n_2 = 8$, use Table A-10.
    H_o: the sequence is random
    H_1: the sequence is not random
    $\alpha = .05$
    C.R. $G \leq 6$
    $G \geq 16$
    calculations:
    $G = 10$
    conclusion:

    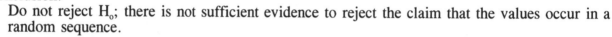

    Do not reject H_o; there is not sufficient evidence to reject the claim that the values occur in a random sequence.

    A lack of randomness would mean there was a pattern. Recognition of that pattern would give the bettor an advantage and put the casino at a disadvantage.

7. Since $n_1 = 10$ and $n_2 = 14$, use Table A-10.
    H_o: the sequence is random
    H_1: the sequence is not random
    $\alpha = .05$
    C.R. $G \leq 7$
    $G \geq 18$
    calculations:
    $G = 4$
    conclusion:

    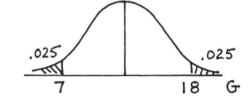

    Reject H_o; there is sufficient evidence to reject the claim that the values occur in a random sequence and to conclude that the sequence is not random (in fact, there are fewer runs than expected by chance -- i.e., like items tend to cluster together).

    The pattern that seems to repeat is 5 true followed by 7 false. Rather than rely on those specific numbers, however, a student might be safer merely to note that the instructor tends to present groups of true statements together and groups of false statements together.

9. Since $n_1 = 17$ and $n_2 = 14$, use Table A-10.
   H_o: the sequence is random
   H_1: the sequence is not random
   $\alpha = .05$
   C.R. $G \leq 10$
       $G \geq 23$

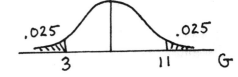

   calculations:
     $G = 14$
   conclusion:
     Do not reject H_o; there is not sufficient evidence to reject the claim that the values occur in a random sequence.
   Yes; it appears that we elect Democrat and Republican candidates in a sequence that is random.

11. Define July 1 to be the middle of the year.
   A A A A A B B B B B B
   Since $n_1 = 6$ and $n_2 = 6$, use Table A-10.
   H_o: dates after and before the middle occur in a random sequence
   H_1: dates after and before the middle do not occur in a random sequence
   $\alpha = .05$
   C.R. $G \leq 3$
       $G \geq 11$
   calculations:
     $G = 2$
   conclusion:
     Reject H_o; there is sufficient evidence to reject the claim that dates after and before the middle of the year occur in a random sequence and to conclude that the sequence is not random (in fact, dates on the same side of the middle tend to occur in groups).

13. Since $n_1 = 21$ and $n_2 = 19$, use the normal approximation.
   $\mu_G = 2n_1n_2/(n_1+n_2) + 1$
       $= 2(21)(19)/40 + 1 = 20.95$
   $\sigma_G^2 = [2n_1n_2(2n_1n_2-n_1-n_2)]/[(n_1+n_2)^2(n_1+n_2-1)]$
       $= [2(21)(19)(758)]/[(40)^2(39)] = 9.694$
   H_o: the sequence is random
   H_1: the sequence is not random
   $\alpha = .05$ [assumed]
   C.R. $z < -z_{.025} = -1.960$
       $z > z_{.025} = 1.960$

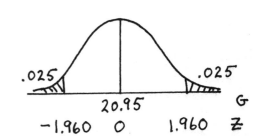

   calculations:
     $G = 7$
     $z_G = (G - \mu_G)/\sigma_G$
       $= (7 - 20.95)/\sqrt{9.694}$
       $= -13.95/3.113$
       $= -4.481$
   conclusion:
     Reject H_o; there is sufficient evidence to reject the claim that the sequence is random and to conclude that the sequence is not random (in fact, there are fewer runs than expected by chance -- i.e., those of like gender tended to be grouped together).

15. The sequence in A's and B's is as follows.

```
B  B  B  A  A  B  A  A  B  A  A  A  A  A  B  B  B  A  B  A
B  A  B  B  B  A  B  B  A  A  A  B  B  A  A  B  B  A  A  B
A  A  B  A  A  B  A  A  B  B  A  A  B  B  A  A  B  A  B  B
A  A  B  B  B  A  A  B  A  B  B  A  B  A  A  B  B  A  A  A
A  A  B  A  B  B  B  A  B  A  B  B  B  B  B  A  B  A  A  A
```

Since $n_1 = 49$ and $n_2 = 51$, use the normal approximation.

$\mu_G = 2n_1n_2/(n_1+n_2) + 1$
$\quad = 2(49)(51)/100 + 1 = 50.98$

$\sigma_G^2 = [2n_1n_2(2n_1n_2-n_1-n_2)]/[(n_1+n_2)^2(n_1+n_2-1)]$
$\quad = [2(49)(51)(4898)]/[(100)^2(99)] = 24.727$

H_o: the sequence is random
H_1: the sequence is not random
$\alpha = .05$ [assumed]
C.R. $z < -z_{.025} = -1.960$
$\quad\;\; z > z_{.025} = 1.960$

calculations:

$G = 54$

$z_G = (G - \mu_G)/\sigma_G$
$\quad = (54 - 50.98)/\sqrt{24.73}$
$\quad = 3.02/4.973$
$\quad = .607$

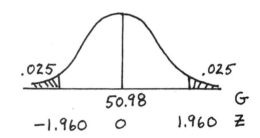

conclusion:

Do not reject H_o; there is not sufficient evidence to reject the claim that the sequence is random.

17. The minimum possible number of runs is $G = 2$ and occurs when all the A's are together and all the B's are together (e.g., A A B B).

The maximum possible number of runs is $G = 4$ and occurs when the A's and B's alternate (e.g., A B A B).

Because the critical region for $n_1=n_2=2$ is

C.R. $G \leq 1$
$\quad\;\; G \geq 6$

the null hypothesis of the sequence being random can never be rejected at the .05 level. Very simply, this means that it is not possible for such a small sample to provide 95% certainty that a non-random phenomenon is occurring.

Review Exercises

1. Below are the scores (in order) for each group.
   claim: the populations have different distributions

Pres	R	Pope	R	Mona	R
0	1	3	4	7	6
1	2	6	5	10	7
2	3	11	8	12	9.5
12	9.5	15	12.5	13	11
15	12.5	18	17.5	17	15.5
16	14	19	19	25	22
17	15.5	25	22	33	25
18	17.5		88.0		96.0
21	20				
25	22				
26	24				
36	26				
	167.0				

$n_1 = 12 \quad R_1 = 167$
$n_2 = 7 \quad R_2 = 88$
$n_3 = 7 \quad R_3 = 96$

$n = \Sigma n = 26 \quad \Sigma R = 351$

check:
$\Sigma R = n(n+1)/2$
$= 26(27)/2$
$= 351$

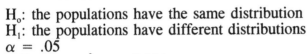

H_o: the populations have the same distribution
H_1: the populations have different distributions
$\alpha = .05$
C.R. $H > \chi^2_{2,.05} = 5.991$
calculations:
$$H = [12/n(n+1)] \cdot [\Sigma(R_i^2/n_i)] - 3(n+1)$$
$$= [12/26(27)] \cdot [(167)^2/12 + (88)^2/7 + (96)^2/7] - 3(27)$$
$$= [.0171] \cdot [4746.9] - 81$$
$$= .144$$
conclusion:
   Do not reject H_o; there is not sufficient evidence to conclude that the populations have different distributions.

2. Let the Winners be group 1.
   claim: the populations have the same distribution

pair	1	2	3	4	5	6	7	8
W-R	12	-5	-2	-2	6	0.5	3.5	0
R	7	-5	-2.5	-2.5	6	1	4	-

n = 7 non-zero ranks
$\Sigma R- = 10.0$
$\Sigma R+ = 18.0$
$\Sigma R = 28.0$ [check: $\Sigma R = n(n+1)/2 = 7(8)/2 = 28$]

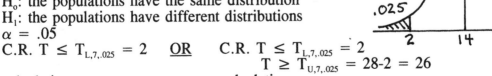

H_o: the populations have the same distribution
H_1: the populations have different distributions
$\alpha = .05$
C.R. $T \leq T_{L,7,.025} = 2$ **OR**   C.R. $T \leq T_{L,7,.025} = 2$
$\qquad\qquad\qquad\qquad\qquad\qquad T \geq T_{U,7,.025} = 28-2 = 26$

calculations: calculations:
   T = 10   T = 18
conclusion:
   Do not reject H_o; there is not sufficient evidence to reject the claim that the populations have the same distribution.

3. Since $n_1 = 6$ and $n_2 = 24$, use the normal approximation.

$$\mu_G = 2n_1n_2/(n_1+n_2) + 1$$
$$= 2(6)(24)/30 + 1$$
$$= 10.6$$
$$\sigma_G^2 = [2n_1n_2(2n_1n_2-n_1-n_2)]/[(n_1+n_2)^2(n_1+n_2-1)]$$
$$= [2(6)(24)(258)]/[(30)^2(29)]$$
$$= 2.847$$

H_o: the sequence is random
H_1: the sequence is not random
$\alpha = .05$
C.R. $z < -z_{.025} = -1.960$
   $z > z_{.025} = 1.960$
calculations:
   $G = 12$
   $z_G = (G - \mu_G)/\sigma_G$
   $= (12 - 10.6)/\sqrt{2.847}$
   $= 1.4/1.687$
   $= .830$

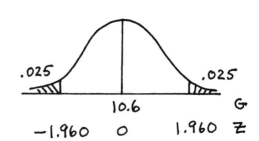

conclusion:
   Do not reject H_o; there is not sufficient evidence to reject the claim that the sequence is random.

NOTE: In light of the "obvious" repeating DAAAA pattern, it's disappointing that the test did not reject randomness and suggest the pattern that every 5th item is a defect. But the test statistic will be the same for any arrangement of 6 D's and 24 A's that starts with D, ends with A, and has no two adjacent D's. In addition, for 6 D's and 24 A's even the maximum $G=13$ (if the arrangement had started with A, ended with A, and had no adjacent D's) would not cause rejection -- i.e., for this n_1 and n_2 the test can detect too few runs but cannot detect too many runs.

4. The following table summarizes the calculations.
   claim: $\rho_s \neq 0$

x	R_x	y	R_y	d	d^2
2.41	1	2	2.5	-1.5	2.25
7.57	4	3	4.5	-0.5	0.25
9.55	7	3	4.5	2.5	6.25
8.82	6	6	8	-2	4
8.72	5	4	6	-1	1
6.96	3	2	2.5	0.5	0.25
6.83	2	1	1	1	1
11.42	8	5	7	1	1
	36		36.0	0.0	16.00

$r_s = 1 - [6(\Sigma d^2)]/[n(n^2-1)]$
$= 1 - [6(16)]/[8(63)]$
$= 1 - .190$
$= .810$

H_o: $\rho_s = 0$
H_1: $\rho_s \neq 0$
$\alpha = .05$
C.R. $r_s < -.738$
   $r_s > .738$
calculations:
   $r_s = .810$

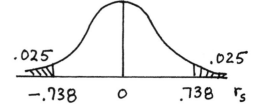

conclusion:
   Reject H_o; there is sufficient evidence to conclude that $\rho_s \neq 0$ (in fact, $\rho_s > 0$).

5. Let MicroAir be group 1.
   claim: median difference = 0

pair	1	2	3	4	5	6	7	8	9	10	11
H-S	+	+	+	+	0	+	+	-	+	+	+

   n = 10 +'s and -'s

   H_o: median difference = 0
   H_1: median difference $\neq$ 0
   α = .05
   C.R. $x \leq x_{L,10,.025} = 1$ **OR**   C.R. $x \leq x_{L,10,.025} = 1$
   $x \geq x_{U,10,.025} = 10-1 = 9$

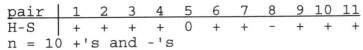

   calculations: calculations:
      x = 1   x = 9

   conclusion:
   Reject H_o; there is sufficient evidence to reject the claim that median difference = 0 and to conclude that median difference $\neq$ 0 (in fact, median difference > 0 -- i.e., the MicroAir times tend to be longer).

   On the basis of this test, Flight Services appears to be the system to adopt. It would also be wise to check the variances. If the typical overall difference between the times is only slight, but one system is much more variable than the other, one may wish to choose the more consistent system.

6. Below are the scores (in order) for each group. The group listed first is considered group 1.
   claim: the populations have the same distribution

Drug	R	Ctrl	R
508	1	549	4
512	2	563	5
516	3	565	6
603	8	598	7
621	9	652	12.5
624	10	668	14
627	11	674	15
652	12.5	676	16
711	18.5	703	17
729	20	711	18.5
747	21	772	22
777	23	789	25
787	24	800	26
	163.0	821	27.5
		821	27.5
		830	29
		837	30
			302.0

   $n_1 = 13$ $\Sigma R_1 = 163$
   $n_2 = 17$ $\Sigma R_2 = 302$

   $n = \Sigma n = 30$ $\Sigma R = 465$

   check: $\Sigma R = n(n+1)/2$
              $= 30(31)/2$
              $= 465$

   $R = \Sigma R_1 = 163$

   $\mu_R = n_1(n+1)/2$
        $= 13(31)/2$
        $= 201.5$

   $\sigma_R^2 = n_1 n_2(n+1)/12$
        $= (13)(17)(31)/12$
        $= 570.9$

H_o: the populations have the same distribution
H_1: the populations have different distributions
$\alpha = .05$
C.R. $z < -z_{.025} = -1.960$
   $z > z_{.025} = 1.960$
calculations:
   $z_R = (R - \mu_R)/\sigma_R$
      $= (163 - 201.5)/\sqrt{570.9}$
      $= -38.5/23.894$
      $= -1.611$

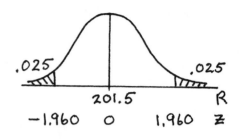

conclusion:
   Do not reject H_o; there is not sufficient evidence to reject the claim that the populations have the same distribution.

7. Below are the scores (in order) for each group.
   claim: the populations have different distributions

1	R	2	R	3	R	4	R
62	4	60	3	64	6	55	1
66	8.5	63	5	73	13.5	57	2
73	13.5	70	10	77	16.5	65	7
74	15	71	11	78	18.5	66	8.5
80	20	77	16.5	82	22.5	72	12
85	26.5	82	22.5	85	26.5	78	18.5
87	28.5	84	25	89	30	81	21
91	31.5	87	28.5	91	31.5	83	24
	147.5		121.5		165.0		94.0

$n_1 = 8$ $R_1 = 147.5$
$n_2 = 8$ $R_2 = 121.5$
$n_3 = 8$ $R_3 = 165.0$
$n_4 = 8$ $R_4 = 94.0$

$n = \Sigma n = 32$ $\Sigma R = 528.0$

check:
$\Sigma R = n(n+1)/2$
   $= 32(33)/2$
   $= 528$

H_o: the populations have the same distribution
H_1: the populations have different distributions
$\alpha = .01$
C.R. $H > \chi^2_{3,.05} = 7.815$
calculations:
   $H = [12/n(n+1)] \cdot [\Sigma(R_i^2/n_i)] - 3(n+1)$
      $= [12/32(33)] \cdot [(147.5)^2/8 + (121.5)^2/8 + (165)^2/8 + (94)^2/8] - 3(33)$
      $= [.0114] \cdot [9072.4] - 99 = 4.096$

conclusion:
   Do not reject H_o; there is not sufficient evidence to conclude that the populations have different distributions.

8. Since $n_1 = 14$ and $n_2 = 16$, use Table A-10.
   claim: the sequence is random

   H_o: the sequence is random
   H_1: the sequence is not random
   $\alpha = .05$
   C.R. $G \leq 10$
      $G \geq 22$
   calculations:
      $G = 11$

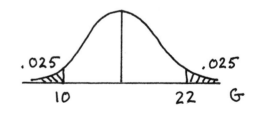

   conclusion:
   Do not reject H_o; there is not sufficient evidence to reject the claim that the sequence is random.

9. Below are the scores (in order) for each group. The group listed first is considered group 1.
   claim: the populations have the same distribution

MO	R	CS	R
23.00	1	29.00	10
24.50	2	30.99	12
24.75	3	32.00	14
26.00	4	32.99	16
27.00	5	33.00	17
27.98	6	33.98	18
27.99	7	33.99	19
28.15	8	34.79	20
29.99	10	35.79	21
29.99	10	37.75	22
31.50	13	38.99	30
32.75	15		192
	84		

$n_1 = 12$ $\Sigma R_1 = 84$
$n_2 = 11$ $\Sigma R_2 = 192$

$n = \Sigma n = 23$ $\Sigma R = 276$

check: $\Sigma R = n(n+1)/2$
$= 23(24)/2$
$= 276$

$R = \Sigma R_1 = 84$

$\mu_R = n_1(n+1)/2$
$= 12(24)/2$
$= 144$
$\sigma_R^2 = n_1 n_2(n+1)/12$
$= (11)(12)(24)/12$
$= 264$

H_o: the populations have the same distribution
H_1: the populations have different distributions
$\alpha = .05$
C.R. $z < -z_{.025} = -1.960$
    $z > z_{.025} = 1.960$
calculations:
   $z_R = (R - \mu_R)/\sigma_R$
      $= (84 - 144)/\sqrt{264}$
      $= -60/16.248$
      $= -3.693$
conclusion:

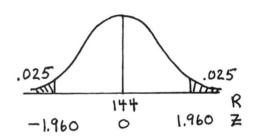

Reject H_o; there is sufficient evidence reject the claim that the populations have the same distribution and to conclude that they have different distributions (in fact, the mail order prices are lower).

10. Let the Pre-training weights be group 1.
    claim: median difference = 0

pair	1	2	3	4	5	6	7	8	9	10
Pr-Po	+	0	0	0	+	+	+	+	0	+

n = 6 +'s and -'s

H_o: median difference = 0
H_1: median difference $\neq 0$
$\alpha = .05$
C.R. $x \leq x_{L,6,.025} = 0$ OR C.R. $x \leq x_{L,6,.025} = 0$
                                      $x \geq x_{U,6,.025} = 6\text{-}0 = 6$
calculations: calculations:
   $x = 0$ $x = 6$
conclusion:
   Reject H_o; there is sufficient evidence to reject the claim that median difference = 0 and to conclude that median difference $\neq 0$ (in fact, the pre-training weights are larger).

11. Let the Pre-training weights be group 1.
    claim: the populations have the same distribution

pair	1	2	3	4	5	6	7	8	9	10
Pr-Po	5	0	0	0	8	1	1	4	0	1
R	5	-	-	-	6	2	2	4	-	2

n = 6 non-zero ranks
$\Sigma R- = 0$
$\Sigma R+ = 21$
$\Sigma R = 21$ [check: $\Sigma R = n(n+1)/2 = 6(7)/2 = 21$]

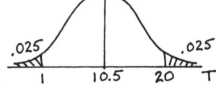

H_o: the populations have the same distribution
H_1: the populations have different distributions
$\alpha = .05$
C.R. $T \leq T_{L,6,.025} = 1$ OR   C.R. $T \leq T_{L,6,.025} = 1$
                                             $T \geq T_{U,6,.025} = 21-1 = 20$
calculations: calculations:
   $T = 0$ $T = 21$
conclusion:
   Reject H_o; there is sufficient evidence to reject the claim that the populations have the same distribution and to conclude that they have different distributions (in fact, the Pre-training weights tend to be larger).

12. The following table summarizes the calculations.

R_x	R_y	d	d^2
4	1	3	9
3	5	-2	4
5	6	-1	1
6	2	4	16
7	7	0	0
8	8	0	0
2	3	-1	1
1	4	-3	9
36	36	0	40

$$r_s = 1 - [6(\Sigma d^2)]/[n(n^2-1)]$$
$$= 1 - [6(40)]/[8(63)]$$
$$= 1 - .476$$
$$= .524$$

H_o: $\rho_s = 0$
H_1: $\rho_s \neq 0$
$\alpha = .05$
C.R. $r_s < -.738$
     $r_s > .738$
calculations:
   $r_s = .524$
conclusion:
   Do not reject H_o; there is not sufficient evidence to reject the claim that $\rho_s = 0$.

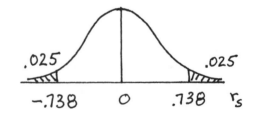

Cumulative Review Exercises

1. a. Since $n_1 = 20$ and $n_2 = 10$, use Table A-10.
   claim: the sequence is random
   H_o: the sequence is random
   H_1: the sequence is not random
   $\alpha = .05$
   C.R. $G \leq 9$
   $\quad\quad G \geq 20$
   calculations:
   $\quad G = 21$
   conclusion:

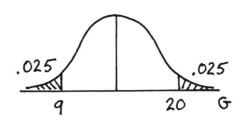

   Reject H_o; there is sufficient evidence to reject the claim that the sequence is random and
   to conclude that it is not random (in fact, there are more runs than expected -- i.e., there
   appears to have been an effort to avoid consecutive occurrences of the same gender).

   b. Let p be the proportion of women: $\hat{p} = x/n = 10/30 = .333$
   claim: $p \neq .50$
   H_o: $p = .50$
   H_1: $p \neq .50$
   $\alpha = .05$
   C.R. $z < -z_{.025} = -1.960$
   $\quad\quad z > z_{.025} = 1.960$
   calculations:
   $\quad z_p = (\hat{p} - \mu_p)/\sigma_p$
   $\quad\quad = (.333 - .50)/\sqrt{(.5)(.5)/30}$
   $\quad\quad = -.167/.0913$
   $\quad\quad = -1.826$
   conclusion:
   Do not reject H_o; there is not enough evidence to conclude that $p \neq .50$.

   c. Let p be the proportion of women: $\hat{p} = x/n = 10/30 = .333$
   $\hat{p} \pm z_{.025} \cdot \sqrt{\hat{p}\hat{q}/n}$
   $.333 \pm 1.960 \cdot \sqrt{(.333)(.667)/30}$
   $.333 \pm .169$
   $.165 < p < .502$

   d. There is not enough evidence to be 95% certain that there was a bias against either gender.
   There is 95% confidence, however, that the sample was not obtained in a random sequence.
   The pollster appears to have deliberately avoided consecutive occurrences of the same
   gender. Since any effort to make the selection anything other than a random sample may
   involve subjective decisions, there is the possibility that the sample may be biased in subtle
   ways and/or not representative of the population.

2. a. Let the After scores be group 1.
   claim: median difference = 0

pair	1	2	3	4	5	6	7	8	9	10
A-B	+	0	-	+	+	+	-	+	+	+

   n = 9 +'s and -'s

   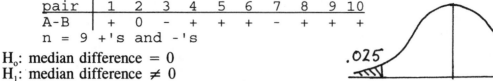

H_o: median difference = 0
H_1: median difference $\neq$ 0
α = .05

C.R. x $\leq$ $x_{L,9,.025}$ = 1   <u>OR</u>   C.R. x $\leq$ $x_{L,9,.025}$ = 1
                                              x $\geq$ $x_{U,9,.025}$ = 9-1 = 8

calculations: calculations:
   x = 2   x = 7

conclusion:
   Do not reject H_o; there is not sufficient evidence to reject the claim that
   median difference = 0.

   b. Let the After scores be group 1.
   claim: the populations have the same distribution

pair	1	2	3	4	5	6	7	8	9	10
A-B	100	0	-10	120	140	110	-30	90	150	150
R	4	-	-1	6	7	5	-2	3	8.5	8.5

   n = 9 non-zero ranks
   $\Sigma R-$ = 3
   $\Sigma R+$ = 42
   ΣR = 45   [check: ΣR = n(n+1)/2 = 9(10)/2 = 45]

H_o: the populations have the same distribution
H_1: the populations have different distributions
α = .05

C.R. T $\leq$ $T_{L,9,.025}$ = 6   <u>OR</u>   C.R. T $\leq$ $T_{L,9,.025}$ = 6
                                              T $\geq$ $T_{U,6,.025}$ = 45-6 = 39

calculations: calculations:
   T = 3   T = 42

conclusion:
   Reject H_o; there is sufficient evidence to reject the claim that the populations have the
   same distribution and to conclude that they have different distributions (in fact, the After
   scores tend to be larger).

c. Let the After scores be group 1.
   original claim: $\mu_d = 0$ [$n \le 30$ and σ_d unknown, use t]
   $d = x_1 - x_2$: 100 0 -10 120 140 110 -30 90 150 150
    $n = 10$
    $\Sigma d = 820$ $\qquad\qquad$ $\overline{d} = 82$
    $\Sigma d^2 = 110200$ $\qquad$ $s_d = 69.09$
   H_o: $\mu_d = 0$
   H_1: $\mu_d \neq 0$
   $\alpha = .05$
   C.R. $t < -t_{9,.025} = -2.262$
    $t > t_{9,.025} = 2.262$
   calculations:

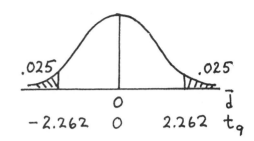

$\qquad t_{\overline{d}} = (\overline{d} - \mu_{\overline{d}})/s_{\overline{d}}$
$\qquad\qquad = (82 - 0)/(69.09/\sqrt{10}\,)$
$\qquad\qquad = 82/21.847$
$\qquad\qquad = 3.753$
   conclusion:
   Reject H_o; there is sufficient evidence to reject the claim that $\mu_d = 0$ and to conclude that $\mu_d \neq 0$ (in fact, $\mu_d > 0$).

d. The test in part (a) failed to reject the hypothesis of no difference.
   The test in part (b) rejected the hypothesis of no difference in a "relatively close" decision.
   The test in part (c) "strongly" rejected the hypothesis of no difference.

   Each test used progressively more information from
   $\qquad$ (a) the sign of the difference
   $\quad$ to (b) the rank of the difference
   $\quad$ to (c) the actual value of the difference.

   In general, the more information a test uses, the better able it is to detect departures from the null hypothesis. These results also support the statement that non-parametric tests [like parts (a) and (b)] lack the sensitivity of parametric tests [like part (c)] and so require stronger evidence before a null hypothesis is rejected.

FINAL NOTE: Congratulations! You have completed statistics -- the course that everybody likes to hate. I hope that this manual has helped to make the course a little more understandable for you. I hope you leave the course with an appreciation of broad principles and not memories of merely manipulating formulas. I wish you well in your continued studies, and that you achieve your full potential wherever your journey of life may lead.